Innovative Multifunctional Nanomaterial for Photocatalysis, Sensing, and Imaging

Azad Kumar
Siddharth University, India

A volume in the Advances in Chemical and Materials Engineering (ACME) Book Series

IGI Global
PUBLISHER of TIMELY KNOWLEDGE

Published in the United States of America by
IGI Global
Engineering Science Reference (an imprint of IGI Global)
701 E. Chocolate Avenue
Hershey PA, USA 17033
Tel: 717-533-8845
Fax: 717-533-8661
E-mail: cust@igi-global.com
Web site: http://www.igi-global.com

Library of Congress Cataloging-in-Publication Data

Names: Kumar, Azad, 1988- editor.
Title: Innovative multifunctional nanomaterial for photocatalysis, sensing, and imaging / edited by Azad Kumar.
Description: Hershey, PA : Engineering Science Reference, [2023] | Includes bibliographical references and index. | Summary: "Innovative Multifunctional Nanomaterial For Photocatalysis, Sensing, and Imaging clarifies the type, chemical composition, and applications of multifunctional nanomaterials in photocatalysis, sensing, and imaging. This book discusses the most current advances in multifunctional materials for photocatalysis, sensing, and imaging in biological systems. Covering topics such as drug development, nanoparticles, and sensing, this premier reference source is an excellent resource for chemists, engineers, professionals, students and educators of higher education, researchers, and academicians"-- Provided by publisher.
Identifiers: LCCN 2023017281 (print) | LCCN 2023017282 (ebook) | ISBN 9781668487433 (h/c) | ISBN 9781668487440 (s/c) | ISBN 9781668487457 (eISBN)
Subjects: LCSH: Optical detectors--Materials. | Image converters--Materials. | Photocatalysis--Materials. | Nanostructured materials.
Classification: LCC TK8360.O67 I5434 2023 (print) | LCC TK8360.O67 (ebook) | DDC 620.1/15--dc23/eng/20230906
LC record available at https://lccn.loc.gov/2023017281
LC ebook record available at https://lccn.loc.gov/2023017282

This book is published in the IGI Global book series Advances in Chemical and Materials Engineering (ACME) (ISSN: 2327-5448; eISSN: 2327-5456)

British Cataloguing in Publication Data
A Cataloguing in Publication record for this book is available from the British Library.

All work contributed to this book is new, previously-unpublished material.
The views expressed in this book are those of the authors, but not necessarily of the publisher.

For electronic access to this publication, please contact: eresources@igi-global.com.

Advances in Chemical and Materials Engineering (ACME) Book Series

ISSN:2327-5448
EISSN:2327-5456

Editor-in-Chief: J. Paulo Davim, University of Aveiro, Portugal

MISSION

The cross disciplinary approach of chemical and materials engineering is rapidly growing as it applies to the study of educational, scientific and industrial research activities by solving complex chemical problems using computational techniques and statistical methods.

The **Advances in Chemical and Materials Engineering (ACME) Book Series** provides research on the recent advances throughout computational and statistical methods of analysis and modeling. This series brings together collaboration between chemists, engineers, statisticians, and computer scientists and offers a wealth of knowledge and useful tools to academics, practitioners, and professionals through high quality publications.

COVERAGE

- Metallic Alloys
- Fracture Mechanics
- Sustainable Materials
- Ductility and Crack-Resistance
- Computational methods
- Heat Treatments
- Electrochemical and Corrosion
- Toxicity of Materials
- Polymers Engineering
- Artificial Intelligence Methods

IGI Global is currently accepting manuscripts for publication within this series. To submit a proposal for a volume in this series, please contact our Acquisition Editors at Acquisitions@igi-global.com or visit: http://www.igi-global.com/publish/.

The Advances in Chemical and Materials Engineering (ACME) Book Series (ISSN 2327-5448) is published by IGI Global, 701 E. Chocolate Avenue, Hershey, PA 17033-1240, USA, www.igi-global.com. This series is composed of titles available for purchase individually; each title is edited to be contextually exclusive from any other title within the series. For pricing and ordering information please visit http://www.igi-global.com/book-series/advances-chemical-materials-engineering/73687. Postmaster: Send all address changes to above address.

Titles in this Series

For a list of additional titles in this series, please visit:
http://www.igi-global.com/book-series/advances-chemical-materials-engineering/73687

Innovations in Materials Chemistry, Physics, and Engineering Research
Eugene de Silva (Virginia Research Institute, USA) and Pramudi Abeydeera (Polymeric Ltd., USA)
Engineering Science Reference • copyright 2023 • 446pp • H/C (ISBN: 9781668468302) • US $270.00 (our price)

Handbook of Research on Advanced Functional Materials for Orthopedic Applications
R. Ranjith (SNS College of Technology, India) and J. Paulo Davim (University of Aveiro, Portugal)
Engineering Science Reference • copyright 2023 • 371pp • H/C (ISBN: 9781668474129) • US $325.00 (our price)

Sustainable Utilization of Nanoparticles and Nanofluids in Engineering Applications
Sampath Boopathi (Mechanical Engineering, Muthayammal Engineering College, India) and J. Paulo Davim (University of Aveiro, Portugal)
Engineering Science Reference • copyright 2023 • 372pp • H/C (ISBN: 9781668491355) • US $270.00 (our price)

Modeling, Characterization, and Processing of Smart Materials
Ajay Kumar (Mechanical Engineering Department, School of Engineering and Technology, JECRC University, India) Parveen Kumar (Mechanical Engineering Department, Rawal Institute of Engineering and Technology, India) Ashish Kumar Srivastava (G.L. Bajaj Institute of Technology and Management, India) and Vikas Goyat (Mechanical Engineering Department, SRM Institute of Science and Technology, India)
Engineering Science Reference • copyright 2023 • 370pp • H/C (ISBN: 9781668492246) • US $270.00 (our price)

Production, Properties, and Applications of Engineered Cementitious Composites
S. Praveenkumar (PSG College of Technology, India) and J. Paulo Davim (University of Aveiro, Portugal)

For an entire list of titles in this series, please visit:
http://www.igi-global.com/book-series/advances-chemical-materials-engineering/73687

IGI Global
PUBLISHER of TIMELY KNOWLEDGE

701 East Chocolate Avenue, Hershey, PA 17033, USA
Tel: 717-533-8845 x100 • Fax: 717-533-8661
E-Mail: cust@igi-global.com • www.igi-global.com

Table of Contents

Detailed Table of Contents

Development of innovative nanomaterials for multifunctional application in photo-catalysis, electro-chemical sensing etc. are gaining extraordinary importance worldwide. In this sense, several metal chalcogenides nano-materials, such as FeS, CdS, ZnS, Fe3O4 etc., nano-materials and their composite have been synthesized and established as effective heterogenous recyclable catalyst for conducting several photochemical and electrochemical reactions. When the nano-materials have the ability to absorb solar light energy, the performance for photo-catalytic reactions has been increased much more. In this chapter it has been elaborated that how the solar energy can be used to drive different organic reactions pathway for green synthesis as well as to develop decomposition of organic pollutants in waste-water treatment. Another interesting function of the same nano-materials in the field of bio-technology has been expounded based on its superior electro-catalytic and bio-sensing activity towards hydrogen peroxide and glucose for determination of diabetic disease.

The present global concern associated with the energy crisis and environmental pollution can be addressed through semiconductor photocatalysis technology by converting solar energy into green fuels. WO3 is considered as a favorable candidate due to its durability in the oxidation reaction, primarily in acidic pH, absorption of the visible photon ($\lambda \leq 480$ nm), cost beneficiary, and less toxicity. However, due to the undesirable position of its conduction band, WO3 suffers from photoexcited charge-carrier recombination and shows lower photo-response for the reduction process. To improve the activity under visible light, various approaches have been implemented, such as band alignment and restriction of charge-carrier recombination. This review summarises the numerous endeavors like manipulation of crystal facets and morphologies, incorporation of dopants, and composite formation through hetero-junctions using various semiconductors.

Photocatalysis is a traditional approach to solve energy crises and environmental pollution control. Graphitic carbon nitride (g-C3N4), a 2-D conjugated polymer, along with different other forms of carbon based components like carbon nano-tubes, carbon quantum dots, carbon nano-sheets, carbon nano-wires etc. acts as nano-mediator and serves as promising candidates with bismuth-based oxides for applications in environmental and energy sectors leading to sustainable development. The latest efforts towards overcoming the limitations for improvement in performances of these materials can be controlled by different factors e.g. surface morphology, band gap, size, crystallinity, charge separation efficiency and rate of the charge transfer etc. Here, we primarily focus on the detailed strategies related to the fabrication of different carbon components incorporated bismuth oxides based nano-composites for applications in sustainable development.

Water scarcity has been exacerbated by environmental pollution caused by increasing global industrialization and urbanization. This has led to continued research for more ecologically safe and cleaner technologies for eliminating pollutants. In the recent past, nano-engineered metal oxides such as titanium oxide (TiO2) have shown promising photocatalytic effect of completely mineralizing organic pollutants to form harmless byproducts. TiO2 is the most frequently used photocatalyst, however its effectiveness is limited to light absorption within the ultra violet region due to its wide band-gap. A redshift by TiO2 has been achieved through tuning of the anatase-rutile phase transitions, doping, surface modifications and creation of multi component heterojunctions. This chapter discusses the synthesis of TiO2–based photocatalysts and photocatalytic activity enhancement of TiO2 through doping and formation of hybrids. Their application in the photo-degradation of dyes and other organic pollutants as well as their limitations and future prospects are also addressed.

Metallic nanoparticles (NPs) have great potential for biological and chemical sensor applications because the surface of the NPs has a sensitive spectral response, and the light signal can be easily monitored by their absorption or scattering. In particular, silver nanoparticles (Ag NPs) are used in various applications due to their advanced physicochemical properties such as their excellent antibacterial, catalytic, and conductivity at the nanoscale. Plant-mediated synthesis methods have come to the fore with the green chemistry approach in the synthesis of NPs. The aim of this chapter is to provide a detailed resource by considering the studies in the literature on the use of Ag NPs produced by green routes mainly in the fields of photocatalysis, sensing, and imaging. Additionally, this chapter aims to determine the importance and potential of plant-mediated synthesized Ag NPs in these applications will be determined and to be a guide for industrial and academic personnel working in this field.

Air contamination and a growing number of harmful gases releasing into the environment is a big issue of present era. Gases like CO, NH3, CO2, SO2, and CH4 can prompt lethal health threats. Gas sensors have attained noteworthy consideration of researchers in monitoring such kind of pollutants. Detection of gases by sensors typically depends upon the sensing properties, like sensitivity, selectivity, response, and recovery time. With the progress of the internet-of-things technology, the uses of gas sensors in the fields of smart homes, wearable gadgets, and smart mobile terminals have developed rapidly. In this chapter, various groups of the sensing materials are outlined thoroughly, comprising gas sensors based on multifunctional materials such as: metals, graphene, cellulose, MXenes, acoustic wave sensors, and catalytic sensors. This chapter also outlines the current accomplishments in the field of gas detection and signifies the current challenges and future outlooks in this field.

Every aspect of civilization, including its growth and sustainability on both political and socioeconomic level, is significantly impacted by access to clean water. As a result, maintaining water supplies and treating wastewater have become major global challenges. Water pollution is caused by a variety of gases, microbes, and other poisons that are added to water during rainfall, running water, etc. Studies into the concurrent removal of several contaminants or coexisting contaminants from water by nanoparticles pave the door for a novel route that is not only affordable but also devoid of secondary contamination. Hence, decontaminating water that has been polluted by hazardous and toxic compounds is a topic that generates considerable interest among researchers. This chapter discusses various types of nanomaterials as photocatalysts that can effectively manage contaminated water. By analyzing earlier studies to investigate the toxic compounds present in water, this chapter also highlights the various water treatment approaches for purification of wastewater.

'Nanomaterials' is a word is made by two words; first is 'nano,' which means very small; and second one is 'material,' which means compound. Nanomaterials are based on nanoscience and nanotechnology. They deal with the synthesis, characterization, and application of nanostructure materials. Nanomaterials atoms size range 10−9m to 10−7m (1 to 100nm). They have interesting and useful behavior. Nano structures are bridges between molecules and infinite bulk systems. Individual nanostructure includes quantum dots, nano wires, and nanotubes, etc. While a collection of nanostructures involves arrays, assemblies, and super lattices of the nanostructures. The uniqueness of nanomaterials structure consists of chemistry of mechanism, dynamics on the basis of nanoscience which is the chemistry (science) only applied for nanomaterials.

Preface

Due to their capacity to combine a variety of properties, including electronic, magnetic, optical, and catalytic, multifunctional nanoparticles are gradually gaining more and more attention. They are highly functioning materials having altered properties that may differ greatly from those of the original materials. Core-to-shell nanoparticles' characteristics can be altered by altering either the constituent materials or the core-to-shell ratio. Multifunctional nanoparticles exhibit the distinctive qualities of the many materials used in combination to satisfy the various application requirements. Functionalization serves a variety of purposes, including surface modification, the capacity to improve functionality, dispersibility, and stability, controlled release of individual nanoparticles, lower material consumption, etc.

One of the most fascinating nanomaterials with intriguing uses is multifunctional nanoparticles (NPs). A crucial building component in nanotechnology is inorganic core nanoparticles covered in organic monolayers. The inorganic core's nanoscopic size gives it special semiconductor, conductive, optical, and magnetic capabilities that are only found in quantum-confined materials. These methods can be used to create fluorescent biomacromolecule tags for activity assays, MRI contrast agents, and parts for nanoscale electronic devices or data storage components.

For use in detection and diagnostic procedures, inorganic NPs have photochemical, photothermal, and magnetic characteristics. Additionally, a range of pharmacological or active agent payloads can be complexed with their stiff outer surface and the strong core. In image contrast and magnetic resonance, magnetic NPs, such as superparamagnetic iron oxide particles, have been utilised to identify tumours. These molecules can be complexed with targeted capabilities that can improve the diagnostic ability of these NPs due to surface modifications including surfactant coating and bioconjugation between the carboxyl and amine groups.

Target-specific delivery via surface ligand decoration, simultaneous achievement of cancer therapeutics and diagnostics, and co-delivery of multiple bioactive(s) with imaging agents are just a few examples of how multifunctional nanoparticles can accomplish multiple goals more effectively than conventional nanoparticles.

These NPs, which can be used for theranostic systems in the form of dendrimers, micelles, liposomes, carbon nanotubes (CNTs), quantum dots (QDs), iron oxide NPs, gold nanoparticles (AuNPs), and silica NPs, are mostly composed of proteins, peptides, lipids, polymers, metals, and their oxides. Optical imaging, magnetic resonance imaging (MRI), ultrasound, computed tomography (CT), positron emission tomography (PET), and single-photon emission computed tomography are some of the imaging modalities.

Therefore, this book is providing the most recent developments, hot topics, challenges, and critical reviews in the field of multi-functional nanomaterials in order to aim the ongoing advancement in the synthesis of multifunctional nanomaterial composites and their applications. In order to highlight the significant research accomplishments in the multi-functional nanomaterials, it is anticipated that this collection of chapters would boost the contributions' high awareness. Additionally, the book focuses on every facet of nanoscience and nanotechnology.

- Development of novel procedures for the synthesis of multifunctional nanomaterial, nanocomposites, carbon based nanomaterials and other nanomaterials by fine-tuning synthesis conditions to get the optimum
- Innovative multifunctional nano materials are being developed for the use as biomarker
- Elaboration of multifunctional nanomaterial can be used for the removal of dyes, hazardous chemicals, and heavy metals;
- High shows the higher selectivity

SOLUTION

The multifunctionality of nanomaterials is highlighted by the principal applications of strongly linked to the Nanomaterials. Applications of nanomaterials in soft tissues, EMI shielding, spintronic, information storage, magnetic data storage, memory device applications, energy harvesting using piezoelectric polymers and nanomultiferroics, nonlinear optical limiting using graphene or ferrite nanoparticles, and even applications in sunscreens, cosmetics, and food packaging will be covered. An important field of contemporary pharmaceutics is the development of efficient processes and materials for targeted transport, controlled release of bioactive chemicals, and diagnostic probes. Treatment times are shortened while unfavorable effects and side effects associated with systemic administration are avoided by effective drug deposition in the target site and controlled activation of therapeutic and imaging agents. Various magnetic nano- and microcapsule kinds are being created right now for MRI, controlled medication release, hyperthermia, and magnetically assisted delivery.

BOOK ORGANIZATION

In this book, we have selected the 8 research and review articles for publication. The chapters in this book reflect a wide range of fundamental and applied research in water and wastewater treatment by the nanomaterials and interdisciplinary subjects. This book is a unique collection of full research papers as well as reviews.

In the first chapter, discuss about the development of innovative nanomaterials for multifunctional application in photo-catalysis, electro-chemical sensing etc. are gaining extraordinary importance worldwide. In this sense, several metal chalcogenides nano-materials, such as FeS, CdS, ZnS, Fe_3O_4 etc. nano-materials and their composite with Prussian blue modified have been synthesized and established as effective heterogonous recyclable catalyst for conducting several photochemical and electrochemical reactions. When the nano-materials have ability to absorb solar light energy, the performance for photo-catalytic reactions has been increased much more. In this chapter it has been elaborated that how the solar energy can be used to drive different organic reactions pathway for green synthesis as well as to develop decomposition of organic pollutants in waste-water treatment. Another interesting function of the same nano-materials in the field of bio-technology has been expounded based on its superior electro-catalytic and bio-sensing activity towards hydrogen peroxide and glucose for determination of diabetic disease.

In the second chapter, the present global concern associated with the energy crisis and environmental pollution can be addressed through semiconductor photocatalysis technology by converting solar energy into green fuels. WO_3 is considered as a favorable candidate due to its durability in the oxidation reaction, primarily in acidic pH, absorption of the visible photon ($\lambda \leq 480$ nm), cost beneficiary, and less toxicity. However, due to the undesirable position of its conduction band, WO_3 suffers from photo excited charge-carrier recombination and shows lower photo-response for the reduction process. To improve the activity under visible light, various approaches have been implemented, such as band alignment and restriction of charge-carrier recombination. This review summarizes the numerous endeavors like manipulation of crystal facets and morphologies, incorporation of dopants, and composite formation through hetero-junctions using various semiconductors.

In the third chapter, discuss about the photocatalysis is a traditional approach to solve energy crises and environmental pollution control. Graphitic carbon nitride (g-C_3N_4), a 2-D conjugated polymer, along with different other forms of carbon based components like carbon nano-tubes, carbon quantum dots, carbon nano-sheets, carbon nano-wires etc. acts as nano-mediator and serves as promising candidates with bismuth-based oxides for applications in environmental and energy sectors leading to sustainable development. The latest efforts towards overcoming the limitations for improvement in performances of these materials can be controlled by different factors e.g. surface morphology, band gap, size, crystallinity, charge separation efficiency and rate of the charge transfer etc. Here, we primarily focus on the detailed strategies related to the fabrication of different carbon components incorporated bismuth oxides based nano-composites for applications in sustainable development.

In the four chapter, Water scarcity has been exacerbated by environmental pollution caused by increasing global industrialization and urbanization. This has led continued research for more ecologically safe and cleaner technologies for eliminating pollutants. In the recent past, nano-engineered metal oxides such as titanium oxide (TiO_2) have shown promising photocatalytic effect of completely mineralizing organic pollutants to form harmless byproducts. TiO_2 is the most frequently used photocatalyst, however its effectiveness is limited to light absorption within the ultra violet region due its wide band-gap. A redshift by TiO_2 has been achieved through tuning of the anatase-rutile phase transitions, doping, surface modifications and creation of multi component heterojunctions. This chapter discusses the synthesis of TiO_2–based photocatalysts and photocatalytic activity enhancement of TiO_2 through doping and formation of hybrids. Their application in the photo-degradation of dyes and other organic pollutants as well as their limitations and future prospects are also addressed.

In the fifth chapter, Metallic nanoparticles (NPs) have great potential for biological and chemical sensor applications because the surface of the NPs has a sensitive spectral response and the light signal can be easily monitored by their absorption or scattering. In particular, silver nanoparticles (Ag NPs) are used in various applications due to their advanced physicochemical properties such as their excellent antibacterial, catalytic, and conductivity at the nanoscale. Plant-mediated synthesis methods have come to the fore with the green chemistry approach in the synthesis of NPs. The aim of this chapter is aimed to provide a detailed resource by considering the studies in the literature on the use of Ag NPs produced by green routes mainly in the fields of photocatalysis, sensing, and imaging. Additionally, this chapter aims to determine the importance and potential of plant-mediated synthesized Ag NPs in these applications will be determined and to be a guide for industrial and academic personnel working in this field.

In the sixth chapter, Air contamination and a growing number of harmful gases releasing into the environment is a big issue of present era. Gases like CO, NH_3, CO_2, SO_2, and CH_4 can prompt lethal health threats. Gas sensors have attained noteworthy consideration of researchers in monitoring such kind of pollutants. Detection of gases by sensors typically depends upon the sensing properties, like sensitivity, selectivity, response and recovery time. With the progress of the Internet-of-Things technology, the uses of gas sensors in the fields of smart homes, wearable gadgets, and smart mobile terminals have developed rapidly In this chapter, various groups of the sensing materials are outlined thoroughly, comprising gas sensors based on multifunctional materials such as; metals, graphene, cellulose, MXenes, acoustic wave sensors and catalytic sensors. This chapter also outlines the current accomplishments in the field of gas detection and signifies the current challenges and future outlooks in this field.

In the seventh chapter, every aspect of civilization, including its growth and sustainability on both political and socioeconomic level, is significantly impacted by access to clean water. As a result, maintaining water supplies and treating wastewater have become major global challenges. Water pollution is caused by a variety of gases, microbes, and other poisons that are added to water during rainfall, running water, etc. Studies into the concurrent removal of several contaminants or coexisting contaminants from water by nanoparticles pave the door for a novel route that is not only affordable but also devoid of secondary contamination. Hence, decontaminating water that has been polluted by hazardous and toxic compounds is a topic that generates a considerable interest among researchers. This review article discusses various types of nanomaterials as a photocatalysts that can effectively manage contaminated water. By analyzing earlier studies to investigate the toxic compounds present in water, this chapter also highlights the various water treatment approaches for purification of wastewater.

In the eighth chapter, discuss about the nanomaterials. ''Nanomaterials'' word is made by two words first one is nano that means very small and second one is material that mean compound. Nanomaterials based on nanoscience and nanotechnology. They deal with the synthesis, characterization and application of nanostructure materials. Nanomaterials atoms size range 10^{-9}m to 10^{-7}m (1 to 100nm). They have interesting and useful behavior. Nano structures are bridge between molecules and infinite bulk systems. Individual nanostructure includes quantum dots, nano wires and nanotubes etc. While collection of nanostructures involves arrays, assemblies and super lattices of the nanostructures. The uniqueness of nanomaterials structure consists of chemistry of mechanism, dynamics on the basis of nanoscience which is the chemistry (science) only applied for nanomaterials.

Azad Kumar
Department of Chemistry, Siddharth University, Kapilvastu, Siddharthnagar, (U.P.), India

Acknowledgment

I am highly thankful to **Dr. Shilpi Srivastav**, Department of Chemistry, Siddharth University, Kapilvastu, Siddharthnagar, (U.P.), India. for their support and guidance during the publication of this book. I am highly thankful to **Dr. Laxman Singh**, Department of Chemistry, Siddharth University, Kapilvastu, Siddharthnagar, (U.P.), India. for their support and guidance during the publication of this book. I would like to express my deep and sincere gratitude to **Victor Shikuku**, Kaimosi Friends University Kenya, **Wilfidah Nyairo**, Maseno University, Kenya, **Dr. Asma Rafiq**, Government, College University Lahore, Pakistan, **Chinmoy Bhattacharya** Department of Chemistry, Indian Institute of Engineering Science & Technology (IIEST), Shibpur, Howrah-711103, West Bengal, for their support and guidance during the publication of this book. I am also thankful to all authors who contribute the chapter for this book. I am also thankful to publisher, IGI Global publisher and their entire team for their invaluable support and encouragement.

I find no way to express in word my deep gratitude and profound reverence to my Parents and all my relatives. This undertaking would not have even been contemplated, let alone be completed, without their prayers, their blessings, their encouragement and the confidence they have showed in me. I recollect all the moments, they have shared with me all these years and waited patiently to see my research this destination in life.

Editorial Advisory Board

Chapter 1
Metal Chalcogenides-Based Multifunctional Nanomaterials for Solar-Light-Driven Photo-Catalysis and Electro-Chemical Sensing

Amit Kumar Dutta
https://orcid.org/0000-0002-1235-766X
Bangabasi Morning College, University of Calcutta, India

ABSTRACT

Development of innovative nanomaterials for multifunctional application in photo-catalysis, electro-chemical sensing etc. are gaining extraordinary importance worldwide. In this sense, several metal chalcogenides nano-materials, such as FeS, CdS, ZnS, Fe_3O_4 *etc., nano-materials and their composite have been synthesized and established as effective heterogenous recyclable catalyst for conducting several photochemical and electrochemical reactions. When the nano-materials have the ability to absorb solar light energy, the performance for photo-catalytic reactions has been increased much more. In this chapter it has been elaborated that how the solar energy can be used to drive different organic reactions pathway for green synthesis as well as to develop decomposition of organic pollutants in waste-water treatment. Another interesting function of the same nano-materials in the field of bio-technology has been expounded based on its superior electro-catalytic and bio-sensing activity towards hydrogen peroxide and glucose for determination of diabetic disease.*

DOI: 10.4018/978-1-6684-8743-3.ch001

INTRODUCTION

As the dimension of the nano-scaled materials so small, new interesting properties have been generated such as interesting chemical and physical properties, large specific surface to volume ratio and high catalyst loading capacity (S. K. Maji, 2014). That is why, several research groups have already paid considerable attention to develop several nano-scaled inorganic materials as multifunctional. Among them, transition-metal chalcogenides-based nanomaterials behaves as multifunctional nanomaterials and are found to have wide applications in the area of catalysis, water splitting, water purification, gas sensing, lithium-ion battery production, biomedical applications, biological imaging etc (L.H. Long, 1999). Most of the chalcogenides are semiconducting in nature and have proper redox potentials, excellent electron transfer ability and outstanding stability in air and solutions (M. R. Hoffmann, 1995). Some nano-composites have been developed based on Prussian blue or Chitosan or Nafion modified metal chalcogenide where Prussian blue or Chitosan or Nafion can enhance the multifunctional efficiency as a whole (A.K. Dutta, 2012).

Photocatalysis is one of the most useful methods where light energy is used to conduct a chemical reaction (Ge *et al.*, 2018). Actually, light energy is the driving force for a chemical reaction. When solar light energy is used in photochemical reaction it will be most useful alternative sustainable energy source as the abundance is virtually unlimited and free (A.K. Dutta, 2012). So, focus needs to be done on improving environmentally friendly, low-energy, and cost-effective technologies for solar-light-driven photo-catalysis. Recently, following naturally occurring photosynthesis process, researchers want to utilise solar light-driven photocatalysis in different organic reactions pathway for green synthesis.

S. Das, 2014 has reported chalcogenide nano-materials, CdS, CuS nanoparticles, as photo-sensitizer to synthesise 2-substituted benzothiazole, benzimidazole under light irradiation at room temperature. Again, Solar light-driven photocatalysis is also now widely used in remediation and restoration of the contaminated water in waste water treatment (Rafiq *et al.*, 2021). Several organic pollutants such as pesticide product, pharmaceutical product, textile effluents, and other organic compound (Hasija *et al.*, 2019) which are highly soluble in water and low-biodegradable, makes severe adverse impacts on life and the environment of the planet. In that sense metal chalcogenides nano-materials show excellent light energy conversion capability due to presence of semiconducting behaviour and direct band gap potential (M. R. Hoffmann, 1995). When the nano-materials are highly porous in nature, they have high catalyst loading capacity and maximum light absorption capability. TiO_2-based materials are well known catalyst and widely has been used for photo-degradation of organic pollutants over the past decade, however, it is catalytically active only under UV irradiation ($\lambda < 400$ nm) because of its wide band gap energy (Eg $\approx$ 3.2

eV) (K. Nakata, 2012). Of late, several non-titania-based materials such as CuO, ZnO, FeS, ZnS etc. (M. Sun, 2008) and even doped hetero-nanostructures Bi_2S_3/BiOCl, Eu-doped Bi_2S_3 (A. Sarkar, 2015) have been found to exhibit solar-light-driven catalytic activity.

Sensing is another important application of nanomaterials in the field of bio-technology for detection and estimation of hydrogen peroxide and glucose levels in foods and blood. Again, substantial amount of hydrogen peroxide is excreted in human urine. Owing to the presence of superior electron transfer ability of the nanomaterials, electrochemical rapid current response has been widely used for sensing application (Rivas *et al.*, 2007). FeS, CuS are one of the most useful electro-catalyst due to presence of Fe^{2+}/Fe^{3+} or Cu^{1+}/Cu^{2+} redox active centres. Again, in the nano-composite with Prussian blue or Chitosan or Nafion modified, which provide more redox active centre, the charge transfer ability and electrochemical behaviour increases much more. Very recent, Y. Cui, 2007 works on various enzymatic sensing techniques and proposed enzyme/protein-modified electrode for determination of human blood glucose level and exposed the entry of different non-enzymatic nano-scaled inorganic materials as hydrogen peroxide and glucose sensor in the bio-medical field. As, the metal chalcogenides nanomaterials are generally chemically and biologically inert, non-toxic, much more stable to biodegradation, and not susceptible to denaturation, have been used as hydrogen peroxide and glucose sensor.

Keeping this in mind, In the subsequent episode, we have enlarged on details about the photo-catalytic activities of several metal chalcogenides nanomaterials and there uses in green organic synthesis, effective photo-active devices for decomposition and remediation of organic pollutants in waste-water treatment. Based on the superior electro-catalytic activities of the nano-material, this chapter also presented how a simple and cheap electrochemical method have been developed for the measurement of hydrogen peroxide and human blood-sugar level.

Nano-Scaled Metal Chalcogenides

As effective heterogeneous photo-catalyst and sensor at a time, the most widely studied nano-sized metal chalcogenides include transition metal sulfides, selenides, oxides such as FeS, CdS, ZnS, CuS, Cu_2S, FeSe, Fe_2O_3, Fe_3O_4 nanoparticles (NPs). Different synthesis methods are available to prepare different shape, size, morphologies even different forms such as nano-particles, nano-tubes etc. that are important factors to affect their catalytic performances.

Iron sulfide (FeS) NPs with different morphologies ie. nanospheres and nanorods have been synthesized from single source precursor complex $[Fe(ACDA)_3]$ [where the ligand is 2-aminocyclopentene-1-dithiocarboxylic acid (HACDA)] through solvothermal way at 120 °C (S. K. Maji *et al.*, 2012) using different nucleophilic solvents Ethylene-glycol (EG), Ethylene-diamine (EN) or Ammonia (NH_3). To improve photo-catalytic performance, the shape size morphology has been controlled during the crystal growth process, where crystal nuclei preferentially grow along a unique direction to produce nano-rods in presence of EN or aggregate into quasi-spherical structures to produce nano-spheres in presence of EG and NH_3. As a result, the band gap energy of different FeS NPs has been automatically tuned in between 2.75 to 3.13 eV which has been measured through UV-Vis absorption spectroscopy.

CdS nano-crystals (S. Das *et al.*, 2013) were prepared by hydrothermal refluxing a cadmium(II) precursor complex with ligand pyridine-2-carboxamide and thiourea inthe presence of excess Ligand.

To prepare another Semiconducting nano-materials, ZnS NPs, an important II–VI semiconductor, S.K. Maji *et al.*, 2011 has chosen the same solvothermal decomposition of precursor $[Zn(ACDA)_2]$ using different nucleophilic solvents. Nanorods (diameter ~6 nm and length ~50 nm) and Nano-sphere (~5 nm) have been deposited in presence of EN (20 ml) and [HDA (15 ml) + TOP (5 ml)], respectively.

Similarly, for synthesis of Cu_2S NPs, $[Cu(ACDC)_2]$ precursor complex has been used and apply same solvothermal way using two nucleophilic solvents, viz. EN and TETA at 110 °C.

Ag_2S quantum dots was obtained using precursor complex Ag(ACDC) through solvothermal way in ethylene glycol solvent at 120° C within 30 min.

γ–Fe_2O_3 NPs (A.K. Dutta *et al.*, 2014) has been prepared from Fe(III) precursor complex $[Fe_3(\mu_3\text{-}O)(O_2CC_6H_5)_6(H_2O)_3]NO_3$ where the complex in solid state just burned inside a quartz tube of a horizontal tubular furnace at 670°C for 1 h.

Iron oxide (Fe_3O_4) NPs have been synthesized from Fe(III) precursor complex single source precursor complex $[Fe_3(\mu_3\text{-}O)(O_2CCH_2Cl)_6(H_2O)_3]NO_3{\cdot}H_2O$ through hydrothermal way using Teflon-lined stainless steel autoclave at 150 °C. Prussian Blue (PB) modified nanocomposites was prepared using above Fe_3O_4 NPs and ultrasonicated in aqueous solution of $K_3[Fe(CN)_6]$ for about 20 min followed by addition of $FeCl_3$ solution. Finally, the resultant mixture was again ultrasonicated for 30 min.

The Powder X-Ray diffraction (XRD) pattern confirmed the formation of pure magnetite (Fe_3O_4) (JCPDS ID. 190629) phase of iron oxide with one major orientation along the (311) plane (Figure 1a). The corresponding XRD of PB–Fe_3O_4 composite shows the presence of a few peaks, denoted by asterisks, in addition to those of Fe_3O_4, due to presence of PB molecules. The transmission electron microscopy (TEM) and high-resolution transmission electron microscopy (HRTEM) of the nano-composite (Figure 1b), confirm the attachment of PB molecules at the corner of the Fe_3O_4 NPs to form a hetero–nanostructure.

Figure 1. (A) XRD pattern (a) Fe_3O_4 and (b) PB–Fe_3O_4 NPs. '' marks denote the peaks corresponding to PB molecule (B) TEM and inset HRTEM image of the PB–Fe_3O_4 hetero–nanostructure*

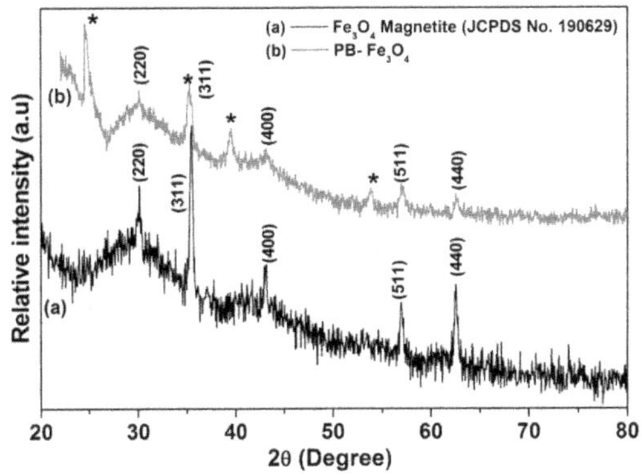

(a)

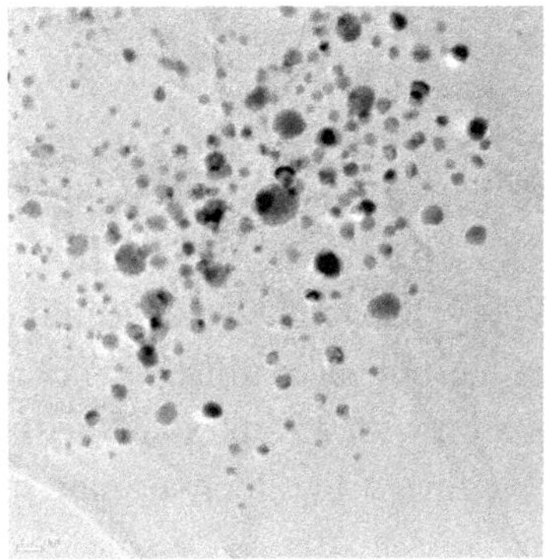

(b)

FeS NPs (A.K. Dutta *et al.*, 2012) has been prepared using hydrothermal decomposition of same Fe(III) precursor complex in presence of thiourea and PVP at 150 °C under nitrogen atmosphere

To synthesize FeSe NPs (A.K. Dutta *et al.*, 2012), an additional compound Na_2SeO_3 has been used along with the above Fe(III) precursor complex followed by same hydrothermal decomposition at same conditions.

Another precursor complex $[CuL_2(H_2O)_2]Cl_2$ [HL = pyridine 2–carboxamide] (A.K. Dutta *et al.*, 2013) has been used as copper source to prepare cupric sulfide (CuS) nanoparticles via solvothermal technique using thiourea at pH=8. Here, the ligand pyridine-2-carboxamide itself acts as capping agent.

Photo-Catalytic Green Synthesis of 2-Substituted Benzothiazoles

We know that visible light energy is the driving force for a chemical reaction. Using photo-catalytic property of nano-materials we can synthesis several organic compound under visible/solar light illumination. Our research group (S. Das, 2014) has synthesised 2-substituted benzothiazole from various aryl and alkyl aldehydes in presence of mesoporous CdS nanosphere under light irradiation at room temperature.

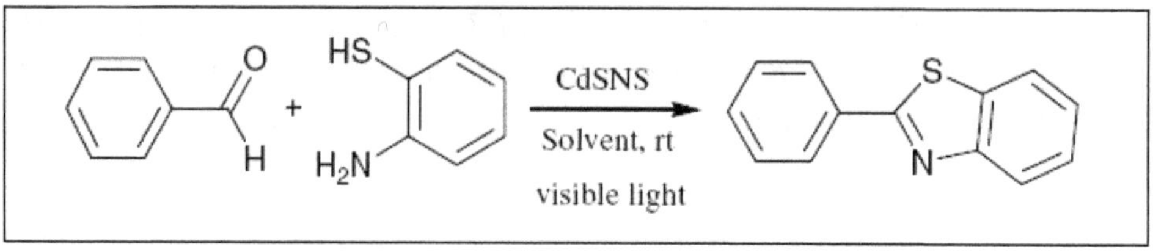

Scheme 1

When benzaldehyde and 2-aminophenol are chosen as starting materials, the reaction was completed within 30 min in methanol as solvent and in presence of 5 mg CdS catalyst with 96% product yield according to scheme 1. To standardise the reaction, if we change the aldehyde part, most of the aldehydes gave excellent yields over 90%.

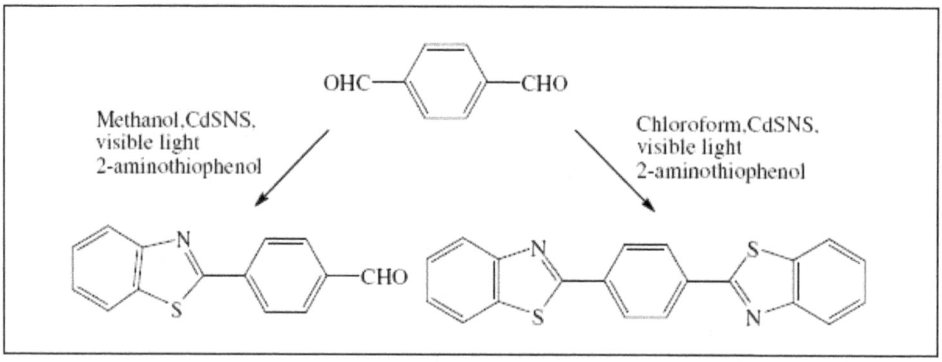

Scheme 2

It can also easily be synthesized di-benzothiazole, 4-benzothiazole-yl-benzaldehyde, from terephthalic aldehyde according to scheme 2. The possible mechanism of this photocatalytic pathway has been described based on radical mechanism which can be initiated by band structure of CdS semiconductor and electron-hole formation upon visible light illumination according to following mechanism (scheme 3). The radical mechanism has been confirmed using radical scavenger, TEMPO (2,2,6,6-tetramethylpiperidinoxyl).

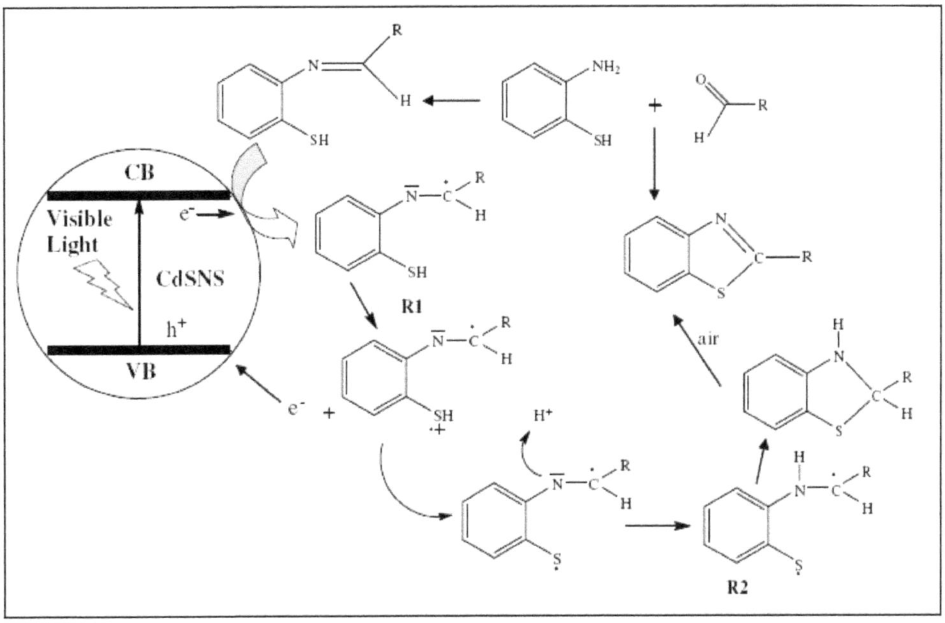

Scheme 3

Photo-Catalytic Degradation of Water Pollutants

To measure the Photo-catalytic activity of the proposed nano-scaled materials, water soluble and non-biodegradable organic pollutants, Rose bengal (RB), methylene blue (MB) dyes and phenol-based pollutants, have been chosen as example because they are commonly used in industries. The catalytic performances have been investigated spectro-photometrically, where absorption curves decrease rapidly with the irradiation time (Figure 2A) catalysed by γ–Fe_2O_3 NPs and the corresponding solution became colourless (Figure 2A inset) within few hours. A comparative performance has been illustrated in Figure 2B with γ–Fe_2O_3 NPs and commercial TiO_2 or commercial tungsten (VI) oxide (WO_3) as an example.

To measure exact time taken for degradation and remediation of polluted water, the decomposition processes have been modelled as a pseudo–first order reaction kinetics expressed by the equation (1)

$$ln(C_0/C_t) = k\,t \qquad (1)$$

where C_0 represents the initial concentration, C_t denotes the concentration at a given reaction time 't', and k is the reaction rate constant. The data have been fitted

Figure 2. (A) UV–Vis spectral changes of RB in aqueous γ–Fe_2O_3 dispersion as a function of irradiation time. (B) Corresponding reaction profile with pseudo–first order kinetics fitting (A.K. Dutta, 2014)

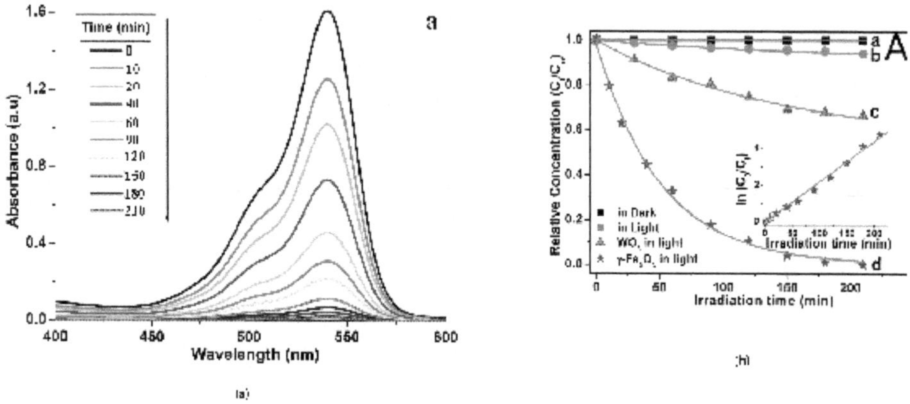

to that kinetic model using a nonlinear least square fitting routine to obtain the rate constant values and summarized in Table 1.

In another studies, the photo–degradation of phenol-based compounds has been carried out using another photo-catalyst, Eu-doped Bi_2S_3 hetero-nanostructure (A. Sarkar, 2015).

Table 1. Comparison of the kinetic parameters for the photo-catalytic activity of the proposed nano-scaled semiconductor nanomaterials

Photo–catalyst	Rate constant with RB (min^{-1})	Rate constant with MB (min^{-1})	Reference
γ–Fe_2O_3	2.15×10^{-2}	8.80×10^{-2}	A.K. Dutta *et al.*, 2014
WO_3	3.3×10^{-3}	4.5×10^{-3}	A.K. Dutta *et al.*, 2014
TiO_2 (Degussa P-25)	1.7×10^{-3}	1.16×10^{-3}	S.K. Maji *et al.*, 2012
CdS (from HDA)	2.2×10^{-2}	–	S.K. Maji *et al.*, 2012
ZnS (from HDA)	2.17×10^{-2}	–	S.K. Maji *et al.*, 2011
FeS (from EG)	–	2.71×10^{-2}	S.K. Maji *et al.*, 2012
FeS	6.02×10^{-2}	5.73×10^{-2}	A.K. Dutta *et al.*, 2012
FeSe	3.39×10^{-2}	2.32×10^{-2}	A.K. Dutta *et al.*, 2012

To understand the mechanism of the photo–catalytic activity of the proposed semiconducting nano-materials, it has been established that during light irradiation, the semiconducting NPs utilizes the incident light energy to excite electron from its valance band (VB) to the conduction band (CB), thus leaving behind hole (Scheme 4). This hole can abstract electron from water bodies to generate hydroxyl radical (OH•) (H. Al–Ekabi, 1989). These radicals actually oxidize or decompose the pollutants.

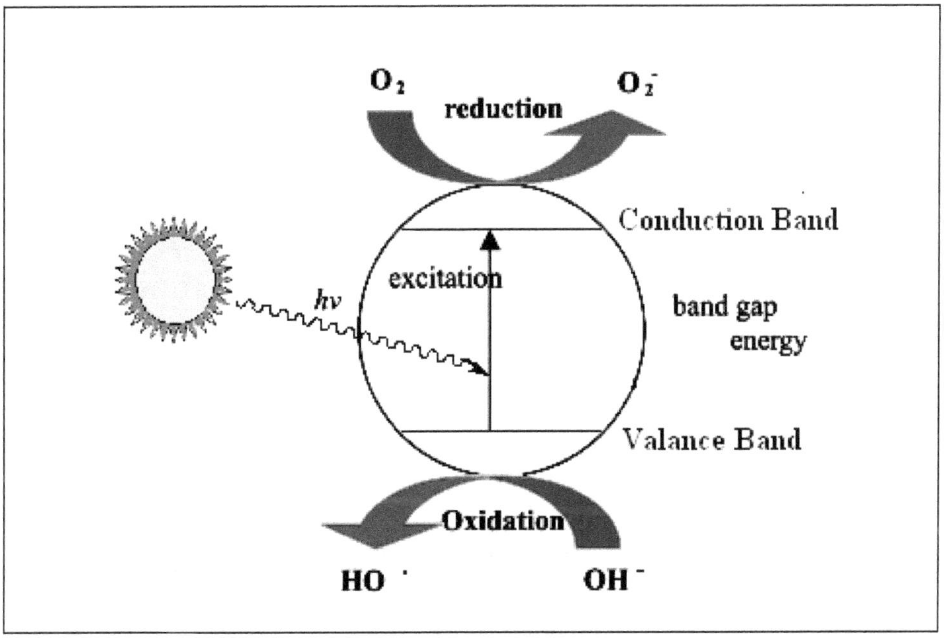

Scheme 4

The overall reactions are presented in equations (2)–(5).

$$\gamma - Fe_2O_3 + hv \rightarrow \gamma - Fe_2O_3\left(e_{CB}^- + h_{VB}^+\right) \quad (2)$$

$$H_2O \rightarrow H^+ + OH^- \quad (3)$$

$$OH^- + h_{VB}^+ \rightarrow OH \quad (4)$$

$$RB / MB + OH \rightarrow Products \quad (5)$$

The aromatic dyes have been further degraded by O_2^- active species, which was generated on the surface of the proposed NPs according to following equations.

$$O_2 + e \rightarrow o_2^- \quad (6)$$

$$O_2^- + 2H^+ + 2e \rightarrow H_2O_2 \quad (7)$$

$$H_2O_2 + e \rightarrow OH^- + OH \quad (8)$$

$$RB / MB + OH \rightarrow Products \quad (9)$$

Electrochemical Sensing

Electrochemical sensing functionality of the proposed nano-materials has been investigated through standard electrochemical set up using three electrode systems.

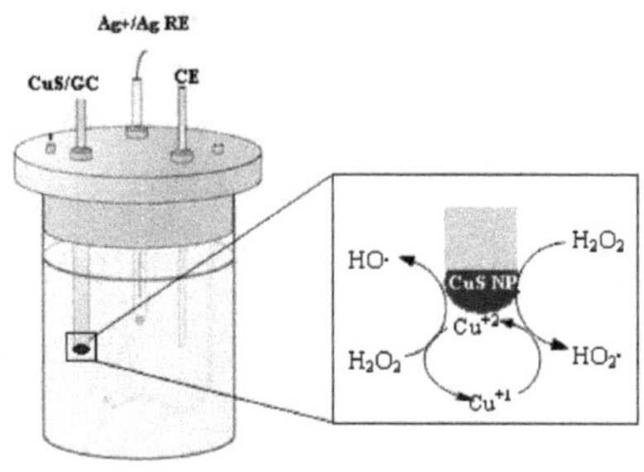

Scheme 5

At first chemically modified electrode like has been constructed through fabrication of nano-materials onto the surface of a glassy carbon electrode (GCE). That NPs/

Figure 3. Cyclic voltammograms (CV) (a) bare GCE and (b) CuS/GCE
(A.K. Dutta, 2014)

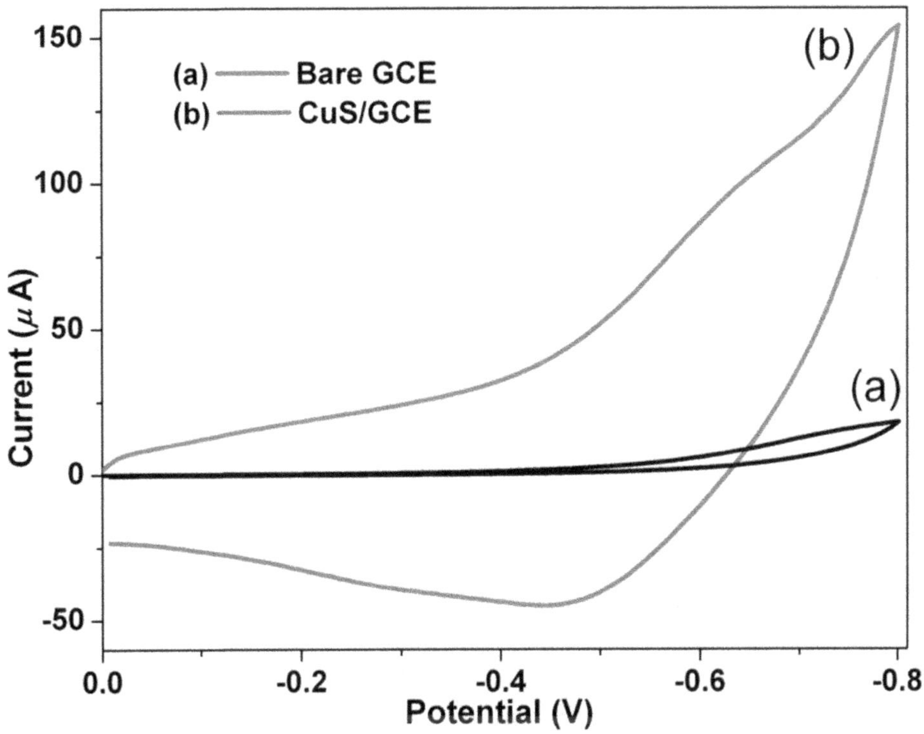

GCE has been used as working electrode along with counter electrode and reference electrode (Standard three electrodes system) in all electrochemical measurements (Scheme 5).

The electro-catalytic activity of the nano-materials have been confirmed through cyclic voltammograms (CV) and Electrochemical impedance spectroscopy (EIS) using the above three electrode cell set up. When the CV and EIS has been investigated with bare GCE and CuS/GCE, the significant current response (Figure 3) and low semicircle diameter in Nyquist diagram confirmed that our propose CuS NPs possess superior electron-transfer ability.

Electrochemical sensing behaviour toward hydrogen peroxide (H_2O_2) has been qualitatively studied by cyclic voltametric measurements with successive addition of H_2O_2 where the reduction peak current of CuS/GCE increases gradually. The quantitative measurement of H_2O_2 has been further carried out through amperometric current–time responses and calibrations (Figure 4). A calibration curve has been constructed based on amperometric current versus H_2O_2 concentration producing a linear range from 10 μM to 1900 μM (Table 2). Using that calibration curve and

Figure 4. Amperometric responses of CuS/GCE in buffer solution (pH 7.0) upon successive addition of Hydrogen peroxide
(A.K. Dutta, 2014)

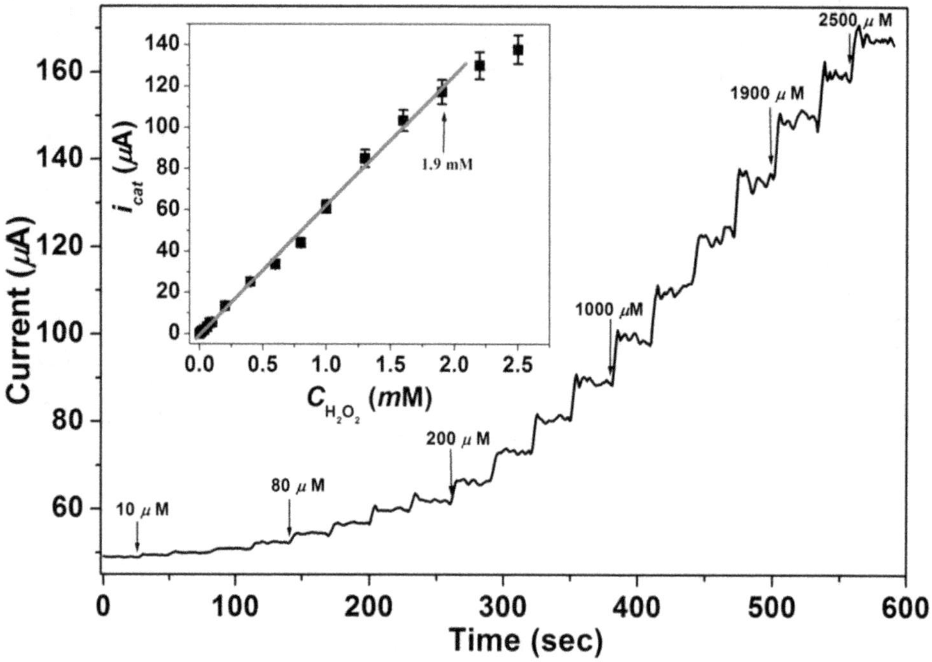

standard addition method, the amount of hydrogen peroxide in real samples such as excreted human urine samples have been measured. The results have been compared with that measured through standard volumetric titration method against standard $KMnO_4$ (Table 3).

Table 2. Comparison of amperometric response parameters towards Hydrogen peroxide using proposed nano-sensors

Nano-sensors	Applied potential (V)	Sensitivity (μA mM^{-1})	Linear range (μM)	Detection limit (μM)
CuS/GC	–0.65	62.93	10–1900	1.1
NCNT/GC	+0.3	24.5	1.76–139	0.37
Cytochromec/Au/C	–0.1	–	10–1000	–
FeS/GC	–0.4	0.85	5 – 140	4.3
FeSe/GC	–0.5 V	19.57	5 – 100	3.0
FeS nanosheet/GC	–0.4V	–	0.5–150	0.092
CdS/GC	–0.6	0.989	1–1900	0.28

Table 3. Determination of hydrogen peroxide in real samples using CuS nano-sensor

Sample	Amperometric method	Reference method ($KMnO_4$ titration)
Urine Sample A	0.128 mM (RSD= 3.2%)	0.136 mM
Urine Sample B	0.175 mM (RSD= 6.14%)	0.193 mM

Similarly, Glucose can be measured using same electrochemical sensing process (S. K. Maji, 2014). Here, an additional compound, biological enzyme glucose-oxidase (GOx), has been used to oxidized glucose to produce H_2O_2 according to following reaction

$$glucose + O_2 \xrightarrow[pH\,7.0]{GOx} H_2O_2 + Gluconic\,acid$$

Then, that H_2O_2 has been employed in the above H_2O_2 electro-catalytic reaction in presence of nano-sensors instead of external H_2O_2 The response plot is linearly correlated to glucose concentration from 10 μM to 3700 μM with a detection limit of 1.2 μM (signal to noise ratio = 3) with the help of Cu_2S nano-sensor (S. K. Maji, 2014).

CONCLUSION

In summary, the proposed nano-scaled materials are widely explored as multi-functional i.e., an efficient photo-catalyst as well as electro-catalyst. They exhibit different catalytic efficiencies based of different size, shape and morphology. The catalytic performances have also been improved by tuning band-gap energy and surface area during synthesis process. Finally, an environmentally friendly, cost-effective technologies have been developed for solar-light-driven photo-catalysis so that most of the solar spectrum and light energies can be utilized in our society for large-scale organic synthesis and waste-water treatment process. Also, non–enzymatic amperometric sensor has been developed based on metal chalcogenide nano-sensors, immobilized on a GC electrode surface for the detection and estimation of hydrogen peroxide and glucose.

ACKNOWLEDGMENT

The Author is grateful to Prof. Bibhutosh Adhikary, Department of Chemistry, Indian Institute of Engineering Science and Technology (IIEST), Shibpur, for helpful discussions. The Author is indebted to UGC, India for financial support through the MRP [PSW–066/15–16 (ERO)]. The author also acknowledges the MHRD (India) for providing instrumental facilities to the Department of Chemistry, IIEST Shibpur and RUSA Scheme of the Department of Higher Education, Government of West Bengal, for providing instrumental facilities to the Department of Chemistry, Bangabasi Morning College, Kolkata.

REFERENCES

Al–Ekabi, H., Serpone, N., Pelizzetti, E., Minero, C., Fox, M. A., & Draper, R. B. (1989). Kinetic studies in heterogeneous photocatalysis. Titania-mediated degradation of 4-chlorophenol in air-equilibrated aqueous media. *Langmuir*, *5*(1), 250–255. doi:10.1021/la00085a048

Barreto, J. C., Smith, G. S., Strobel, N. H. P., McQuillin, P. A., & Miller, T. A. (1995). Terephthalic acid: A dosimeter for the detection of hydroxyl radicals in vitro. *Life Sciences*, *56*(4), 89–96. doi:10.1016/0024-3205(94)00925-2 PMID:7823778

Cui, Y., Barford, J. P., & Renneberg, R. (2007). Development of an interference-free biosensor for glucose-6-phosphate using a bienzyme-based Clark-type electrode. *Sensors and Actuators. B, Chemical*, *123*(2), 696–700. doi:10.1016/j.snb.2006.10.004

Das, S., Samanta, S., Maji, S. K., Samanta, P. K., Dutta, A. K., Srivastava, D. N., Adhikary, B., & Biswas, P. (2013). Visible-light-driven synthesis of 2-substituted benzothiazoles using CdS Tet. *Lett.*, *54*(3), 1090–1096. doi:10.1016/j.tetlet.2012.12.044

Dutta, A. K., Maji, S. K., & Adhikary, B. (2014). γ–Fe_2O_3 nanoparticles: An easily recoverable effective photo-catalyst for the degradation of rose bengal and methylene blue dyes in the waste-water treatment plant. *Materials Research Bulletin*, *49*, 28–34. doi:10.1016/j.materresbull.2013.08.024

Dutta, A. K., Maji, S. K., Srivastava, D. N., Mondal, A., Biswas, P., Paul, P., & Adhikary, B. (2012). Synthesis of FeS and FeSe Nanoparticles from a Single Source Precursor: A Study of Their Photocatalytic Activity, Peroxidase-Like Behavior, and Electrochemical Sensing of H_2O_2. *ACS Applied Materials & Interfaces*, *4*(4), 1919–1927. doi:10.1021/am300408r PMID:22448782

Ge, L., Peng, Z., Wang, W., Tan, F., Wang, X., Su, B., Qiao, X., & Wong, P. K. (2018). g-C_3N_4/MgO nanosheets: Light-independent, metal-poisoning-free catalysts for the activation of hydrogen peroxide to degrade organics. *Journal of Materials Chemistry. A, Materials for Energy and Sustainability*, *6*(34), 16421–16429. doi:10.1039/C8TA05488F

Hasija, V., Raizada, P., Sudhaik, A., Sharma, K., Kumar, A., Singh, P., Jonnalagadda, S. B., & Thakur, V. K. (2019). Recent advances in noble metal free doped graphitic carbon nitride based nanohybrids for photocatalysis of organic contaminants in water: A review. *Applied Materials Today*, *15*, 494–524. doi:10.1016/j.apmt.2019.04.003

Hoffmann, M. R., Martin, S. T., Choi, W., & Bahemann, D. W. (1995). Environmental Applications of Semiconductor Photocatalysis. *Chemical Reviews*, *95*(1), 69–96. doi:10.1021/cr00033a004

Long, L. H., Evans, P. J., & Halliwell, B. (1999). Hydrogen Peroxide in Human Urine: Implications for Antioxidant Defense and Redox Regulation. *Biochemical and Biophysical Research Communications*, *262*(3), 605–609. doi:10.1006/bbrc.1999.1263 PMID:10471371

Maji, S. K., Dutta, A. K., Bhadu, G. R., Paul, P., Mondal, A., & Adhikary, B. (2013). A novel amperometric biosensor for hydrogen peroxide and glucose based on cuprous sulfide nanoplates. *Journal of Materials Chemistry. B, Materials for Biology and Medicine*, *1*(33), 4127–4134. doi:10.1039/c3tb20846j PMID:32260965

Maji, S. K., Dutta, A. K., Biswas, P., Srivastava, D. N., Paul, P., Mondal, A., & Adhikary, B. (2012). Synthesis and characterization of FeS nanoparticles and their photocatalytic, electrocatalytic and biomimic peroxidase behavior. *Applied Catalysis A, General*, *419*, 170–177. doi:10.1016/j.apcata.2012.01.025

Maji, S. K., Dutta, A. K., Srivastava, D. N., Paul, P., Mondal, A., & Adhikary, B. (2011). Effective photocatalytic degradation of organic pollutant by ZnS nanocrystals synthesized via thermal decomposition of single-source precursor. *Polyhedron*, *30*(15), 2493–2498. doi:10.1016/j.poly.2011.06.029

Maji, S. K., Dutta, A. K., Srivastava, D. N., Paul, P., Mondal, A., & Adhikary, B. (2012). Peroxidase-like behavior, amperometric biosensing of hydrogen peroxide and photocatalytic activity by cadmium sulfide nanoparticles. *Journal of Molecular Catalysis A Chemical*, *358*, 1–9. doi:10.1016/j.molcata.2012.03.007

Maji, S. K., Sreejith, S., Mandal, A. K., Dutta, A. K., & Zhao, Y. (2014). Synthesis of Ag_2S quantum dots by a single-source precursor: An efficient electrode material for rapid detection of phenol. *Analytical Methods*, *6*(7), 2059–2064. doi:10.1039/c3ay42264j

Nakata, K., & Fujishima, A. (2012). TiO_2 photocatalysis: Design and applications. J. Photochem. Photobiol. C Photochem. *Journal of Photochemistry and Photobiology C, Photochemistry Reviews*, *13*(3), 169–189. doi:10.1016/j.jphotochemrev.2012.06.001

Rafiq, A., Ikram, M., Ali, S., Niaz, F., Khan, M., Khan, Q., & Maqbool, M. (2021). Photocatalytic degradation of dyes using semiconductor photocatalysts to clean industrial water pollution. *Journal of Industrial and Engineering Chemistry*, *97*, 111–128. doi:10.1016/j.jiec.2021.02.017

Rivas, G. A., Rubianes, M. D., Rodriguez, M. C., Ferreyra, N. F., Luque, G. L., Pedano, M. L., Miscoria, S. A., & Parrado, C. (2007). Carbon nanotubes for electrochemical biosensing. *Talanta*, *74*(3), 291–307. doi:10.1016/j.talanta.2007.10.013 PMID:18371643

Sarkar, A., Ghosh, A. B., Saha, N., Dutta, A. K., Srivastava, D. N., Paul, P., & Adhikary, B. (2015). Enhanced photocatalytic activity of Eu-doped Bi_2S_3 nanoflowers for degradation of organic pollutants under visible light illumination. *Catalysis Science & Technology*, *5*(8), 4055–4063. doi:10.1039/C5CY00473J

Sun, M., Li, D., Li, W., Chen, Y., Chen, Z., He, Y., & Fu, X. (2008). New Photocatalyst, Sb_2S_3, for Degradation of Methyl Orange under Visible-Light Irradiation. *The Journal of Physical Chemistry C*, *112*(46), 18076–18081. doi:10.1021/jp806496d

Chapter 2
Advancement of Photoelectrochemical Properties of WO3-Supported Nano-Materials

Debasish Ray
Indian Institute of Engineering Science and Technology, Shibpur, India

Sangeeta Ghosh
Indian Institute of Engineering Science and Technology, Shibpur, India

Swarnendu Baduri
Indian Institute of Engineering Science and Technology, Shibpur, India

Himanshu Sekhar Sahoo
https://orcid.org/0009-0000-3328-4466
Indian Institute of Engineering Science and Technology, Shibpur, India

Chinmoy Bhattacharya
https://orcid.org/0000-0003-2370-7108
Indian Institute of Engineering Science and Technology, Shibpur, India

ABSTRACT

The present global concern associated with the energy crisis and environmental pollution can be addressed through semiconductor photocatalysis technology by converting solar energy into green fuels. WO_3 is considered as a favorable candidate due to its durability in the oxidation reaction, primarily in acidic pH, absorption of the visible photon ($\lambda \leq 480$ nm), cost beneficiary, and less toxicity. However, due to the undesirable position of its conduction band, WO_3 suffers from photoexcited charge-carrier recombination and shows lower photo-response for the reduction

DOI: 10.4018/978-1-6684-8743-3.ch002

process. To improve the activity under visible light, various approaches have been implemented, such as band alignment and restriction of charge-carrier recombination. This review summarises the numerous endeavors like manipulation of crystal facets and morphologies, incorporation of dopants, and composite formation through hetero-junctions using various semiconductors.

INTRODUCTION

Due to the high escalating rate of the global population, global energy utilization is supposed to be expanding to cope with their technical evolution in succeeding years. To meet the global energy demand currently, the utmost proportion (~ 80%) of the energy source is extracted from finite resources (fossil fuel) such as coal, natural gas, crude oil and petroleum. The unrestrained uses of fossil fuels as energy sources depleting global storage are non-sustainable, leading to energy crises and various types of pollution observed majorly in drastic climatic degradation in the form of global warming. Hence, to tackle these two global issues, society should have to look for new energy sources which are renewable and eco-friendly, to fulfill global energy requirements with minimal cost and environmental consequences. In the present scenario, the contribution from renewable sources such as solar, geothermal, ocean, biomass, hydropower, and wind is about 89 EJ/annum (about 17% of main power dispenses). Nevertheless, a major proportion of these renewable sources comes up with the employment of conventional biomass and developed hydropower plants, which is insignificant from a sustainability perspective. In this concern, an alternative feasible power source is essential to satisfy energy requirements in an environmentally benign fashion. Hydrogen is thought to be a flawless energy conveyor for the future as it has the potential of high energy density, cleanness, renewable and truly carbon-free. Meanwhile, hydrogen (H_2) is empowered with its oxygen (O_2, a thermodynamically convenient reaction); energy is liberated in addition to water vapour (H_2O) as a by-product (in accordance with the half-cell reaction as $4H^+ + 4e^- + O_2 \rightarrow 2H_2O$ and $2H^+ + 2e^- \rightarrow H_2$) that conquered its environment attentive. In addition to this, hydrogen is also feasible to put in the catalytic transformation of CO_2 towards different chemicals and energy-harnessing sources for alleviating climate change and ocean acidification. Especially, hydrogen is conceivable as an energy prosperous reagent for the exothermic production of hydrocarbons utilizing environmental CO_2 as raw material or even methanol, and methane. Beyond this, hydrogen simultaneously comes with the advantage of diverse storage compatibility as an energy source, such as in gaseous form or in liquefied state and some get trapped in metallic hydrides as well, owing superior higher heating value (HHV) and lower heating value (LHV) in comparison to the various practiced fuel sources.

Figure 1. Schematic representation of currently used various routes for H_2 production

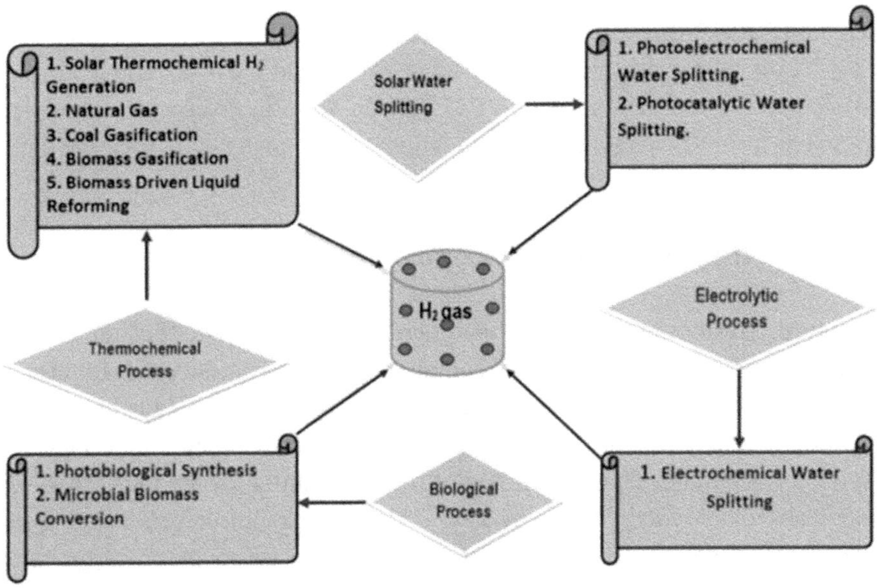

Not only in the form of combustion but also hydrogen can be exploited in fuel cells to produce energy in a chemical reaction fashion. Hence, it may be productively coupled in automobiles and domestics with the ease of portable power sources and substantially various applications. Owing to the aforementioned qualities, hydrogen may be an excellent candidate for the next-generation energy source with environmental concerns. At present, various approaches come into practice (e.g., photo-assisted water splitting, electrolysis of water, hydrocarbon reforming, cracking of petroleum, biological processes, thermal water splitting, etc.), as described in Figure 1, to generate molecular hydrogen (H_2) from various hydrogen-containing materials.

Among the techniques mentioned above, the most governed one in economically favoured fashion for the production of molecular hydrogen is the industrially steam reforming from hydrocarbon (say, methane) and the water-gas shift reaction stands next to it, whereas only 4% is contributed from water, electrolysis. The matter of concern is that the steam reforming of hydrocarbon generates hydrogen along with the mass production of environmental contaminant CO_2 gas; therefore, it is required to transit towards an alternative approach to produce hydrogen in a cleaner and efficient way. Therefore, the production of hydrogen emerges from the energy and material sources, which are cost-effective and eco-friendly with the endorsement of a truly carbon-free manner followed by minimum health clash. The limitless solar energy among the various renewable energy sources can serve as a key to fulfilling the energy requirement without further environmental impairment. It is calculated that to fulfill human society's energy requirement, only a fractional portion (~0.015%) of solar energy outstretched to the earth's surface is sufficient. So, to take into account the limitless solar energy, come up with the liberty to harness it in a convenient form of energy (chemical energy, bond energy) by fabricating suitable photoactive materials.

Photoelectrochemical Water Splitting

To capture and store the limitless solar energy in carbon neutral fashion to prevent the production of any environmental contaminant, and to produce the simplest form of energy hydrogen (H_2) and environmental benign oxygen (O_2) ($H_2O+h\nu \rightarrow 2H_2+O_2$) by splitting water, photoelectrochemical water splitting is classified among the auspicious technique. The expectation of the community for the foundation of the future's sustainable energy economy, such as hydrogen, should have a cleaner essence, possessed by the solar water splitting technology. In a precise manner, the fructification of hydrogen with the assistance of chemical methods through electrolysis of water utilizing solar energy by designing the PEC device is considered to be useful. A concise chronicle of events concerning the exploitation of some inorganic semiconductors for photo-assisted water splitting is adequately demonstrated by Osterloh et al in their report. In 1955, the foremost monstrance step in the domain of Photoelectrochemistry was reported by Brattain et al by exploiting the electrochemical investigation of electrolytic junction of germanium succeeded by the crucial research of Gerischer, Pleskov, and Marcus co-worker. On the other hand, the efficacious operation of water photoelectrolysis in 1971 was brought about by Fujishima and Honda by fabricating PEC cells with TiO_2 nanoparticle semiconductors under illuminating conditions. The momentous work of Fujishima and Honda triggered a tremendous number of researchers towards the domain of water photoelectrolysis, followed by the groups led by Gratzel, Bard, and Nozik, which has produced numerous reports of great impact on this field. Succeeded by the aforementioned immense works, enormous development has been accomplished in the current time to appraise the conceivable and promising candidate to lead photoelectrochemical water splitting more efficiently. The reported articles display that for hydrogen production through PEC, inorganic metal oxide-based semiconductors in the role of photoanode stand out to be well well-equipped materials as they have better durability, electrical properties, and ease of tailoring. Figure 2 shows the conventional PEC water splitting mechanism supported by p-type and n-type semiconductor-based photoelectrodes. The two distinguished halves illustrate in Figure 3 (a) and (b) the electrodics phenomenon,

Figure 2. Illustration of photoelectrochemical solar water splitting reaction in one-step photoexcitation system

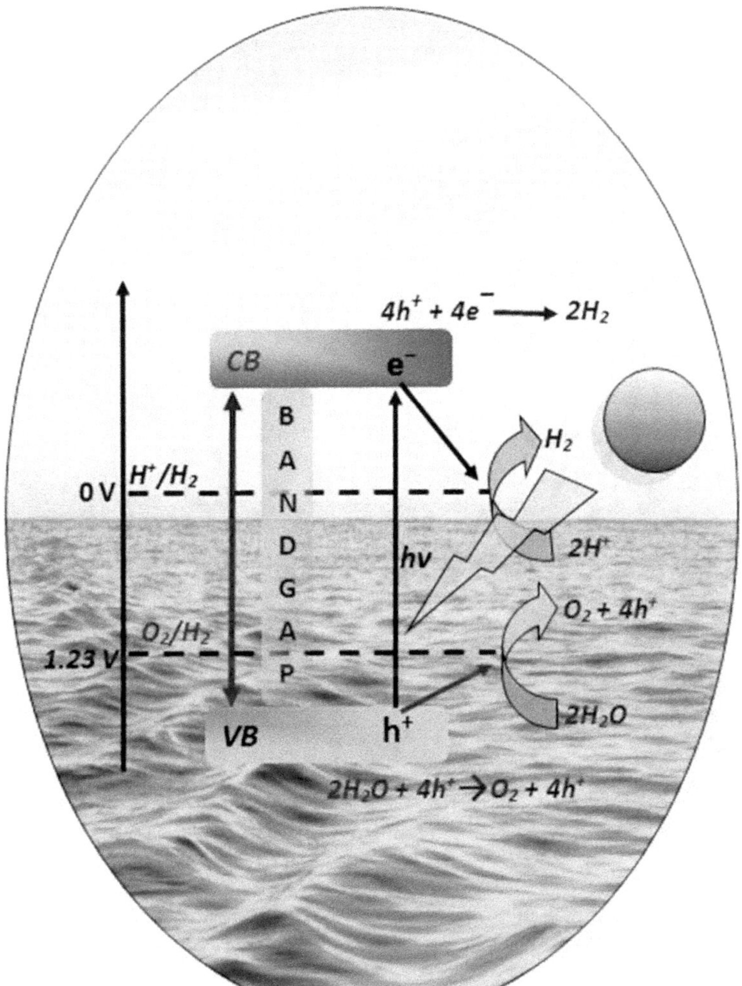

i.e., the evolution of H_2 at photocathode by reducing H_2O and the evolution of O_2 at the photoanode by oxidizing H_2O as displayed in the below equations.

Anodic Oxidation reaction: $2\ H_2O\ (l) \rightarrow O_2\ (g) + 4H^+(aq) + 4e^-$...........(1)

Cathodic Reduction reaction: $2H^+(aq) + 2e^- \rightarrow H_2\ (g)$.............(2)

On illumination, the photoactive material absorbs photon energy greater than its band gap energy and creates excitons (i.e. the photogenerated electron/hole pair). If the semiconductor is of n-type, the photogenerated hole moves to the semiconductor-

Figure 3. Mechanistic approach for Photoelectrochemical a) water oxidation reaction using n-type semiconductor photoanode and b) water reduction reaction with a p-type semiconductor photocathode

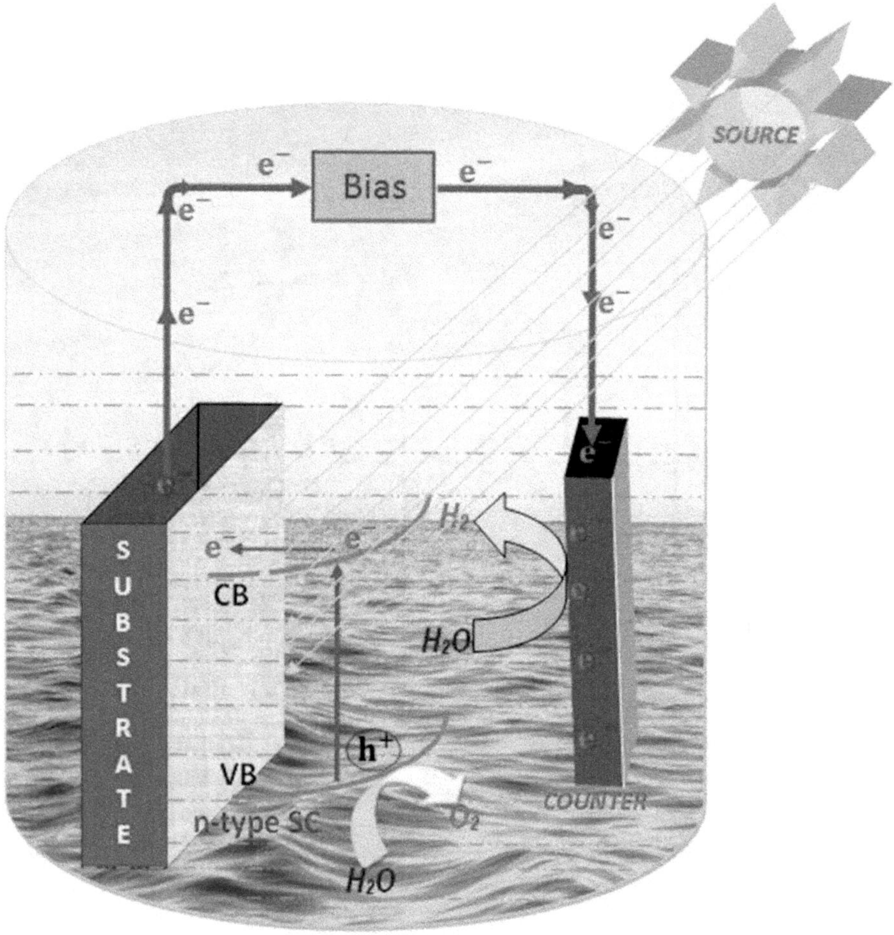

electrolyte interface to oxidize the water to O_2, and the electrons are dragged to the counter electrode (generally Pt in most occasions) by means of external bias to form H_2. But if the semiconductor is of p-type, due to the downward band bending of the interface, the excited electrons are transferred to the semiconductor/electrolyte interface to reduce the H_2O to H_2, and the holes are dragged to the counter electrode to form O_2. As aforementioned, the PEC water splitting technique had a few conveniences over photocatalytic water splitting, such as (i) the produced products H_2 and O_2 are need not be separated as they produced in different chambers; (ii) the recombination of excitons (photogenerated electron/hole pair) are suppressed by self or externally applied bias hence improve the migration of the photogenerated charge to the interface; (iii) conducting substrates were utilized to grow the films, so in future, it can help in mass production for industrialization; (iv) similar to the photocatalytic water splitting it does not demand any external power consumption for stirring. The monstrous efforts have been employed in the cultivation of a convincing model for accomplishing greater efficacy in water splitting, although they had some challenges. The photo-driven water splitting is an uphill reaction with a positive Gibbs free energy of 237 kJ/mol. To acquire such an amount of energy, the semiconductor should have appropriate band gap energy with proper band alignment to absorb photon (visible light) of sufficient energy to bring out the redox (water oxidation and reduction) reaction. To absorb an adequate part of the visible region of the solar spectrum, the band gap of the photoelectrodes must be small (~1.5 eV). The valance band of the semiconductor should be more positive than the water oxidation potential (1.23 V against NHE at zero pH) to oxidize water into oxygen (H_2O/O_2), and the conduction band of the semiconductor should be more negative than the water reduction potential (0 V against NHE at zero pH) to reduce water (H^+/H_2) into hydrogen. Despite that, numerous semiconductors based on inorganic metal oxide are employed for the complete water splitting, although they do not own the requirement, like desired band gap with proper band alignment and sufficient durability. Additionally, in PEC methodology, the system experiences energy loss accompanied by the migration and the recombination of photogenerated excitons (electron/hole pair) in the semiconductor/electrolyte interface or in the bulk solution. Subsequently, for efficacious water splitting, the migration of photogenerated charge, efficient charge separation, and desired band gap with proper band alignment are desirable. Furthermore, the other factors taken into account for the efficiency of overall water splitting are low Debye length of the generated charge, better charge absorption through customized surface morphology, size and shape of the semiconductor via involvement of quantum effects, the position of the flat band, the effect of pH, different ions present in the system and charge transfer processes, etc. Hence, the properties of the photoactive electrodes that own the efficiency of the PEC water splitting for harvesting limitless solar energy for water splitting.

Photoelectrochemical Water Splitting Using WO_3

After the pioneering work of Fujishima and Honda employing TiO_2 nanoparticles numerous no of photocatalysts (both p and n-type) have been tried for photo-assisted water splitting, including main group and transition metals, such as silicon-based oxides and carbides, carbon nitride, transition metal-based oxide, chalcogenides, metal nitrides, respectively. Due to the high chemical durability about neutral and slight acidic pH, including enough earth-abundance with no toxicity with the legacy of tuneable composition made, the tungsten-based oxides (n-type semiconductor) are one of the potential aspirants for the solar light-driven water splitting. The crystal of tungsten-based oxide embodied of perovskite unit and is classified among the most enchanting prospects for photo-assisted water splitting due to its well-positioned band gap energy (2.5-2.8 eV). This allows it to absorb ~12% of the visible region of the solar spectrum with a favourable hole diffusion length of (~150 nm) in contrast with the α-Fe_2O_3 in a few nanometres, including superior charge transporting (12 $cm^2s^{-1}V^{-1}$) property pertaining to TiO_2 (0.3 $cm^2s^{-1}V^{-1}$). Tungsten oxide consolidates the position of the valance band as the valence band is more positive than the water oxidation potential, but as the coin flip to the position of the conduction band minima, it's not able to reduce water to hydrogen as the conduction band minima is not deep enough situated to overcome the reduction potential. Consequently tungsten-based oxides in photo-assisted water splitting only produces $O_{2,}$ not H_2. Hence, to lead the overall water splitting employing tungsten oxide as photoanode, it should be united along with a suitable photocathode for the photo-assisted water reduction for the complete fabrication of the PEC device; otherwise, the tungsten oxide photoelectrode is coupled with the counter electrode with an externally applied bias to drive the electron to counter electrode to accomplish the water splitting to generate both O_2 and H_2 The upward band bending of the electrode/electrolyte interface made the tungsten oxide a suitable choice to exploit as a photoanode. It has already reported that the theoretical maxima of solar to hydrogen conversion efficiency for tungsten oxide is about ~4.5% as a PEC tendon device. It is noteworthy that the greater durability of tungsten oxide in low pH conquers its utility; since at that situation, H_2 production is more efficient. Typically, due to small band gap of the tungsten oxide, in contrast to TiO_2, provides the former the privilege to absorb more in the visible region of the solar spectrum. On the other hand, it portrays a higher stationary photocurrent than the Fe_2O_3 owing to enhanced IPCE (incident photon to current conversion efficiency). Consequently, tungsten oxide is considered more appropriate than Fe_2O_3 and TiO_2 for the application of photoelectrochemical water splitting. As the coin has two faces, the high recombination rate of photogenerated excitons (electron/hole pair) limits tungsten oxide to confront the exigency of practical utilization.

WO_3 Crystal Structure

Typically, the crystalline arrangement of tungsten oxide corresponds to perovskite ABO_3 type in three-dimensional spaces arranged with WO_6 octahedra sharing the corners and edges. Generally, the tungsten oxide exhibits divergence from regular perovskite ABO_3 is observed by virtue of defacement associated with the antiferroelectric dislocation of the W atom and reciprocated twisting of the oxygen octahedral. Consequently, the deformation leads to five stable crystal structures at various ranges of temperature. In the temperature range of –180°C to 900°C, WO_3 distinctly epitomized five different stable crystalline phases, which are inherited with a certain temperature span. At low-temperature range −180°C to −43°C monoclinic-II (ε-WO_3) phase is the dominant followed by the triclinic (δ-WO_3) phase in between −43°C to17°C. With a further rise in temperature from 17°C to 330°C, monoclinic-I with (γ-WO_3) phase dominated over the other phases, whereas more elevation stabilizes the orthorhombic (β-WO_3) phase from 330°C to 740°C, and at high temperature from 740°C to 900°C tetragonal (α-WO_3) is the predominating phase. Since at room temperature, the γ-phase is superior over the others due to better stability. The distinguished band structure corresponding to the various unit cells have been described. The dislocation from the primary octahedral, bringing out the robust enhancement in compression, results in fluctuation of the electronic band gap of the tungsten oxide semiconductor. Hence, the semiconductor band gap energy is inherited from crystallographic arrangement, which ranges from 2.4 eV to 1.6 eV for monoclinic reformatted to normal perovskite-type cubic crystal, respectively. These twisting not only govern the electronic band structure but also have eminent command on the defect states (in between bulk and surface). The monoclinic structure is observed to have a greater degree of durability and prompt for photoelectrochemical water splitting application than the other crystal structures. Additionally, the crystal facet had the same impact on the action of photoelectrochemical water splitting, such as the crystal structure. The needed over potential for water oxidation is owned by the orientation crystal facet over the surface of tungsten oxide. It is already documented that the over-potential of different facets of tungsten oxides for photo-oxidation of water are 1.05 V, 1.10 V, and 1.04 V for (002), (020), and (200) facets, respectively. Typically, in the last decade, it was observed experimentally, in fact, that illumination of light on the (002) crystal phase reflects improved photocatalytic O_2 generation and CO_2 reduction to methane (CH_4). Additionally, tungsten oxides (002) crystal phase is found to have a greater affinity towards the adsorption and deterioration of organic contaminants. The aforementioned enhanced photocatalytic activity as photoanode of tungsten oxide (002) crystal facet upon light illumination chronicled due to expedite photo-induced charge carrier separation and constrains the generation of peroxo moieties commanding the increased photocurrent and improved photo-stability. The free energy diagrams related to various steps for water photo-oxidation onWO_3 photoanode involving the two different facets (200) and (002) have been recorded at constant potential of 1.23 V. It is evident that the overall requirement of free energy for the (200) crystal phase is about 1.62 eV to carry out the desired process ($H_2O \rightarrow O_2$) whereas the (002) phase demands only 1.49 eV for the same. If we go through the step-wise free energy change for different spices formed throughout in the course of the photo water oxidation reaction, the free energy change for the moiety OH* for (002) is about 0.90 eV while for (200) crystal facet, it is only 0.51 eV suggests the generation of OH* is more propitious from (200) phase than the (002). Furthermore, the generation of O* species in the next step from the OH* moiety is more favourable for the (002) phase than that of the (200) in terms of free energy change. Consequently, the (200) crystal facet is overloaded with the more favorable OH* moiety in the course of the reaction, causing the depletion of photoelectrochemical activity and stability of the (200) facet of the tungsten oxide photoanode. The first breakthrough employing WO_3 as a photoanode and Pt as an auxiliary electrode against SCE as the reference electrode to tandem PEC cell was fathered in 1976 by Hodes et al. After that, numerous attempts emerged to upgrade the efficiency of photoelectrochemical water oxidation. The steady enhancement in publications throughout the last decade is evident that the researcher was attracted towards the photoelectrochemical water splitting application using tungsten metal-based oxides. Hence, it turns out to be crucial, to sum up the noteworthy endeavor carried out so far in the documented form on the exploration of WO_3 for the photoelectrochemical and photocatalytic application. In this article, we have summarized the implementation of tungsten-based oxide thin films documented a few years ago in favor of photoelectrochemical water splitting. It emerged from the literature review that the PEC performance of micro and nanostructured tungsten-based oxides as photoanode is strongly influenced by the crystal size, crystal phase, porosity, orientation of crystals, thickness of the tandem film and states of defects in lattice, etc. Hence, this article emerges the immense attempt carried out in the direction of tungsten oxide photo-assisted water splitting on the various subjects such as morphology, and various routes to shape up the fabrication of photoanode.

Different Morphology of WO_3

Morphology plays an essential role in elevating the efficacy of tungsten-based oxide as a photoanode for PEC application. A significant part of the recently reported article on tungsten-based oxide as a photoanode for photoelectrochemical application centralizes the morphology of thin films. Various groups documented numerous types of morphology regarding tungsten-based oxide, revealing that advantages of one phase over another. Notably, nanostructures such as 1D and 2D with high aspect ratios, such as nanotubes (NTs), nanorods (NRs) are observed to be predominant as these nanostructures enjoy the maximum absorption of illuminated light along the axis due to the orthogonal directory, including the advantage of charge migration over shortened radius, paramount the enhanced photoelectrochemical water splitting application due to both effective light absorption and quick charge transfer process. On the other hand, the porous morphology empowers the excitons (photogenerated electron/hole pair) to be assembled across comparatively concise space orthogonally with the depressed charge carrier collection span, which can impart high external and internal quantum yield for accumulation of excitons (photogenerated electron/hole pair). For the WO_3 photoanode, heaps of morphologies were documented by employing various fabricating technology. Some of the morphology has been discuued below to highlight the photoelectrochemical property of WO_3.

1-D Morphology

The favored owned by the one dimensional (1D) morphologies such as nanowires (NWs), nanorods (NRs), an array of branched nanowires (NWs) /nanorods (NRs), nanoneedles, 1D nanostructure with the hierarchical framework, nanoblade with grass-like appearance and tree like hyper-branched, over the other morphologies for photoelectrochemical water splitting application is that, the former invites more light absorption through the orthogonal direction and command electrical footprint for photo-excited charge carrier along the nanostructured array equipped with adequate charge dissolution throughout the diameter. Consequently, the large interfacial contacting surface for the one-dimensional nanostructured-based tungsten oxide as photoanode in comparison to the bulk tungsten oxide leads to enhanced redox active sites and boosts the charge dissolution efficiency and light harvesting characteristics as well.

2-D Morphology

The two-dimensional morphologies such as nanoplate, nanosheets and nanoflake are decorated tungsten oxide photoelectrodes having the advantage of lesser grain boundaries that are supposed to suppress the hole diffusion length to facilitate the charge separation along with the enhanced photo-generated carrier lifetime. These textures also had the advantage of high surface area and adequate film thickness for excellent light harnessing due to the indirect electronic promotion in WO_3 photoanode. The high efficacy of diffusion of photogenerated holes for these unique nanoplate, nanosheets, and nanoflake textured tungsten oxide towards the interfacial region of photoanode/ electrolyte with the concise under-layer that inhibits the dark current. The above-mentioned morphologies of WO_3 photoanode can easily be acquired by featuring the simple hydrothermal technique. Other methods, such as anodization, sputtering and dealloying of Fe-W alloy, were also documented to fabricate such unique morphology for WO_3.

Morphology

Nanorod

To fabricate the well-directed WO_3 nanorod, an elementary and expeditious strategy such as the hydrothermal technique was adopted by the Kalanur group. The merit of the synthesis lies that it is free from any structure-directing agent or seed layering for such specific orientation; only polyperoxo tungstic acid (PTA) was enough to assist the orientation. It is observed that the morphology and the crystalline phase were tuned by varying the reaction conditions like concentration of precursor solution, duration, and temperature of post calcination along with the hydrothermal treatment timing. The phase transition was noticed on heat treatment of nanorod WO_3 over the conducting substrate to obtain a stable monoclinic phase from a less stable orthorhombic phase followed by hexagonal. Among the aforementioned phases, the appreciable photocurrent density of 2.26 mA cm^{-2} was reported for the optimized more stable monoclinic phase of the tungsten oxide nanorod array under the 1.5G illumination along with the highest IPCE of 35% at (λ= 400 nm) at an applied bias of 1.23 V against RHE in presence 0.5M Na_2SO_4 as supporting electrolyte, whereas the other two phases orthorhombic and hexagonal WO_3 nanorod shows inadequate (photocurrent < 10μA cm^{-2}) response. Nevertheless, at lower wavelength, such as 350 nm, the IPCE value reached up to 90%. The amplified photo response of the vertically aligned WO_3 array was favored by the stable monoclinic phase, as featured the controlled assistance to the thin film over conducting substrate.

Nonowire Morphology

Colloid WO_3 nanowire was successfully synthesized by Goncalves et al. in nonhydrolytic media employing WCl_6 and oleyl alcohol as a source of metal and oxygen, respectively. In this condition, the obtained WO_3 nanowire had poor stability with uncontrolled wire length. To tackle both issues, oleylamine was strategically introduced to the colloidal matrix. The instability of the colloid nanowires WO_3 obtained from the WCl_6 is due to the Brønsted acidity associated with its surface sites. Hence, the introduced oleylamine serves as an organic base and is strongly combined with the surface of nanowire WO_3 to govern the growth of the nanoparticles along with the stable colloidal solution. The variable temperature was probed to command the particle size and the phase of the tungsten oxide nanomaterial. The photoanode was fabricated by employing colloidal deposition in toluene through sintering. The sintering of the thin film leads to the change in morphology towards the porous structure. The XRD pattern exploits that the fabricated photoanode at 240°C well index with orthorhombic $WO_3.H_2O$ with greater extent of (002), (020), and (200) crystal facet, among which the preference toward the (200) facet is noticed than the other two facet. The above mentioned facet of tungsten oxide was reported to facilitate the excel charge transfer from the photoactive surface to the water leading to the suppression of exciton (electron-hole pair) recombination and drive towards the improved photocurrent which is in accordance with experimental and theoretical reports carried out over the monoclinic phase of the said photoanode in the domain of photoelectrochemical and photocatalysis. Among the various synthesized nanowire, the orthorhombic WO_3 shows an excellent photocurrent density of 1.96 mA cm^{-2} at an applied potential of 1.23 V against NHE under the illumination of standard 1.5 G and achieve the highest IPCE of 75% at 360 nm under similar condition. Nevertheless, the photoelectrode shows long-term durability. The notable fact is that the smaller grain-sized hexagonal phase shows poor photocatalytic activity than the larger-sized orthorhombic WO_3 nanoparticle due to less grain boundary defect and high porosity of the fabricated electrode.

Rao et al. employed flame vapour deposition (FVD) technique to obtain WO_3 nanotubes and nanowires with high aspect ratios. The course of the reaction was designed in such a way to achieve well-defined uniformed and dense packed tungsten oxide nanowires as well as single and multi-shell nanotubes over the conducting FTO substrate by tuning the substrate temperature in the span of 530°C to 700°C. In comparison to the nanotubes developed from other techniques such as hydrothermal and chemical vapour deposition (CVD), the nanotubes WO_3 photoelectrode obtained from FVD consist of greater areal number density. In between the nanowires and nanotubes grown by the FVD technique, the nanotubes portrayed the foremost PEC behavior consequence of the greater contact area with the FTO along with the large areal number density supports the better light absorption along with the fluent generation and migration of the photo-induced excitons. The optimized nanotube WO_3 photocatalyst was fabricated at 560°C with a growth timing of three hours, showing the saturated photocurrent density of 2.2 mA cm^{-2} under 1.5G illumination in the presence of 0.5 M H_2SO_4 as supporting electrolyte at an applied potential of 2 V with the highest ABPE (applied bias photon to electron efficiency) of 0.21%. The excellence in the PEC performance of the FVD-grown nanotubes and nanowires over the other 1D texture is attributed to their potential toward light absorption by orthogonalizing the light capturing along the long axes in conjugation with the facile charge carrier collection adopting the shortest route.

Branched Nanorod Arrays

Zhang et al. reported the branched nanorod array of WO_3 nanoparticles obtained by the contemporary solvothermal method. During the reaction, ammonium oxalate serves as a capping reagent and it is reported that the synergistic interaction among the ions present {(oxalate as anion ($C_2O_4^{2-}$) and ammonium as cation (NH_4^+)} leads to the formation of branched nanorod morphology. The hydrothermally obtained nanoparticles were first heated in the N_2 atmosphere, followed by heating in the air as the direct annealing in air condition sparks the distortion and crumbles the branched structure. The samples obtained from the annealing only in the presence of N_2 shows poor photocatalytic property as the sample contains NH_4^+ ion, which limits the photo-activity. In contrast, the annealing followed by an oxidizing environment eliminates the NH_4^+ ion. The branched nanorods exhibit enhanced photocatalytic performance than the unbranched one, synthesized employing the similar condition. The branched nanorod exhibits the highest photocurrent density of 1.3 mA cm^{-2} against the applied bias of 1.3 V vs Ag/AgCl. The enhancement in the photoelectrochemical efficacy was attributed to the improved light trapping characteristics along with the larger surface area provided by the branched nanorods photoanode,promotes hole migration at the interface of WO_3/electrolyte. Liu et al. also synthesized the porous branched nanowire WO_3 through one-step hydrothermal method, but the photocatalytic response of the nanorod was not stand in comparison to the photocurrent documented above.

Nano-grassblades

Jin et al. documented the two-step hydrothermal technique to obtain the nano-grassblade like morphology of tungsten oxide thin film onto the conducting substrate. In the initial step, the nanoseed was decorated on the conducting substrate, followed by the nanograss blade generation in the next one, the presence of urea controlls the generation desired morphology of WO_3. The seed layering obtained in the first step was reported to be prominent and play a crucial role in achieving the desired morphology of WO_3 nanoparticles as the seed layer made the reliable adhesion between the substrate and the thin film. Additionally, the growth time for the later step had control over the orientation, morphology, and the exposed facet of the ultimate WO_3 nanoparticles, with no such influence on their optical property. The obtained nanoparticle owned a large aspect ratio oriented perpendicular to the substrate with a single crystalline monoclinic phase. The highest photocurrent density of 1.2 mA cm^{-2} under the illumination of standard 1.5G at a potential of 1.2 V against SCE with a lower onset potential of –0.1V vs. SCE at neutral pH (maintained by 0.1 M KH_2PO_4/K_2HPO_4 buffer solution) for the nanoparticle obtained by second step of hydrothermal treatment for 2 hr. The exhibited enhanced photocatalytic activity of the optimized photoanode is due to the alternation of the specific surface area supporting the enhancement of the interfacial region between WO_3/electrolyte for solar water splitting.

Hierarchical Architectures Consisting of 1D Nanostructures

To achieve the hierarchical architecture composed of 1D nanoparticles, Biswas et al. employed a simple hydrothermal technique for the application of PEC water splitting. The synthesis was carried out using ammonium meta-tungstate as a metal precursor in the presence of different acids like HCl, H_2SO_4, and HNO_3. These acids play the significant character in achieving hierarchical architecture. The irregular sphere-like morphology was obtained without these acids under a similar synthetic condition. The single crystalline phase of 1D WO_3 nanoparticle with flower-like micro arrangements was grown along the preferential direction of (001) facet with the repetition of 1D WO_3 building blocks. The highest photocurrent density is shown by the optimized hierarchical architectures of 1.6 mA cm^{-2} against the applied bias of 1.2 V against Ag/AgCl in 0.5 (M) sulphuric acid solution with the highest IPCE (at λ= 400 nm) of 45% under the similar condition. The advancement in the photo response activity of the hierarchical architecture of tungsten oxide photoanode is featured in the facilitated charge separation through the excellent charge migration among the well-crystalline 1D building blocks.

Hierarchical Tree-Like/Hyperbranched

The hierarchical tree-like nanostructure supported by the decent porosity assists in efficient internal scattering, along with the higher degree of light absorption leading to improved charge collection. Shin et al. employed the modern pulsed laser deposition technology to obtain a particular hierarchical tree-like nanostructured WO_3 photoanode with a greater porosity. The partial pressure of oxygen during the deposition plays a crucial role in tuning the degree of porosity. The XRD analysis shows that the porous hierarchical tree, like WO_3 thin film, had a preferred orientation of growth along the (010) direction. The improved photocurrent density of 1.8 mA cm^{-2} under the illumination of standard 1.5G at a potential of 1.23 V against RHE with the exceeded IPCE value of 70% in the swept range of 350 nm to 400 nm for porous hierarchical tree like WO_3 photoanode is manifested due to the inheritance of greater surface area coupled with the enriched light absorption property. Balandeh et al. also employed the aforementioned technology to fabricate a WO_3 thin film developed with arrayed hyperbranched facilitated with a high aspect ratio (equivalent to the morphology achieved by Shin et al.) The photo response of such specially designed hyperbranched photoanodes manifests a low onset potential of 0.4 V against RHE while they attained the saturation at an applied bias of just 0.8 V (vs. RHE) with the highest photocurrent density value of 2.2 mA cm^{-2}. This photo response is comparable with the response documented by Shin et al. with analogous morphology. Nukui et al. synthesised an interesting Quasi-1D hyperbranched WO_3

thin film having nanotree like morphology by using a simple hydrothermal method using tungsten metal as substrate in acidic media in the presence of nitric acid, oxalic acid, and rubidium sulphate for 30 min at 150°C. The obtained thin film was annealed at 973K to get well crystalline WO_3 with reduced oxygen defects. Nukui et al. compared the oxygen evolution behavior of synthesized hexagonal nanotree like WO_3 photoanode with the dense morphological WO_3 photocalyst in proximity of sacrificial reducing agent. They reported a high quantum efficiency of 3.2% under visible light irradiation with the oxygen generation rate of 9μmol per hour. This advancement in the photo response ability of specially designed WO_3 nanofilms is driven by the selective directional growth of long-range ordered hyperbranched morphology, which facile improved photogenerated charge injection with the higher extent of charge migration.

Doping Of Tungsten Oxide

Doping has been documented as a remarkably efficient technique to remodel the semiconductor adsorption edge to enhance photocatalytic activity at less energized radiation, suggesting a red shift towards the visible region. The fundamental consequence of the foreign element inclusion during the doping is either to modulate the band edge position of the semiconductor or the incorporation of foreign element, alternates the electronic structure of the photocatalyst. Such refinement assists in suppressing the recombination of excitons and enhances charge carrier separation; these, in turn, may lead to betterment of interfacial (electrolyte/ photocatalyst) characteristics. In connection with the tungsten oxide semiconductor, it has already been observed that the inclusion of noble metal, transition metal, non-metal, and rare earth metals, can shrink the band gap energy and impede the generation of recombination site of excitons (e^-/h^+ pair), hence leading to the betterment of photocatalytic performance under illuminating situation.

Transition metal ion doping

The comparable ionic size of the transition metal ion with the semiconductor made them an ultimate choice as impurity or doped. These dopants have the potential to substitute the W^{6+}-ion from the WO_3 lattice, reshuffling the crystalline structure and diminishing the band gap. Feng et al. reported the fabrication of Ti^{4+}-ion incorporated WO_3 nanocuboid employing a simple hydrothermal technique. The lattice structure remains intact unless the amount of dopant reaches the molar ratio up to 10%. Further addition of the Ti^{4+}ion leads to the phase transformation, and a separate tetragonal TiO_2 starts to appear.

Additionally, they confirmed that the efficiency of photo-assisted degradation of RhB for @10% Ti^{4+} ion is higher than that of the @2% and @5% doped Ti^{4+} ion. The incorporation of Ti^{4+} ion as a foreign element creates energy levels in between the band edges, i.e. valence band and conduction band, either by electron accepting or donating. In 2015, Song et al. illustrated that effectual donation is more convenient for the ion having a compatible size. Their group employed a carbon sphere as a template to fabricate Fe-incorporated nanostructured tungsten trioxide with a routine precipitation approach and annealed for 6 hrs at 400°C. The XRD analysis confirmed that no subsidiary phase has been detected as the iron content achieved up to 8%. The UV-vis spectroscopy suggests that the inclusion of iron ion as an impurity develops an impurity level top of the valence band, as the ionic radius of the Fe^{3+} (0.64Å) ion is comparable to the parental W^{6+} (0.62Å) ion can easily fit into the WO_3 crystal and therefore shrinkage the band gap. facilitate the deterioration of organic contaminants such as RhB and methyl red upon visible light illumination (420nm). Meanwhile, the transition metals such as Co^{2+} ion and Nb^{5+} ion of different ionic radii are chosen for inclusion into the WO_3 lattice, and comparable performance was observed. The crystal lattice was considerably distorted with the inclusion of both ions. These deformities prompted supplementary defect in the crystal lattice, for instance, oxygen vacancies, that facilitates the generation of reactive species on the interface and improved photo response.

Tungsten Oxide Nanostructures and Nanocomposites for Photoelectrochemical Water Splitting Forming Composites or Heterojunctions

Pure WO_3 is limited only to absorb some extent of the solar spectrum (~12%) with the sluggish interfacial (electrode/electrolyte) charge migration in addition to the prompt exciton (electron-hole pair) recombination. Therefore, it is vital to address these loopholes to enhance the photoelectrochemical property of pure WO_3. These can be discoursed by coupling with another suitable semiconductor to form a heterojunction. Coupling two n-type semiconductors having proper band edge alignment, i.e., likewise VB and CB energies, enables one to construct a type-II heterojunction which is enriched with the privilege to localize the wave function of electron and holes in the shell or core material, henceforth recuperate the charge separation. On the other hand, to facilitate adequate charge separation, a Z-scheme or p-n junction was fabricated by conjugating an n-type semiconductor and a p-type semiconductor. Till date, numerous of semiconductors was put together with WO_3 to fabricate a heterojunction, for example, $WO_3/NiWO_4$, $WO_3/ZnWO_4$, WO_3/TiO_2, $WO_3/CuWO_4$, NiO/WO_3, WO_3/Bi_2S_3, $WO_3/BiVO_4$, WO_3/Bi_2MoO_6 and $WO_3/\alpha\text{-}Fe_2O_3$ to escalate the photoelectrochemical and photocatalytic performance of the

WO_3 photoelectrode. Apart from binary heterojunction, some ternary heterojunction structures were also documented, like $WO_3/Sb_2S_3/rGO$, in which rGO is employed as the conductive layer to improve the charge transfer kinetics.

All are aware that the electron transportation in the heterojunction material is the direct consequence of the fermi level and band edge position. For the context of tungsten oxide, the efficient way to advance the photoelectrochemical performance is by constructing a type II heterojunction by conjugating WO_3 with other semiconductors having narrower band gap energy. Typically, the photoexcited electrons perhaps easily migrate to the more negative CB of the WO_3 from the more positive CB of the secondary semiconductor. On the other hand, the photogenerated holes at the VB of the tungsten trioxide will start to move towards the more positive CB of the coupled semiconductor having narrower band gap energy (Figure.4). Unlike

Figure 4. Simplified representation of type II heterojunction

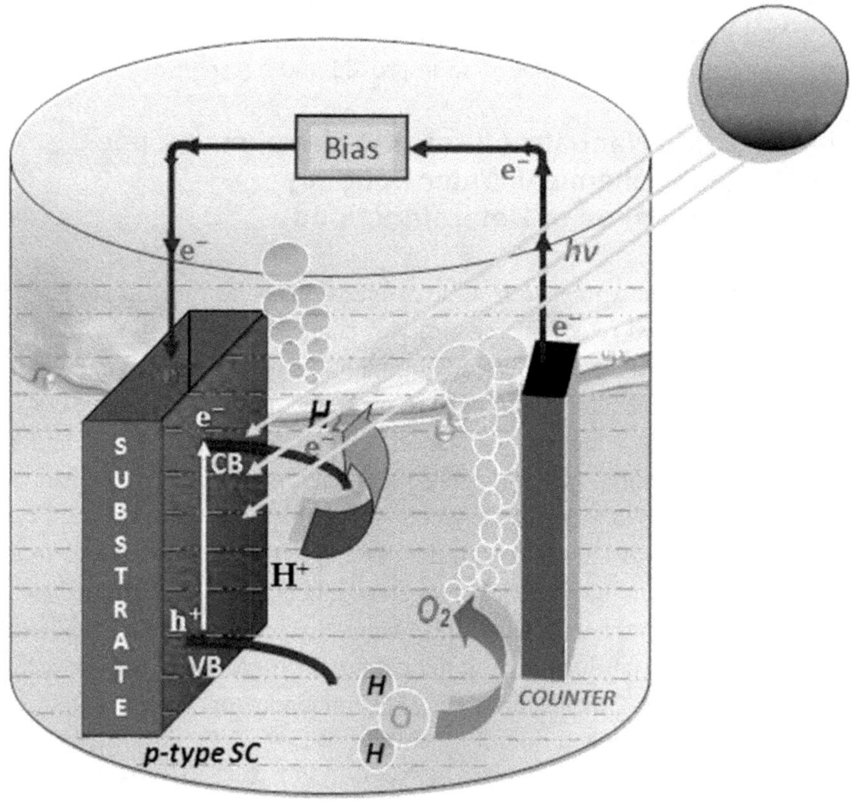

Figure 5. Illustration of photogenerated electron-hole pair separation and transfer through the p-n junction band alignment

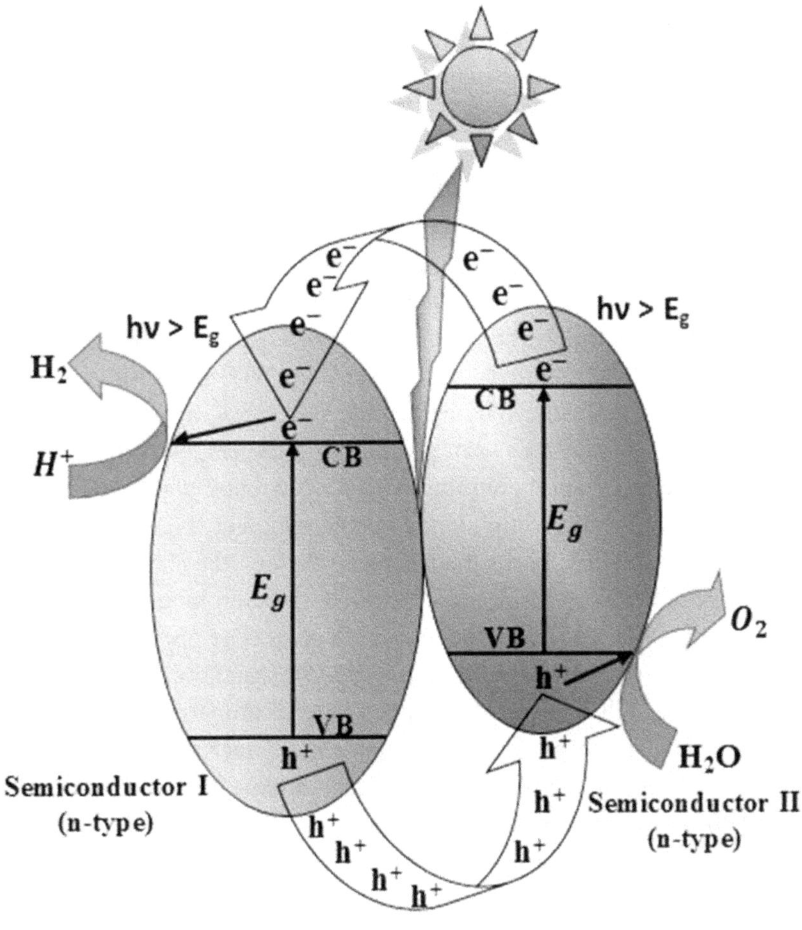

the n-type semiconductor, in which excess electrons engaged to drive the charge carrier while in p-type semiconductors the holes are present in excess compared to the electron to serve for the said carrier. Consequently, in Z-scheme or p-n type heterojunction, (Figure 5), the photogenerated electron at the CB of n-type WO_3 is readily available to recombine with the photoexcited holes originated at VB of the p-type semiconductor. Hence, it is essential to address the optimization of oxidizing and reducing ability in monitoring the PEC.

WO_3/Chalcogenide Heterojunction

In the class of numerous semiconductor materials, metal sulphides were frequently honored as photosensitizers due to their narrower band gap energy for diverse wide bandgap semiconductors. Since majorly metal sulphides have more negative band edge positions of VB and CB than the WO_3, respectively, fastening the WO_3 with metal sulphides to fabricate heterojunction effectively boosts the PEC activity. At first, Zhang et al. documented WO_3/Sb_2S_3 heterojunction photoelectrode. Three types of pristine WO_3 were fabricated by their group (WO_3 microcrystals, WO_3 nanorod, and WO_3 nanoplates), and they coupled these WO_3 with the Sb_2S_3 to get the desired heterojunction via simple hydrothermal technique. The band gap energy of the fabricated semiconductors was evaluated by UV-vis spectroscopy and found to be 2.75 eV for pristine WO_3 microcrystals, 2.82 eV for WO_3 nanorods, and 2.96 eV for WO_3 nanoplates, respectively. Whereas for the WO_3/Sb_2S_3 heterojunction, the band gap energies reduces to 2.28 eV, 2.50 eV, and 2.75 eV, respectively. This reduction in the band gap energy upon coupling with narrow band gap Sb_2S_3 suggests that the advancement in the absorption in the visible region of the solar spectrum could significantly raise the PEC performance. Nevertheless, the charge carrier property at the interface of the WO_3/Sb_2S_3 gets improved for heterojunction due to the facile separation of the photogenerated electron-hole pairs. The photocurrent density for the coupled photoelectrodes WO_3/Sb_2S_3 microcrystals, nanorod and nanoplates were enhanced by 398%, 308%, and 258%, respectively, in comparison to the pristine WO_3 microcrystals, WO_3 nanorod and WO_3 nanoplates. Afterward, Wang et al. lodge a magnificent and efficacious WO_3/Sb_2S_3 heterojunction. Their group fabricates the photoanode by coupling the 3D WO_3 nanoplates together with the Sb_2S_3 nanorods to form a heterojunction employing a simple hydrothermal technique followed by chemical bath deposition (CBD). In accordance with the UV-vis absorption spectra, the WO_3/Sb_2S_3 heterojunction shows improved visible light absorption in comparison to the pristine WO_3. Furthermore, the seed layer supported WO_3/Sb_2S_3 heterojunction shows far better absorption in the visible region of the solar spectrum. Conveniently, compared to the CB band edge position of the pristine WO_3, the CB lies at more negative for the Sb_2S_3, and with stunning contact virtue, the photogenerated electrons can easily migrate from the CB of the Sb_2S_3 to the more positive CB of WO_3, on the other hand, the photo-generated holes are moving swiftly from VB of the WO_3 to more negative VB of the Sb_2S_3. The aforementioned methodology is momentous to intensify the charge transfer efficacy and improve the carrier density with the reduction of charge recombination, hence tailoring the considerable improvement in PEC performance and durability. The seed layer supported WO_3/Sb_2S_3 composites display a remarkably high photocurrent density (5.95 mA cm^{-2} vs RHE at 0.9 V), which is 35 times greater compared to the bare WO_3 (0.17 mA cm^{-2} vs RHE at 0.9

V). For seed layer-supported WO_3/Sb_2S_3 composites, the highest photo-conversion efficiency was observed to be 1.70% (against RHE at 0.76V), which is just 0.03% (vs RHE at 0.92 V) for the bare WO_3.

WO_3/Metal Oxide Heterojunction

As the coin has two faces beside the small band gap energy of metal sulphide semiconductors, their applications as photoelectrode for water splitting were limited by photo-corrosion. Hence, it becomes essential to evolve a novel reduced bandgap semiconductor owing to great photo-corrosion endurance. In this regard, metal oxides can serve as an alternative to the metal sulphide as a photocatalyst to couple with WO_3 to create heterojunction, as they are fairly stable under numerous reaction conditions. In this context, Zhang et al. fabricated a WO_3/Cu_2O heterojunction employing the electrodeposition technique. The physiochemical characterization SEM and TEM revealed the rod-like morphology with the carpet-like layer on the surface of deposited WO_3. To exploit the impact on PEC water splitting behavior they fabricate the heterojunction with a series of variable deposition times. Based on variable composition, they conclude that the highest photocurrent efficiency was portrayed by the sample having a deposition time of 240 s, which was ascribed to the adequate proportion of Cu_2O nanoparticles, which attributed to both the facile charge transfer along the interface and enhanced visible light absorption. The limited fraction of Cu_2O is proficient in generating a lot of interface defects on the WO_3 surface, which intensifies the recombination rate of excitons (photo-generated electron-hole pairs). On the other hand, if the amount of deposited Cu_2O is in excess, it stimulates the lattice scattering and subsequently diminishes the charge carrier concentration, leading to decreased PEC efficiency. Another photoanode heterojunction of $CuWO_4/WO_3$ was fabricated by Zhan et al. and showed that the photoelectrochemical response of the WO_3 has escalated on conjugating with $CuWO_4$ due to the effectual migration of photogenerated charge carrier at the interface of $CuWO_4/WO_3$ heterojunction, prolongation towards the visible light response, and enhanced electron migration properties. Ying et al. decorated the WO_3 photoelectrode with both exfoliated and unexfoliated Bi_2MoO_6 nanosheets to create WO_3/Bi_2MoO_6 heterojunction. It is observed that the WO_3 electrode covered with the exfoliated Bi_2MoO_6 nanosheet shows the highest photocatalytic performance with the excellent photocurrent density of 2.2 mA cm^{-2} at an external bias of 0.8 V against SCE whereas the pristine WO_3 and unexfoliated Bi_2MoO_6 nanosheets shows 1.1 mA cm^{-2} and 1.5 mA cm^{-2}, respectively, under the same condition. This enhancement can be featured in the proper band edge alignment between the WO_3 and Bi_2MoO_6 that alleviates the charge migration with enhanced electron-hole separation. The WO_3 covered with uniformly exfoliated Bi_2MoO_6 nanosheet also favored the high electric mobility.

$WO_3/BiVO_4$ Heterojunction

The coupling of $BiVO_4$ with the WO_3 photoanode to form heterojunction was found to be the most enchanting photocatalyst to investigate over other materials. The enthusiasm lies in coupling $BiVO_4$ with the WO_3 as it advances the charge migration at the interface of $BiVO_4/WO_3$ heterojunction in addition to the improved light absorption ability. Since the band gap energy of $BiVO_4$ lies in the order of 2.4 eV, it serve as an encouraging photoanode with the privilege of fair visible light absorbing proficiency. Not only that, the band position of CB is located just adjacent to the thermodynamic hydrogen evolution potential. However the performance of $BiVO_4$ is limited due to its short hole diffusion length along with the firm recombination of photogenerated charge carriers. It is interesting to note that as the WO_3 surface is coupled with the $BiVO_4$ to create the $BiVO_4/WO_3$ heterojunction, the photoexcited electrons started to flow from the CB of $BiVO_4$ (+0.02V vs RHE) to the more positive CB of WO_3 (+0.41 V vs RHE). On the other hand, the reverse will observed for the hole, i.e., the hole started to stream from VB of the WO_3 to the VB of the $BiVO_4$ because of adjacent band edge alignment among $BiVO_4$ and WO_3. This technique had the potential to sort out the aforementioned limitation due to the recombination of the photogenerated charge carrier by adequately promoting the photogenerated excitons (electron-hole). In general, the $BiVO_4$ acts as a primary light absorber, whereas WO_3 serves as an electron conductor in the $BiVO_4/WO_3$ heterojunction system. The collaborative effect significantly suppresses the recombination rate of photoexcited electron and hole pair leading to the enhancement of the photoelectric conversion performance. Additionally, functioning as a convenient oxygen evolution catalyst, conjugation of $BiVO_4$ manages to inhibit the generation of peroxo species, which in turn extend the durability by hindering the connection between WO_3 and corrosive electrolyte. The first attempt was made by Chatchai et al. to couple $BiVO_4$ with the $WO_{3.}$ Afterward, an extensive no of efforts were made to enhance the PEC performance of $BiVO_4/WO_3$ heterojunction photoelectrode. Lee et al. reported that exploiting immensely ordered 1D WO_3 nanofilm with elevated aspect ratio and active area is an efficient way to enhance the PEC performance of $BiVO_4/WO_3$ heterojunction. Their group fabricated a set of WO_3 nanorods and coupled them with the $BiVO_4$ to form a $BiVO_4/WO_3$ heterojunction. The heterojunction photoanode displays much better photocatalytic efficiency than the pristine $WO_{3.}$ The heterojunction photoelectrode developed by employing facile physical deposition and calcinations technique may lead to the generation of grain boundaries limited its photocatalytic performance, as the grain boundaries enhance the impedance at the interface of the $BiVO_4/WO_3$ heterojunction film. Hence, an in situ fabrication technique was schemed to achieve a high standard of $BiVO_4/WO_3$ heterojunction. The first in situ synthesis of $BiVO_4/WO_3$ heterojunction was documented by Iqbal et al. by practicing simple hydrothermal technique to get single phased nanorods like $BiVO_4/WO_3$heterojunction. In above-mentioned heterojunctions, it was reported that $BiVO_4$ ingrained hooked to the nanorods of $WO_{3.}$ The significantly enhanced photo-response was recorded for the typical nanorod like $BiVO_4/WO_3$heterojunction photoanode. Liu et al. reported the conversion of Bi_2WO_6/WO_3 heterojunction to $BiVO_4/WO_3$ by treating the film hydrothermally with V_2O_5 in an acidic medium. They conclude that the $BiVO_4/WO_3$heterojunction photoanode displays the highest photoelectrochemical performance in comparison to pure WO_3, pure $BiVO_4$, $Bi_2WO_6/WO_{3.}$ $BiVO_4/WO_3$heterojunction fabricated by simple physical deposition. The photoelectrochemical water splitting efficiency is also altered by the change in the proportion of $BiVO_4$to the WO_3in the heterostructure. Seo et al. fabricated the $BiVO_4/WO_3$heterostructure with different proportions of $BiVO_4$ and WO_3 and accomplished the most favorable proportion of the constituents. They observed that when $BiVO_4$appeared at low concentration, the WO_3served majorly as photon absorbing semiconductor; consequently the photoexcited holes were promptly able to recombine at the interface of the heterostructure. On the other hand, when the thickness of the $BiVO_4$layer exceeded 900 nm, the PEC performance will again diminished as the photoexcited excitons are not utilized potentially due to the limitation of charge carrier mobility of $BiVO_4$. Hence, heterostructure with an adequate ratio of both $BiVO_4$and WO_3 will demonstrate a superior PEC activity. In recent times, PEC activity has been improved by synthesizing a special type of nanostructured heterojunction. The specially designed nanostructured core-shell has gained preferential attention due to the potential to maximize each separate processes that impact the PEC activity. Due to the enhanced interfacial and surface area of core-shell structure of $BiVO_4/WO_3$heterojunction, it takes advantage to produce more photoexcitons along with the facile separation of photogenerated carriers. Due to the well-organized cavity embedded framework of core-shell nanostructure, they owned the liberty to absorb multiple reflections from the light source, equipped it to utilize the solar photon more beneficially to advance photocatalytic performance. Hence, it is imperative to formulate an optimized fraction of $BiVO_4/WO_3$core-shell heterostructure to improve the PEC activity. In this context, Rao et al. reported in situ W-doped $BiVO_4/WO_3$core-shell photoanode for improved PEC activity by the combination of drop casting and flame vapour deposition. The supporting nanorod WO_3layer was fabricated by spin coating and followed by flame vapour deposition to maintain $W_{18}O_{49}$ substiochiometry.

The $BiVO_4$is coated by a simple drop casting technique, and reported that W was in situ doped during annealing due to the intimate connection of $WO_3$and $BiVO_{4.}$ The advancement in PEC activity is observed due to the synchronous effect of momentous light absorption and charge segregation. In the core-shell, $BiVO_4$majorly serves as light absorbing material to initiate the charge generation, and WO_3 serve as the charge carrier. Another specially designed yolk-shell like WO_3 was reported by Jin et al. [106]. The obtained yolk-shell WO_3 by means of $PbWO_4$ intermediated $PbWO_4$ was synthesised by the facile hydrothermal technique followed acid treatment to get the desired morphology. The $WO_3/BiVO_4$ heterostructure was achieved by the simple drop-casting technique onto the WO_3 layer and inside of WO_3 and followed by calcinations. The hollowWO_3, spherical yolk-shell WO_3 structure, and yolk-shell $WO_3/BiVO_4$ heterojunctions morphology of the fabricated photoanodes were confirmed by the SEM images. The intimate contact of $BiVO_4$ with the yolk-shell WO_3 to form the $WO_3/BiVO_4$ heterojunction was confirmed HR-TEM and corresponding energy-dispersive X-ray (EDX). The improved PEC performance of yolk-shell WO_3 in compare to the hollow WO_3 was featured in the yolk-shell structure through the efficient charge carrier separation and diminished recombination rate of photogenerated charge carriers. They also documented a decrease in the on-set potential and convenient rise in photocurrent density upon the addition of oxygen evolution reaction (OER) co-catalyst such as FeOOH/NiOOH on the surface of the $WO_3/BiVO_4$ composite. In the technically fabricated yolk-shell $WO_3/BiVO_4/$ OER composite heterojunction, the same role is taken by both the semiconductors to form a typical type-II heterojunction. The photoexcited electrons are easily injected to the more negative CB of the WO_3 from the CB of $BiVO_4$ while in the next photogenerated holes at the VB of the tungsten trioxide will start to move towards the VB of the $BiVO_{4.}$

OUTLOOK AND SUMMARY

The green energy generation by mean of photoelectrochemical water splitting admiringly represent an endurable approach to transform limitless solar energy into bond energy and in this context tungsten based oxides are till now belonging as eminent photoelectrode material for the PEC technology. The photo assisted water oxidation is favourable for tungsten oxide due to the proper alignment of valence band of tungsten oxide, laying lower to the water oxidation potential while the reduction is not supportive due higher positioning of conduction band than the water reduction potential. By the understanding of band alignment tungsten oxide based semiconductors, they are best implemented as photoanode materials in the fabrication of PEC device to perform complete water splitting with an application of external bias. The photoelectrodes complied together with tandems the PEC device are thought to be forthcoming destiny towards green energy. With passes of time, the nanostructured tungsten oxides exhibit noteworthy performance. Due to this a huge no of research articles were documented in literature on the PEC action of tungsten oxide and improvement in PEC performance of tungsten oxides. In recent article, we are addressed the features that manipulate the PEC activity of tungsten oxide thin films by emphasizing the erstwhile attempts. This review discussed the various aspects of WO_3 addressing its crystal facet, morphologies, different ways for improvement of their catalytic performance as photoanode material such as impurity (inclusion of foreign element as dopant), conjugate its with different low band gap semiconductor to form heterojunction to improve charge transfer process etc. As discussed, the proper alignment of the band edges, the electron hole separation and fluent migration of the photogenerated charge carriers are the fundamental characteristics in the crafting of WO_3 photolelectrode for efficient PEC water splitting. Appreciably significant section of the recent review associated to WO_3 PEC water splitting is emphasised on acquiring the distinct morphology to obtained enhanced surface area whereas the specialised architecture help in transportation of photogenerated charge carrriers. Taking these points of interest into consideration of WO_3 photoanodes having various morphologies ranging from 1D to 3D have been fabricated like as nanowires, nanoplates, nanorods, nanosheets, nanoflakes, tree like nanobranched, nano grass blade etc. The recent article reflect that, although accomplishing significant advancement in WO_3 water oxidation performance, still some windows are open for additional encouragement. Some boundaries, just like long time durability of the photoanodes and escalation of the coupled photoanode with the PEC tools is still to encounter, hence channelizing the experimentation in this direction likely to be feasible in respect to the capitalization of the PEC engineering for the renewable fabrication.

ACKNOWLEDGEMENT

The author DR sincerely acknowledges the institute fellowship of IIEST, Shibpur. This work was also financially supported by Department of Science and Technology, Government of West Bengal, India, (Sanction Order No. 1084 (Sanc.)ST/P/S&T/4G-1/2013,dt.08-01-2015) to the Department of Chemistry, IIEST, Shibpur.

REFERENCES

Ahmad, H., Kamarudin, S. K., Minggu, L. J., & Kassim, M. (2015). Hydrogen from photo-catalytic water splitting process: A review. *Renewable & Sustainable Energy Reviews*, *43*, 599–610. doi:10.1016/j.rser.2014.10.101

Ahmed, M., & Xinxin, G. (2016). A review of metal oxynitrides for photocatalysis. *Inorganic Chemistry Frontiers*, *3*(5), 578–590. doi:10.1039/C5QI00202H

Baek, J. H., Kim, B. J., Han, G. S., Hwang, S. W., Kim, D. R., Cho, I. S., & Jung, H. S. (2017). $BiVO_4/WO_3/SnO_2$ double-heterojunction photoanode with enhanced charge separation and visible-transparency for bias-free solar water-splitting with a perovskite solar cell. *ACS Applied Materials & Interfaces*, *9*(2), 1479–1486. doi:10.1021/acsami.6b12782 PMID:27989115

Balandeh, M., Mezzetti, A., Tacca, A., Leonardi, S., Marra, G., Divitini, G., Ducati, C., Meda, L., & Di Fonzo, F. (2015). Quasi-1D hyperbranched WO_3 nanostructures for low-voltage photoelectrochemical water splitting. *Journal of Materials Chemistry. A, Materials for Energy and Sustainability*, *3*(11), 6110–6117. doi:10.1039/C4TA06786J

Bard, A. J. (1979). Photoelectrochemistry and heterogeneous photo-catalysis at semiconductors. *Journal of Photochemistry*, *10*(1), 59–75. doi:10.1016/0047-2670(79)80037-4

Biswas, S. K., & Baeg, J. O. (2013). A facile one-step synthesis of single crystalline hierarchical WO_3 with enhanced activity for photoelectrochemical solar water oxidation. *International Journal of Hydrogen Energy*, *38*(8), 3177–3188. doi:10.1016/j.ijhydene.2012.12.114

Brattain, W. H., & Garrett, C. G. B. (1955). Experiments on the interface between germanium and an electrolyte. *The Bell System Technical Journal*, *34*(1), 129–176. doi:10.1002/j.1538-7305.1955.tb03766.x

Bullett, D. W. (1983). Bulk and surface electron states in WO_3 and tungsten bronzes. *Journal of Physics. C. Solid State Physics*, *16*(11), 2197–2207. doi:10.1088/0022-3719/16/11/022

Cahen, D., Hodes, G., & Manassen, J. (1976). Tungsten trioxide as a photoanode for a photoelectrochemical cell (PEC). *Nature*, 260–312.

Chae, S. Y., Jung, H. J., Jeon, H. S., Min, B. K., Hwang, Y. J., & Joo, O. S. (2014). Morphology control of one-dimensional heterojunctions for highly efficient photoanodes used for solar water splitting. *Journal of Materials Chemistry. A, Materials for Energy and Sustainability*, *2*(29), 11408–11416. doi:10.1039/c4ta00702f

Chae, S. Y., Lee, C. S., Jung, H., Joo, O. S., Min, B. K., Kim, J. H., & Hwang, Y. J. (2017). Insight into charge separation in $WO_3/BiVO_4$ heterojunction for solar water splitting. *ACS Applied Materials & Interfaces*, *9*(23), 19780–19790. doi:10.1021/acsami.7b02486 PMID:28530789

Chandra, D., Saito, K., Yui, T., & Yagi, M. (2013). Crystallization of tungsten trioxide having small mesopores: Highly efficient photoanode for visible-light-driven water oxidation. *Angewandte Chemie International Edition*, *52*(48), 12606–12609. doi:10.1002/anie.201306004 PMID:24123708

Chatchai, P., Murakami, Y., Kishioka, S. Y., Nosaka, A. Y., & Nosaka, Y. (2009). Efficient photocatalytic activity of water oxidation over $WO_3/BiVO_4$ composite under visible light irradiation. *Electrochimica Acta*, *54*(3), 1147–1152. doi:10.1016/j.electacta.2008.08.058

Chen, Q., Li, J., Zhou, B., Long, M., Chen, H., Liu, Y., Cai, W., & Shangguan, W. (2012). Preparation of well-aligned WO_3 nanoflake arrays vertically grown on tungsten substrate as photoanode for photoelectrochemical water splitting. *Electrochemistry Communications*, *20*, 153–156. doi:10.1016/j.elecom.2012.03.043

Chen, X., Zhang, Z., Chi, L., Nair, A. K., Shangguan, W., & Jiang, Z. (2016). Recent advances in visible light driven photoelectrochemical water splitting: Catalyst nanostructures and reaction systems. *Nano-Micro Letters*, *8*(1), 1–12. doi:10.100740820-015-0063-3 PMID:30464988

Chen, Z., Ning, G., Ma, G., Meng, Q., Zhang, Y., Gao, J., Jin, M., Chen, Z., Yuan, M., Wang, X., Liu, J.-M., & Zhou, G. (2017). Effective silicon nanowire arrays/WO_3 core/shell photoelectrode for neutral pH water splitting. *Nanotechnology*, *28*(27), 275401. doi:10.1088/1361-6528/aa749d PMID:28531092

Chiarello, G. L., Aguirre, M. H., & Selli, E. (2010). Hydrogen production by photocatalytic steam reforming of methanol on noble metal modified TiO_2. *Journal of Catalysis*, *273*(2), 182–190. doi:10.1016/j.jcat.2010.05.012

Chiarello, G. L., Bernareggi, M., Pedroni, M., Magni, M., Pietralunga, S. M., Tagliaferri, A., Vassallo, E., & Selli, E. (2017). Enhanced photopromoted electron transfer over a bilayer WO_3 n-n heterojunction prepared by RF diode sputtering. *Journal of Materials Chemistry. A, Materials for Energy and Sustainability*, *5*(25), 12977–12989. doi:10.1039/C7TA03887A

Choi, Y. W., Kim, S., Seong, M., Yoo, H., & Choi, J. (2015). NH_4-doped anodic WO_3 prepared through anodization and subsequent NH_4OH treatment for water splitting. *Applied Surface Science*, *324*, 414–418. doi:10.1016/j.apsusc.2014.10.059

Di Valentin, C., & Pacchioni, G. (2014). Spectroscopic properties of doped and defective semiconducting oxides from hybrid density functional calculations. *Accounts of Chemical Research*, *47*(11), 3233–3241. doi:10.1021/ar4002944 PMID:24828320

Diehl, R., Brandt, G., & Saije, E. (1978). The crystal structure of triclinic WO_3. *Acta Crystallographica. Section B, Structural Crystallography and Crystal Chemistry, 34*(4), 1105–1111. doi:10.1107/S0567740878005014

Duonghong, D., Borgarello, E., & Graetzel, M. (1981). Dynamics of light-induced water cleavage in colloidal systems. *Journal of the American Chemical Society, 103*(16), 4685–4690. doi:10.1021/ja00406a004

Fan, X., Gao, B., Wang, T., Huang, X., Gong, H., Xue, H., Guo, H., Song, L., Xia, W., & He, J. (2016). Layered double hydroxide modified WO_3 nanorod arrays for enhanced photoelectrochemical water splitting. *Applied Catalysis A, General, 528*, 52–58. doi:10.1016/j.apcata.2016.09.014

Feng, C., Wang, S., & Geng, B. (2014). Ti(IV) doped WO_3 nanocuboids: Fabrication and enhanced visible-light-driven photocatalytic performance. *Nanoscale, 3*(9), 3695. doi:10.1039/c1nr10460h PMID:21785781

Fujishima, A., Honda, K. (1972). Electrochemical photolysis of water at a semiconductor electrode. *Nature, 238*, 238037a0.

Gerischer, H. (1966). Electrochemical behavior of semiconductors under illumination. *Journal of the Electrochemical Society, 113*(11), 1174–1182. doi:10.1149/1.2423779

Gonçalves, R. H., Leite, L. D. T., & Leite, E. R. (2012). Colloidal WO_3 nanowires as a versatile route to prepare a photoanode for solar water splitting. *ChemSusChem, 5*(12), 2341–2347. doi:10.1002/cssc.201200484 PMID:23139181

Gondal, M. A., Suliman, M. A., Dastageer, M. A., Chuah, G. K., Basheer, C., Yang, D., & Suwaiyan, A. (2016). Visible light photocatalytic degradation of herbicide (Atrazine) using surface plasmon resonance induced in mesoporous Ag-WO_3/SBA-15 composite. *Journal of Molecular Catalysis A Chemical, 425*, 208–216. doi:10.1016/j.molcata.2016.10.015

Grätzel, M. (2005). Solar energy conversion by dye-sensitized photovoltaic cells. *Inorganic Chemistry, 44*(20), 6841–6851. doi:10.1021/ic0508371 PMID:16180840

Guo, Y., Quan, X., Lu, N., Zhao, H., & Chen, S. (2007). High photocatalytic capability of self-assembled nanoporous WO_3 with preferential orientation of (002) planes. *Environmental Science & Technology, 41*(12), 4422–4427. doi:10.1021/es062546c PMID:17626446

Hains, A. W., Liang, Z., Woodhouse, M. A., & Gregg, B. A. (2010). Molecular semiconductors in organic photovoltaic cells. *Chemical Reviews, 110*(11), 6689–6735. doi:10.1021/cr9002984 PMID:20184362

Huang, F., Yan, A., & Zhao, H. (2016). Influences of doping on photocatalytic properties of TiO_2 photocatalyst. In W. Cao (Ed.), *Semiconductor photocatalysis - materials, mechanisms and applications* (1st ed., p. 51). Intech Open. doi:10.5772/63234

Huang, Z. F., Song, J., Pan, L., Zhang, X., Wang, L., & Zou, J.-J. (2015). Tungsten oxides for photocatalysis, electrochemistry, and phototherapy. *Advanced Materials*, *27*(36), 5309–5327. doi:10.1002/adma.201501217 PMID:26287959

Ibrahim, A. A. M., Khan, I., Iqbal, N., & Qurashi, A. (2017). Facile synthesis of tungsten oxide-bismuth vanadate nanoflakes as photoanode material for solar water splitting. *International Journal of Hydrogen Energy*, *42*(5), 3423–3430. doi:10.1016/j.ijhydene.2016.09.095

Iqbal, N., Khan, I., Yamani, Z. H. A., & Qurashi, A. (2017). A facile one-step strategy for in-situ fabrication of WO_3-$BiVO_4$ nanoarrays for solar-driven photoelectrochemical water splitting applications. *Solar Energy*, *144*, 604–611. doi:10.1016/j.solener.2017.01.057

Izumi, Y. (2013). Recent advances in the photocatalytic conversion of carbon dioxide to fuels with water and/or hydrogen using solar energy and beyond. *Coordination Chemistry Reviews*, *257*(1), 171–186. doi:10.1016/j.ccr.2012.04.018

Jacobsson, T. J., Fjällström, V., Edoff, M., & Edvinsson, T. (2014). Sustainable solar hydrogen production: From photoelectrochemical cells to PV-electrolyzers and back again. *Energy & Environmental Science*, *7*(7), 2056–2070. doi:10.1039/C4EE00754A

Jewell, J., Vinichenko, V., McCollum, D. (2016). Comparison and interactions between the long-term pursuit of energy independence and climate policies. *Nat Energy*, *1*, 201673.

Jiang, C., Moniz, S. J., Wang, A., Zhang, T., & Tang, J. (2017). Photoelectrochemical devices for solar water splitting—Materials and challenges. *Chemical Society Reviews*, *46*(15), 4645–4660. doi:10.1039/C6CS00306K PMID:28644493

Jiao, Y., Zheng, Y., Jaroniec, M., & Qiao, S. Z. (2014). Origin of the electrocatalytic oxygen reduction activity of graphene-based catalysts: A roadmap to achieve the best performance. *Journal of the American Chemical Society*, *136*(11), 4394–4403. doi:10.1021/ja500432h PMID:24580116

Jin, B. J., Jung, E., Ma, M., Kim, S., Zhang, K., Kim, J. I., Son, Y., & Park, J. H. (2018). Solution-processed yolk-shell-shaped WO_3/$BiVO_4$ heterojunction photoelectrodes for efficient solar water splitting. *Journal of Materials Chemistry. A, Materials for Energy and Sustainability*, *6*(6), 2585–2592. doi:10.1039/C7TA08452H

Jin, T., Xu, D., Diao, P., He, W., Wang, H., & Liao, S. (2016). Tailored preparation of WO_3 nano-grassblades on FTO substrate for photoelectrochemical water splitting. *CrystEngComm, 18*(36), 6798–6808. doi:10.1039/C6CE01186A

Kalantar-zadeh, K., Vijayaraghavan, A., Ham, M. H., Zheng, H., Breedon, M., & Strano, M. S. (2010). Synthesis of atomically thin WO_3 sheets from hydrated tungsten trioxide. *Chemistry of Materials, 22*(19), 5660–5666. doi:10.1021/cm1019603

Kalanur, S. S., Hwang, Y. J., Chae, S. Y., & Joo, O. S. (2013). Facile growth of aligned WO_3 nanorods on FTO substrate for enhanced photoanodic water oxidation activity. *Journal of Materials Chemistry. A, Materials for Energy and Sustainability, 1*(10), 3479–3488. doi:10.1039/c3ta01175e

Kalanur, S. S., Yoo, I. H., Eom, K., & Seo, H. (2018). Enhancement of photoelectrochemical water splitting response of WO_3 by means of Bi doping. *Journal of Catalysis, 357*, 127–137. doi:10.1016/j.jcat.2017.11.012

Kalanur, S. S., Yoo, I. H., Park, J., & Seo, H. (2017). Insights into the electronic bands of WO_3/$BiVO_4$/TiO_2, revealing high solar water splitting efficiency. *Journal of Materials Chemistry. A, Materials for Energy and Sustainability, 5*(4), 1455–1461. doi:10.1039/C6TA07592D

Kehl, W. L., Hay, R. G., & Wahl, D. (1952). The structure of tetragonaltungsten trioxide. *Journal of Applied Physics, 23*(2), 212–215. doi:10.1063/1.1702176

Kenney, M. J., Gong, M., Li, Y., Wu, J. Z., Feng, J., Lanza, M., & Dai, H. (2013). High-performance silicon photoanodes passivated with ultrathin nickel films for water oxidation. *Science, 342*(6160), 836–840. doi:10.1126cience.1241327 PMID:24233719

Kwong, W. L., Lee, C. C., & Messinger, J. (2016). Transparent nanoparticulate FeOOH improves the performance of a WO_3 photoanode in a tandem water-splitting device. *The Journal of Physical Chemistry C, 120*(20), 10941–10950. doi:10.1021/acs.jpcc.6b02432

Lai, C. W., & Sreekantan, S. (2013). Preparation of hybrid WO_3-TiO_2 nanotube photoelectrodes using anodization and wet impregnation: Improved water-splitting hydrogen generation performance. *International Journal of Hydrogen Energy, 38*(5), 2156–2166. doi:10.1016/j.ijhydene.2012.12.025

Lang, X., Chen, X., & Zhao, J. (2013). Heterogeneous visible light photocatalysis for selective organic transformations. *Chemical Society Reviews, 43*(1), 473–486. doi:10.1039/C3CS60188A PMID:24162830

Lee, M. G., Kim, D. H., Sohn, W., Moon, C. W., Park, H., Lee, S., & Jang, H. W. (2016). Conformally coated $BiVO_4$ nanodots on porosity-controlled WO_3 nanorods as highly efficient type II heterojunction photoanodes for water oxidation. *Nano Energy*, *28*, 250–260. doi:10.1016/j.nanoen.2016.08.046

Leonard, K. C., Nam, K. M., Lee, H. C., Kang, S. H., Park, H. S., & Bard, A. J. (2013). $ZnWO_4/WO_3$ composite for improving photoelectrochemical water oxidation. *The Journal of Physical Chemistry C*, *117*(31), 15901–15910. doi:10.1021/jp403506q

Li, G., Shrotriya, V., Huang, J. (2005). High-efficiency solution processable polymer photovoltaic cells by self-organization of polymer blends. Nat Mater 4:nmat, 1500.

Li, W., Da, P., Zhang, Y., Wang, Y., Lin, X., Gong, X., & Zheng, G. (2014). WO_3 nanoflakes for enhanced photoelectrochemical conversion. *ACS Nano*, *8*(11), 11770–11777. doi:10.1021/nn5053684 PMID:25347213

Li, W., Liu, C., Yang, Y., Li, J., Chen, Q., & Liu, F. (2012). Plate like WO_3 from hydrothermal RF sputtered tungsten thin films for photoelectrochemical water oxidation. *Materials Letters*, *84*, 41–43. doi:10.1016/j.matlet.2012.06.022

Li, W., Zhan, F., Li, J., Liu, C., Yang, Y., Li, Y., & Chen, Q. (2015). Enhancing photoelectrochemical water splitting by aluminum-doped plate-like WO_3 electrodes. *Electrochimica Acta*, *160*, 57–63. doi:10.1016/j.electacta.2015.01.095

Li, Y. G., Feng, J., Li, H. J., Wei, X. L., Wang, R. R., & Zhou, A. N. (2016). Photoelectrochemical splitting of natural sea water with a-Fe_2O_3/WO_3 nanorod arrays. *International Journal of Hydrogen Energy*, *41*(7), 4096–4105. doi:10.1016/j.ijhydene.2016.01.027

Liew, S. L., Zhang, Z., Goh, T. W. G., Subramanian, G. S., Seng, H. L. D., Hor, T. S. A., Luo, H. K., & Chi, D. Z. (2014). Yb-doped WO_3 photocatalysts for water oxidation with visible light. *International Journal of Hydrogen Energy*, *39*(9), 4291–4298. doi:10.1016/j.ijhydene.2013.12.204

Lin, H. S., & Lin, L. Y. (2017). Improving visible-light responses and electric conductivities by Incorporating Sb_2S_3 and reduced grapheme oxide in a WO_3 nanoplate array for photoelectrochemical water oxidation. *Electrochimica Acta*, *252*, 235–244. doi:10.1016/j.electacta.2017.09.007

Liu, C. J., Li, Y. M., Li, W. Z., Zhu, J., Li, J., Chen, Q. Y., & Yang, Y. H. (2014). CdS quantum dots sensitized plate-like WO_3 photoelectrodes with a TiO_2 buffer-layer. *Materials Letters*, *120*, 170–173. doi:10.1016/j.matlet.2014.01.053

Liu, C. J., Yang, Y. H., Chen, J., Li, S., Li, W. Z., & Tang, X. D. (2017). An in-situ transformation approach for fabrication of $BiVO_4/WO_3$ heterojunction photoanode with high photoelectrochemical activity. *Chemical Engineering Journal*, *326*, 603–611. doi:10.1016/j.cej.2017.05.179

Liu, C. J., Yang, Y. H., Li, W. Z., Li, J., Li, Y. M., & Chen, Q. Y. (2016). In situ synthesis of Bi_2S_3 sensitized WO_3 nanoplate arrays with less interfacial defects and enhanced photoelectrochemical performance. *Scientific Reports*, *6*(1), 23451. doi:10.1038rep23451 PMID:26988275

Liu, X., Wang, F., & Wang, Q. (2012). Nanostructure-based WO_3 photoanodes for photoelectrochemical water splitting. *Physical Chemistry Chemical Physics*, *14*(22), 7894–7911. doi:10.1039/c2cp40976c PMID:22534756

Liu, Y., Zhao, L., Su, J., Li, M., & Guo, L. (2015). Fabrication and properties of a branched $(NH_4)_xWO_3$ nanowire array film and a porous WO_3 nanorod array film. *ACS Applied Materials & Interfaces*, *7*(6), 3532–3538. doi:10.1021/am507230t PMID:25623076

Ma, Y., Jia, Y. L., Wang, L., Yang, M., Bi, Y. P., & Qia, Y. X. (2016). Exfoliated thin Bi_2MoO_6 nanosheets supported on WO_3 electrode for enhanced photoelectrochemical water splitting. *Applied Surface Science*, *390*, 399–405. doi:10.1016/j.apsusc.2016.08.116

Mali, M. G., Yoon, H., Kim, M., Swihart, M. T., Al-Deyab, S. S., & Yoon, S. S. (2015). Electrosprayed heterojunction $WO_3/BiVO_4$ films with nano textured pillar structure for enhanced photoelectrochemical water splitting. *Applied Physics Letters*, *106*(15), 151603. doi:10.1063/1.4918583

Marcus, R. A. (1964). Chemical and electrochemical electron-transfer theory. *Annual Review of Physical Chemistry*, *15*(1), 155–196. doi:10.1146/annurev.pc.15.100164.001103

Martinez, S. C., Hernández, S., & Russo, N. (2015). $BiVO_4$ as photocatalyst for solar fuels production through water splitting: A short review. *Applied Catalysis A, General*, *504*, 158–170. doi:10.1016/j.apcata.2014.11.044

Mehmood, F., Iqbal, J., Jan, T., Gul, A., Mansoor, Q., & Faryal, R. (2017). Structural, photoluminescence, electrical, anti-cancer and visible light driven photocatalytic characteristics of Co doped WO_3 nanoplates. *Vibrational Spectroscopy*, *93*, 78–89. doi:10.1016/j.vibspec.2017.09.005

Mi, Q., Zhanaidarova, A., Brunschwig, B. S., Gray, H. B., & Lewis, N. S. (2012). A quantitative assessment of the competition between water and anion oxidation at WO_3 photoanodes in acidic aqueous electrolytes. *Energy & Environmental Science*, *5*(2), 5694–5700. doi:10.1039/c2ee02929d

Migas, D. B., Shaposhnikov, V. L., Rodin, V. N., & Borisenko, V. E. (2010). Tungsten oxides. I: Effects of oxygen vacancies and doping on electronic and optical properties of different phases of WO_3. *Journal of Applied Physics*, *108*(9), 093713. doi:10.1063/1.3505688

Mohamed, A. M., Shaban, S. A., El Sayed, H. A., Alanadouli, B. E., & Allam, N. K. (2016). Morphology photoactivity relationship: WO_3 nanostructured films for solar hydrogen production. *International Journal of Hydrogen Energy*, *41*(2), 866–872. doi:10.1016/j.ijhydene.2015.09.108

Mu, W., Xie, X., Li, X., Zhang, R., Yu, Q., Lv, K., Wei, H., & Jian, Y. (2014). Characterizations of Nb-doped WO_3 nanomaterials and their enhanced photocatalytic performance. *RSC Advances*, *4*(68), 36064. doi:10.1039/C4RA04080E

Navarro, Y. R. M. (2009). Water splitting on semiconductor catalysts under visible-light irradiation. *ChemSusChem*, *2*(6), 471–485. doi:10.1002/cssc.200900018 PMID:19536754

Ng, K. H., Minggua, L. J., Mark-Lee, W. F., Arifin, K., Jumali, M. H. H., & Kassim, M. B. (2018). A new method for the fabrication of a bilayer WO_3/Fe_2O_3 photoelectrode for enhanced photoelectrochemical performance. *Materials Research Bulletin*, *98*, 47–52. doi:10.1016/j.materresbull.2017.04.019

Nozik, A. J. (1978). Photoelectrochemistry: Applications to solar energy conversion. *Annual Review of Physical Chemistry*, *29*(1), 189–222. doi:10.1146/annurev.pc.29.100178.001201

Nukui, Y., Srinivasan, N., Shoji, S., Atarashi, D., Sakai, E., & Miyauchi, M. (2015). Vertically aligned hexagonal WO_3 nanotree electrode for photoelectrochemical water oxidation. *Chemical Physics Letters*, *635*, 306–311. doi:10.1016/j.cplett.2015.07.006

Osterloh, F. E. (2013). Inorganic nanostructures for photoelectrochemical and photocatalytic water splitting. *Chemical Society Reviews*, *42*(6), 2294–2320. doi:10.1039/C2CS35266D PMID:23072874

Palmstrom, A. F., Santra, P. K., & Bent, S. F. (2015). Atomic layer deposition in nanostructured photovoltaics: Tuning optical, electronic and surface properties. *Nanoscale*, *7*(29), 12266–12283. doi:10.1039/C5NR02080H PMID:26147328

Pleskov, Y. V., & Gurevich, Y. Y. (1986). *Semiconductor Photoelectrochemistry*. Consultants Bureau. doi:10.1007/978-1-4684-9078-7

Porosoff, M. D., Yan, B., & Chen, J. G. (2016). Catalytic reduction of CO_2 by H_2 for synthesis of CO, methanol and hydrocarbons: Challenges and opportunities. *Energy & Environmental Science*, *9*(1), 62–73. doi:10.1039/C5EE02657A

Qi, X. P., She, G. W., Liu, Y. Y., Mua, L. X., & Shi, W. S. (2012). Electrochemical synthesis of CdS/ZnO nanotube arrays with excellent photoelectrochemical properties. *Chemical Communications*, *48*(2), 242–244. doi:10.1039/C1CC15674H PMID:22083259

Qin, D. D., Tao, C. L., Friesen, S. A., Wang, T.-H., Varghese, O. K., Bao, N.-Z., Yang, Z.-Y., Mallouk, T. E., & Grimes, C. A. (2011). Dense layers of vertically oriented WO_3 crystals as anodes for photoelectrochemical water oxidation. *Chemical Communications*, *48*(5), 729–731. doi:10.1039/C1CC15691H PMID:22117192

Rao, P. M., Cai, L. L., Liu, C., Cho, I. S., Lee, C. H., Weisse, J. M., Yang, P. D., & Zheng, X. L. (2014). Simultaneously efficient light absorption and charge separation in WO_3/$BiVO_4$ core/shell nanowire photoanode for photoelectrochemical water oxidation. *Nano Letters*, *14*(2), 1099–1105. doi:10.1021/nl500022z PMID:24437363

Rao, P. M., Cho, I. S., & Zheng, X. (2013). Flame synthesis of WO_3 nanotubes and nanowires for efficient photoelectrochemical water-splitting. *Proceedings of the Combustion Institute*, *34*(2), 2187–2195. doi:10.1016/j.proci.2012.06.122

Rodríguez-Pérez, M., Chacón, C., Palacios-González, E., Rodríguez-Gattorno, G., & Oskam, G. (2014). Photoelectrochemical water oxidation at electrophoretically deposited WO_3 films as a function of crystal structure and morphology. *Electrochimica Acta*, *140*, 320–331. doi:10.1016/j.electacta.2014.03.022

Saito, R., Miseki, Y., & Sayama, K. (2012). Highly efficient photoelectrochemical water splitting using a thin film photoanode of $BiVO_4$/SnO_2/WO_3 multi-composite in a carbonate electrolyte. *Chemical Communications (Cambridge)*, *48*(32), 3833–3835. doi:10.1039/c2cc30713h PMID:22437791

Salje, E. (1976). Structural phase transitions in the system WO_3-$NaWO_3$. *Ferroelectrics*, *12*(1), 215–217. doi:10.1080/00150197608244166

Salje, E. (1977). The orthorhombic phase of WO_3. *Acta Crystallographica. Section B, Structural Crystallography and Crystal Chemistry*, *33*(2), 574–577. doi:10.1107/S0567740877004130

Seabold, J. A., & Choi, K. S. (2011). Effect of a cobalt-based oxygen evolution catalyst on the stability and the selectivity of photo-oxidation reactions of a WO_3 photoanode. *Chemistry of Materials*, *23*(5), 1105–1112. doi:10.1021/cm1019469

Seo, H., Park, G., Oh, K. H., Kang, S. H., Lee, H. C., Cho, S. K., & Nam, K. M. (2017). Analysis of charge separation processes in WO_3-$BiVO_4$ composite for efficient photoelectrochemical water oxidation. *Journal of Electroanalytical Chemistry (Lausanne, Switzerland)*, *789*, 17–23. doi:10.1016/j.jelechem.2017.02.021

Shin, S., Han, H. S., Kim, J. S., Park, I. J., Lee, M. H., Hong, K. S., & Cho, I. S. (2015). A tree-like nanoporous WO_3 photoanode with enhanced charge transport efficiency for photoelectrochemical water oxidation. *Journal of Materials Chemistry. A, Materials for Energy and Sustainability*, *3*(24), 12920–12926. doi:10.1039/C5TA00823A

Song, H., Li, Y., Lou, Z., Xiao, M., Hu, L., Ye, Z., & Zhu, L. (2015). Synthesis of Fe-doped WO_3 nanostructures with high visible-light-driven photocatalytic activities. *Applied Catalysis B: Environmental*, *166–167*, 112–120. doi:10.1016/j.apcatb.2014.11.020

Su, J. Z., Guo, L. J., Bao, N. Z., & Grimes, C. A. (2011). Nanostructured WO_3/$BiVO_4$ heterojunction films for efficient photoelectrochemical water splitting. *Nano Letters*, *11*(5), 1928–1933. doi:10.1021/nl2000743 PMID:21513345

Su, J. Z., Zhang, T., & Wang, L. (2017). Engineered WO_3 nanorods for conformal growth of WO_3/$BiVO_4$ core-shell heterojunction towards efficient photoelectrochemical water oxidation. *Journal of Materials Science Materials in Electronics*, *28*(5), 4481–4491. doi:10.100710854-016-6082-0

Tanisaki, S. (1960). Crystal structure of monoclinic tungsten trioxideat room temperature. *Journal of the Physical Society of Japan*, *15*(4), 573–581. doi:10.1143/JPSJ.15.573

Tee, S. Y., Win, K. Y., & Teo, W. S. (2017). Recent progress in energy-driven water splitting. *Advancement of Science*, 1600337. PMID:28546906

Valdés, Á., & Kroes, G. J. (2009). First principles study of the photo oxidation of water on tungsten trioxide (WO_3). *The Journal of Chemical Physics*, *130*(11), 114701. doi:10.1063/1.3088845 PMID:19317549

Waisman, H., Rozenberg, J., Sassi, O., & Hourcade, J. C. (2012). Peakoil profiles through the lens of a general equilibrium assessment. *Energy Policy*, *48*, 744–753. doi:10.1016/j.enpol.2012.06.005

Walter, M. G., Warren, E. L., McKone, J. R., Boettcher, S. W., Mi, Q., Santori, E. A., & Lewis, N. S. (2010). Solar water splitting cells. *Chemical Reviews*, *110*(11), 6446–6473. doi:10.1021/cr1002326 PMID:21062097

Wang, D. P., Bassi, P. S., Qi, H., Zhao, X., Gurudayal, Wong, L., Xu, R., Sritharan, T., & Chen, Z. (2016). Improved charge separation in $WO_3/CuWO_4$ composite photoanodes for photoelectrochemical water oxidation. *Materials (Basel)*, *9*(5), 348. doi:10.3390/ma9050348 PMID:28773473

Wang, F., Di Valentin, C., & Pacchioni, G. (2012). Doping of WO_3 for photocatalytic water splitting: Hints from density functional theory. *The Journal of Physical Chemistry C*, *116*(16), 8901–8909. doi:10.1021/jp300867j

Wang, G., Ling, Y., Wang, H., Yang, X., Wang, C., Zhang, J. Z., & Li, Y. (2012). Hydrogen-treated WO_3 nanoflakes show enhanced photostability. *Energy & Environmental Science*, *5*(3), 6180–6187. doi:10.1039/c2ee03158b

Wang, N., Zhu, J., Zheng, X., Xiong, F., Huang, B., Shi, J., & Li, C. (2015). A facile two-step method for fabrication of plate-like WO_3 photoanode under mild conditions. *Faraday Discussions*, *176*, 185–197. doi:10.1039/C4FD00139G PMID:25406337

Wang, S., Chen, H., Gao, G., Butburee, T., Lyu, M., Thaweesak, S., Yun, J.-H., Du, A., Liu, G., & Wang, L. (2016). Synergistic crystal facet engineering and structural control of WO_3 films exhibiting unprecedented photoelectrochemical performance. *Nano Energy*, *24*, 94–102. doi:10.1016/j.nanoen.2016.04.010

Wang, S. C., Chen, P., Bai, Y., Yun, J. H., Liu, G., & Wang, L. (2018). New $BiVO_4$ dual photoanodes with enriched oxygen vacancies for efficient solar-driven water splitting. *Advanced Materials*, *20*(20), 1800486. doi:10.1002/adma.201800486 PMID:29602201

Wang, S. C., Chen, P., Yun, J. H., Hu, Y. X., & Wang, L. Z. (2017). An electrochemically treated $BiVO_4$ photoanode for efficient photoelectrochemical water splitting. *Angewandte Chemie International Edition*, *56*(29), 8500–8504. doi:10.1002/anie.201703491 PMID:28516511

Wang, Y. D., Tian, W., Chen, L., Cao, F. R., Guo, J., & Li, L. (2017). Three-Dimensional WO_3 nanoplate/Bi_2S_3 nanorod heterojunction as a highly efficient photoanode for improved photoelectrochemical water splitting. *ACS Applied Materials & Interfaces*, *46*(46), 40235–40243. doi:10.1021/acsami.7b11510 PMID:29067799

Wenzhang, L., Jie, L., Xuan, W., & Qiyuan, C. (2012). Preparation and water-splitting photocatalytic behavior of S-doped WO_3. *Applied Surface Science*, *263*, 157–162. doi:10.1016/j.apsusc.2012.09.021

Wijs, G. A., & de. (1999). Anomalous behavior of the semiconducting gap in WO_3 from first principles calculations. *Physical Review B: Condensed Matter*, *59*(4), 2684–2693. doi:10.1103/PhysRevB.59.2684

Woodward, P. M., Sleight, A. W., & Vogt, T. (1995). Structure refinement of triclinic tungsten trioxide. *Journal of Physics and Chemistry of Solids*, *56*(10), 1305–1315. doi:10.1016/0022-3697(95)00063-1

Wu, P. D., Liu, Z. F., Chen, D., Zhou, M., & Wei, J. D. (2018). Flake-like NiO/WO_3 p-n heterojunction photocathode for photoelectrochemical water splitting. *Applied Surface Science*, *440*, 1101–1106. doi:10.1016/j.apsusc.2018.01.292

Xiao, Q., & Gao, L. (2013). One-step hydrothermal synthesis of C, W-codoped mesoporous TiO_2 with enhanced visible light photocatalytic activity. *Journal of Alloys and Compounds*, *551*, 286–292. doi:10.1016/j.jallcom.2012.10.040

Xie, Y. P., Liu, G., Yin, L., & Cheng, H. M. (2012). Crystal facet dependent photocatalytic oxidation and reduction reactivity of monoclinic WO_3 for solar energy conversion. *Journal of Materials Chemistry*, *22*(14), 6746–6751. doi:10.1039/c2jm16178h

Yan, A., Xie, C., Huang, F., Zhang, S. S., & Zhang, S. S. (2014). An efficient method to modulate the structure, morphology and properties of WO_3 through niobium doping. *Journal of Alloys and Compounds*, *610*, 132–137. doi:10.1016/j.jallcom.2014.04.188

Yang, J., Li, W., Li, J., Sun, D., & Chen, Q. (2012). Hydrothermal synthesis and photoelectrochemical properties of vertically aligned tungsten trioxide (hydrate) plate-like arrays fabricated directly on FTO substrates. *Journal of Materials Chemistry*, *22*(34), 17744–17752. doi:10.1039/c2jm33199c

Zeng, Q. Y., Li, J. H., Li, L. S., Bai, J., Xia, L. G., & Zhou, B. X. (2017). Synthesis of WO_3/$BiVO_4$ photoanode using a reaction of bismuth nitrate with peroxovanadate on WO_3 film for efficient photoelectrocatalytic water splitting and organic pollutant degradation. *Applied Catalysis B: Environmental*, *217*, 21–29. doi:10.1016/j.apcatb.2017.05.072

Zhan, F. Q., Li, J., Li, W. Z., Liu, Y. S., Xie, R. R., Yang, Y. H., Li, Y. M., & Chen, Q. Y. (2015). In situ formation of $CuWO_4$/WO_3 heterojunction plates array films with enhanced photoelectrochemical properties. *International Journal of Hydrogen Energy*, *40*(20), 6512–6520. doi:10.1016/j.ijhydene.2015.03.131

Zhang, H. F., Zhou, W. W., Yang, Y. P., & Cheng, C. W. (2017). 3D WO_3/$BiVO_4$/cobalt phosphate composites inverse opal photoanode for efficient photoelectrochemical water splitting. *Small*, *13*(16), 1603840. doi:10.1002mll.201603840 PMID:28165199

Zhang, J., Ling, Y., Gao, W., Wang, S., & Li, J. (2013). Enhanced photoelectrochemical water splitting on novel nanoflake WO_3 electrodes by dealloying of amorphous Fe–W alloys. *Journal of Materials Chemistry. A, Materials for Energy and Sustainability*, *1*(36), 10677–10685. doi:10.1039/c3ta12273e

Zhang, J., Liu, Z. H., & Liu, Z. F. (2016). Novel WO_3/Sb_2S_3 heterojunction photocatalyst based on WO_3 of different morphologies for enhanced efficiency in photoelectrochemical water splitting. *ACS Applied Materials & Interfaces*, *8*(15), 9684–9691. doi:10.1021/acsami.6b00429 PMID:27032422

Zhang, J., Ma, H. P., & Liu, Z. F. (2017). Highly efficient photocatalyst based on all oxides WO_3/Cu_2O heterojunction for photoelectrochemical water splitting. *Applied Catalysis B: Environmental*, *201*, 84–91. doi:10.1016/j.apcatb.2016.08.025

Zhang, K., Shi, X. J., Kima, J. K., & Park, J. H. (2012). Photoelectrochemical cells with tungsten trioxide/Mo-doped $BiVO_4$ bilayers. *Physical Chemistry Chemical Physics*, *14*(31), 11119–11124. doi:10.1039/c2cp40991g PMID:22772604

Zhang, T., Wang, L., Su, J., & Guo, L. (2016). Branched tungsten oxide nanorod arrays synthesized by controlled phase transformation for solar water oxidation. *ChemCatChem*, *8*(12), 2119–2127. doi:10.1002/cctc.201600267

Zhao, W., Wang, Z., Shen, X., Li, J., Xu, C., & Gan, Z. (2012). Hydrogen generation via photoelectrocatalytic water splitting using a tungsten tri oxide catalyst under visible light irradiation. *International Journal of Hydrogen Energy*, *37*(1), 908–915. doi:10.1016/j.ijhydene.2011.03.161

Zhao, Z., Butburee, T., Lyv, M., Peerakiatkhajohn, P., Wang, S., Wang, L., & Zheng, H. (2016). Etching treatment of vertical WO_3 nanoplates as a photoanode for enhanced photoelectrochemical performance. *RSC Advances*, *6*(72), 68204–68210. doi:10.1039/C6RA11750C

Zhou, Y. G., Zhang, L. Y., Lin, L. H., Wygant, B. R., Liu, Y., Zhu, Y., Zheng, Y. B., Mullins, C. B., Zhao, Y., Zhang, X. H., & Yu, G. H. (2017). Highly efficient photoelectrochemical water splitting from hierarchical $WO_3/BiVO_4$ nanoporous sphere arrays. *Nano Letters*, *17*(12), 8012–8017. doi:10.1021/acs.nanolett.7b04626 PMID:29185764

Zhu, J., Li, W., Li, J., Li, Y., Hu, H., & Yang, Y. (2013). Photoelectrochemical activity of $NiWO_4/WO_3$ heterojunction photoanode under visible light irradiation. *Electrochimica Acta*, *112*, 191–198. doi:10.1016/j.electacta.2013.08.146

Chapter 3
Recent Progress on Carbon-Modified Bismuth-Based Oxides for Applications in Photo-Assisted Devices

Sangeeta Ghosh
Indian Institute of Engineering Science and Technology, Shibpur, India

Aditya Poddar
Indian Institute of Engineering Science and Technology, Shibpur, India

Himanshu Sekhar Sahoo
https://orcid.org/0009-0000-3328-4466
Indian Institute of Engineering Science and Technology, Shibpur, India

Swarnendu Baduri
Indian Institute of Engineering Science and Technology, Shibpur, India

Debasish Ray
Indian Institute of Engineering Science and Technology, Shibpur, India

Sukumar Kundu
Indian Institute of Engineering Science and Technology, Shibpur, India

Jitendra Kumar Singh
Hanyang University, South Korea

Chinmoy Bhattacharya
https://orcid.org/0000-0003-2370-7108
Indian Institute of Engineering Science and Technology, Shibpur, India

ABSTRACT

Photocatalysis is a traditional approach to solve energy crises and environmental pollution control. Graphitic carbon nitride (g-C_3N_4), a 2-D conjugated polymer, along with different other forms of carbon based components like carbon nano-tubes, carbon quantum dots, carbon nano-sheets, carbon nano-wires etc. acts as nano-mediator and serves as promising candidates with bismuth-based oxides for applications in

DOI: 10.4018/978-1-6684-8743-3.ch003

environmental and energy sectors leading to sustainable development. The latest efforts towards overcoming the limitations for improvement in performances of these materials can be controlled by different factors e.g. surface morphology, band gap, size, crystallinity, charge separation efficiency and rate of the charge transfer etc. Here, we primarily focus on the detailed strategies related to the fabrication of different carbon components incorporated bismuth oxides based nano-composites for applications in sustainable development.

INTRODUCTION

Since the industrial revolution, the world's demands for energy have steadily increased, and to meet this challenge, clean, renewable resources must be explored to the fullest. Apart from the non-conventional sources of energy and to compete with fossil fuels, a cost-effective means of storing converted energy must be found. To harness the limitless solar energy in carbon neutral fashion without producing any environmental contaminant, the generation of the purest form of chemical energy, i.e. hydrogen (H_2) and environmentally benign oxygen (O_2) is achieved through photoelectrochemical (PEC) water splitting (Water $\underset{(Semiconductor/\,photon)}{\overset{hole}{\rightarrow}}$ Hydrogen+ Oxygen) and this is classified among the auspicious technique. The expectation of the community for the foundation of the future's sustainable energy economy, such as hydrogen, should have cleaner essence possessed by the solar water splitting technology. PEC cells have been proven as a distinguished source to tackle this issue by taking complete advantage of solar energy in meeting sustainable energy goals through an efficient and effective means of hydrogen production that serves as a fuel. A concise chronicle of events concerning the exploitation of some inorganic semiconductors for photo-assisted water splitting is adequately demonstrated by Osterloh et al. in their report. In 1955 it was first reported in the domain of photoelectrochemistry by Brattain et al.to explore the electrochemical investigation of electrolytic junction of germanium, succeeded by the crucial research of Heinz Gerischer, Rüdiger Memming, Arthur J. Nozik, Allen J. Bard, John O' Mara Bockris, Laurence Peter, Michael Grätzel, Helmut Tributsch, Akira Fujishima, Kenichi Honda, Krishnan Rajeshwar, Thomas J. Meyer, Prashant V. Kamat, Nathan S. Lewis, Frank E. Osterloh, Akihiko Kudo, Kyoung-Shin Choi and many others. In this regard, the efficacious operation of water photoelectrolysis in 1971 was brought about by Fujishima and Honda through fabricating PEC cell using TiO_2 nanoparticle semiconductors under illuminating conditions. The momentous work of Fujishima and Honda triggered a tremendous number of researchers towards the domain of water photoelectrolysis leading to the publications of numerous reports with significant impact on this field. Succeeded

Figure 1. Oxygen evolution and hydrogen evolution reaction for overall solar water splitting

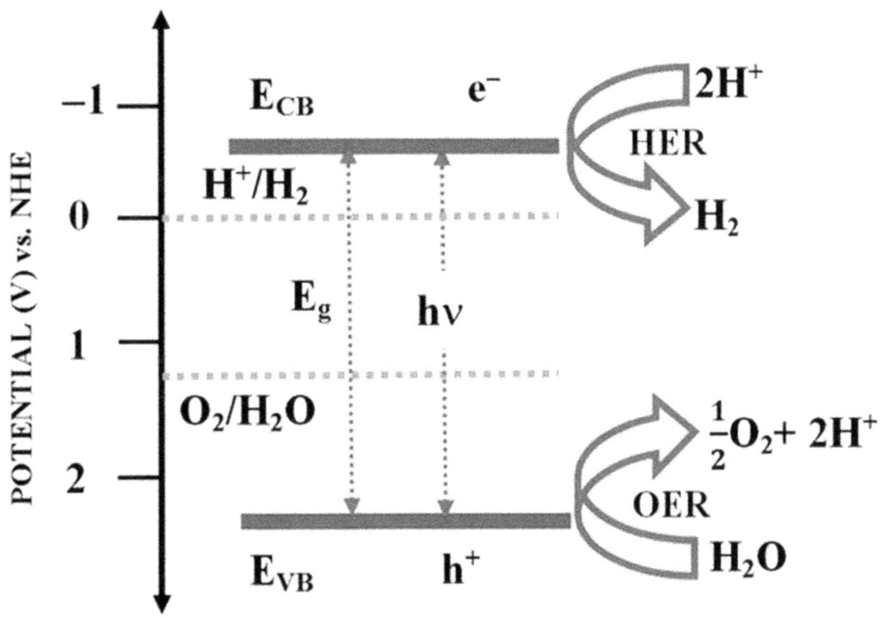

by the aforementioned immense works an enormous development has been accomplished in the current time to appraise the conceivable and promising candidate to lead photoelectrochemical water splitting more efficiently. The reported articles demonstrated the production of H_2/O_2 using different semiconductors in the role of photoelectrodes and out to be well-equipped materials as they owned better durability, electrical properties & ease of tailoring. The two distinguished halves illustrate the electrodics phenomenon, the evolution of H_2 at photocathode by reducing H_2O and the evolution of O_2 at photoanode/counter electrode by oxidizing H_2O as displayed in the following equations and also depicted in the Fig. 1.

Anodic Oxidation Reaction: $2\,H_2O\,(l) \rightarrow O_2\,(g) + 4H^+(aq) + 4e^-$

Cathodic Reduction Reaction: $2H^+(aq) + 2e^- \rightarrow H_2\,(g)$

In the above mentioned diagram (Fig. 1), on illumination, the photoactive material absorbs photon energy greater than its band gap energy and creates excitons (i.e. the photo-generated electron/hole pair). If the semiconductor is of n-type, the photo-generated hole moves to the semiconductor-electrolyte interface to oxidize the water to O_2 molecule and the photo-generated electrons are dragged to the counter electrode (generally Pt) through external bias to reduce water to H_2. On the other hand, if the semiconductor is of p-type, due to downward band bending at the semiconductor/electrolyte interface the excited electrons are transferred towards the semiconductor/electrolyte interface to reduce the water molecule to H_2 and the photo-generated holes are guided to the counter electrode through external bias to oxidize water to O_2. The PEC water splitting technique possesses few conveniences over the photocatalytic water splitting such as (i) the produced products H_2 and O_2 are needed not be separate as they produced in different chambers; (ii) the recombination of excitons (photo-generated electron/hole pair) are suppressed by self or externally applied bias hence improve the migration of the photo-generated charge carriers to the interface; (iii) conducting substrates were utilized to grow the films, so in future it can be helpful in mass production for industrialization; (iv) more importantly like the photocatalytic water splitting it does not demand any external power consumption for stirring. Though they had some challenges, monstrous efforts have been employed to cultivate a convincing model for greater efficacy in water splitting.

The photo-driven water splitting is an uphill reaction with positive Gibbs free energy of 237 kJ/mol. To acquire such an amount of energy, the semiconductor should have appropriate band gap energy with proper band alignment so that they can absorb photon (visible light) of suitable wavelength to bring out the redox (water oxidation and reduction) reactions. The band gap of the photoelectrodes must be small (~1.5 eV) to absorb an adequate part of the visible region of the solar spectrum and the valance band (VB) of the semiconductor should be more positive than the water oxidation potential (1.23 V against NHE at 0 pH) to oxidize $H_2O \rightarrow O_2$ (H_2O/O_2) whereas the conduction band (CB) of the semiconductor should be more negative than the water reduction potential (0 V against NHE at zero pH) to form hydrogen. Considering the same, numerous semiconductors based on inorganic metal oxides have been employed for the complete water splitting process, although they don't fulfill the requirement like desired band gap with proper band alignment and sufficient durability. Additionally, in PEC methodology, the system experiences energy loss accompanied by the migration and the recombination of photo-generated excitons (electron/hole pair) in the semiconductor/electrolyte interface or the bulk semiconductor. Therefore, for efficacious water splitting, charge carrier separation after illumination followed by the faster migration of photo-generated charges, along with desired band gap with proper band alignment is desirable. Furthermore, the other factors to be taken into account are low Debye length for the photo-generated charge, better charge absorption through customized surface morphology, size and shape of the semiconductor particles via involvement of quantum effects, the position of the flat band, the impact of pH, different ions present in the system and charge transport processes etc. Hence the properties of the photoactive electrodes to be judiciously selected that owns the efficiency of the PEC water splitting process for harvesting solar energy into a usable form of chemical energy.

After the pioneering work of Fujishima and Honda employing TiO_2 nanoparticles, several n-type and p-type photocatalysts have been explored for photo-assisted water splitting process, including main group and transition metals, such as silicon-based oxides and carbides, carbon nitride, transition metal-based oxide, chalcogenides, metal nitrides etc. Due to the high chemical durability in neutral and slightly acidic pH, including enough earth-abundance with no toxicity with the legacy of tunable composition, the bismuth-based oxides (generally n-type semiconductivity) are one of the potential aspirants for the solar light-driven water splitting. The crystal of bismuth-based oxide embodied of perovskite unit and is classified among the most enchanting prospect for photo-assisted water splitting due to its well-positioned band gap energy (2.5-2.8 eV), allowing it to absorb nearly 12% of the visible region of the solar spectrum with a favourable hole diffusion length of (~150 nm), in contrast with the α-Fe_2O_3 in few nanometres, including superior charge transporting (12 $cm^2s^{-1}V^{-1}$) property pertaining to TiO_2 (0.03 $cm^2s^{-1}V^{-1}$). In agreement with the water splitting, the conduction band minima of the semiconductor should be more negative than the water reduction H^+/H_2 potential i.e. 0 V against NHE at zero pH; on the other hand, the valance band maxima should be more positive than the water oxidation H_2O/O_2 potential i.e. 1.23 V against NHE at pH 0. Bismuth oxide consolidates the position of the valance band as the value is more positive than the water oxidation potential, but as the coin flip to the position of the conduction band minima, it's not able to reduce water to H_2as it is not situated deep enough to overcome the reduction potential. Consequently, bismuth-based oxides in photo-assisted water splitting only produce O_2 not H_2. Hence to lead the overall water splitting employing bismuth oxide as photoanode it should be united along with a suitable photocathode for the photo-assisted water reduction for the complete fabrication of the PEC device otherwise, the bismuth oxide photoelectrode is coupled with the counter electrode with an externally applied bias to drive the electron to counter electrode to accomplish the water splitting process. The upward band bending of the electrode/electrolyte interface made the bismuth oxide the suitable choice to exploit as a photoanode. It's already documented that the theoretical maxima of solar to hydrogen conversion efficiency for bismuth vanadate are about ~4.6% as a PEC tandem device. Bismuth vanadate is a well-known n-type semiconductor photocatalyst with a direct band gap of 2.48 eV. It absorbs significant portions of visible light and is also used as stable, neutral, nontoxic and relatively cheap electrode material. It's noteworthy that the greater durability of bismuth vanadate in low pH conquers its utility since; H_2 production is more efficient in that situation. The valence band of the Bi_2O_3 photocatalyst is composed of mainly a combination of the Bi^{3+} 6s orbital on top of the O^{2-} 2p orbital. Typically due to relatively smaller band gap than TiO_2, bismuth based oxides are capable of absorbing more in the visible region of the solar spectrum. On the other hand; it portrays a higher stationary photocurrent than Fe_2O_3 owing to enhanced

IPCE (incident photon to current conversion efficiency). Consequently, bismuth oxide is thought to be more appropriate compared to Fe_2O_3 and TiO_2 for applying PEC water splitting.

In this aspect, further modifications craved a new path where carbon-based nano-materials have played a major role to enhance the photocatalytic properties. There are three types of carbon nano-tubes (CNTs): single-walled nano-tube (SWCNT), double-walled nano-tube (DWCNT) and multi-walled nano-tube (MWCNT), which have been extensively used after their discovery for their promising physical, chemical and mechanical capabilities. These CNTs have gained popularity amongst researchers as it enhance the intrinsic properties of the semiconductors attached to them it in higher scale. Similarly, modifications are also made with cost-effective carbon quantum dots (QDs) as their toxicity is very low, i.e. well compatible with the ecosystem and possesses high stability in an aqueous environment with a larger surface area. These properties enhance photocatalytic abilities of many semiconductors drastically. Other carbon-based nano-materials like carbon nano-sheets, carbon nano-wires have also been used to improve photocatalytic abilities of the pristine semiconductor materials by increasing the surface area, altering morphology design or any other intrinsic properties. Now a days lot of efforts have been dedicated to construct different carbon modified Bi-based metal oxide composite photocatalysts for better photo-induced properties depending on some fundamental factors. This review, therefore, summarizes the progress of Bi-based photocatalysts via the incorporation of C-based nano-materials and their energy and environmental sector applications.

Different Morphologies of g-C_3N_4

Carbon Nano-tubes (CNTs)

In recent times, CNTs have been getting substantial research attention because of their prior physicochemical as well as electronic and mechanical properties. They have high thermo-chemical stability and strong mechanical properties i.e. significant elasticity and tensile strength makes them suitable for various potential applications, like, different sensors, electro-catalysts, catalysts, conversion of energy and remediation in environment. Moet al. synthesized g-CN nano-tubes a through self-templating approach and reported it using melamine (ML) as the only precursor first time. This technique is completely free of either acid or alkali and the slow in-situ conversion method is used to convert a portion of melamine (ML) to cyanuric acid. g-CN nano-tubes with a high number of N-defects are obtained to pyrolyze the intact portion of ML and the converted part together. Strong Van der Walls bonding is present in CNT with various definite adsorption sites to adsorb molecules. CNTs are very good electron acceptors; when oxides of metal are decorated with it; nanocomposite formation occurs for both PEC water splitting and photo-catalysis process. To improve PEC performance significantly, CNTs are coupled with CdS and TiO_2. The utilization of carbon nano-tubes (CNT) as a charge collector in PEC systems or dye-sensitized solar cells (DSSC) conceptually could be more modern. As a result, it improves the devices performance on different light absorber nanostructures, e.g. Fe_2O_3, TiO_2 and ZnO. CNTs form a compact layer that decreases transmission of photons with a lower energy of band gap which is situated at the top of PEC thin film.

Carbon nano-sheets

The photocatalytic performance of nano-sheets is much better than bulk g-CN. In photo-degradation and water splitting process, nano-sheets of g-CN show improved activity than bulk g-CN. Nano-sheets have large surface area ratio with high crystallinity which are some acute properties needed to perform photo-driven reactions. It has been reported that the meso-porous form of g-CN nano-tubes has a BET surface area of 38.6 $m^2\ g^{-1}$ and under a –0.2V potential bias its photocurrent is 4.2 times higher than the bulk material. During solar-assisted hydrogen evolution reactions, the nano-sheets have demonstrated extended light absorption abilities and achieve better photo-catalytic performance.

Carbon Nano-wires

Nano-wires are the 1-D rod-like structure of g-CN which has recently received a lot of scientific attention due to higher surface area and volume ratios with more active sites are present to carry out reactions. Oh et al. synthesized g-CN nano-wires composite to develop the surface area of meso-porous carbon spheres to impregnate it with meso-porous carbon spheres and a three dimensional composite is formed.

Due to the increased surface area, the composite exhibited better electrochemical properties than that of graphene, carbon nanotubes and activated carbon based materials. For example, using conducting material poly(3,4-ethylene-dioxy-thiophene): poly(4-styrenesulfonate), i.e. PEDOT:PSS Tang et al. fabricated g-CN nano-wires based composite comprising nano-wire as substrate. This three dimensional resultant material is recyclable, flexible and possessing significant capacitive property. Xie et al. fabricated g-CN nano-wires via solvothermal technique to increase the surface area after poly-condensation reaction of pretreatment of cyanuric acid where melamine(ML) precursor is also fused at 500°C for 2 hrs. This designed nano-wires thus achieved an improved decomposition of methylene blue (MB) if visible active light is exposed as the number of active sites to occur reaction increases in nano-structured composite.

Quantum dots composite

g-CN quantum dots (i.e. 0-D structures) have been synthesized via different routes and has received attention of the researchers due to their unique properties e.g. physicochemical stability, greater surface area, appropriate band gap and bio-compatibility. Lin et al. first obtained bulk g-CN through thermal treatment of ML at ~550ºC temperature followed by etching in a mixture of concentrated HNO_3 and H_2SO_4. For environmental and biological applications, g-CNQDs are used as a fluorescent probe along with their thermal stability, nontoxic nature, water-solubility and bio-compatibility.

3-D g-C_3N_4

Due to the increased surface area of graphitic carbon nitride, there is also several drawback of it. To diminish the drawback Liang et al. designed a macro-porous 3D g-CN by converting ML sponge wisely as a template and also performed through a single-step polymerization reaction of ML sponge dipped in urea and the final product was immediately cut into different shapes of 3D arrangement comprising g-CN. However, urea and ML are the g-CN precursors to form a macroscopic 3D g-CN which has increased photo-catalytic activity for hydrogen evolution. The meso-porous 3D g-CN was fabricated by Wang et al. The in presence of silica template through optimization of the self-polymerization reaction condition and has reported that the performance of the material for photo-catalytic hydrogen production is almost 6.5 times greater than bulk g-CN. The 3D frameworks afford an increased activity in light absorption and for carrying out suitable photo-induced reactions more active sites are also present. Luo et al. prepared 3D g-CN which is 29.5 times superior to pristine g-CN indicating an immense photo-catalytic hydrogen formation process.

From Bulk to various morphologies of g-C_3N_4

Wang et al. synthesized g-CN in bulk, nano-ribbons, nano-sheets and quantum dots in a favorable method. Bulk g-CN was fabricated through thermal etching of urea and bulk g-CN was heated to obtain g-CN nano-sheets. These nano-sheets were treated with acid to achieve nano-rods and these nano-rods were subsequently cut to obtain the quantum dots through the hydrothermal method, presented in Fig. 2.

Reduced Graphene Oxide (RGO)

RGO is an atomically thin sheet of carbon atoms. It is obtained from GO via reduction reactions e.g. chemical reduction, electrochemical reduction and thermal reduction.

Figure 2. Schematic representation of processes to obtain different morphologies of g-C_3N_4 from the bulk

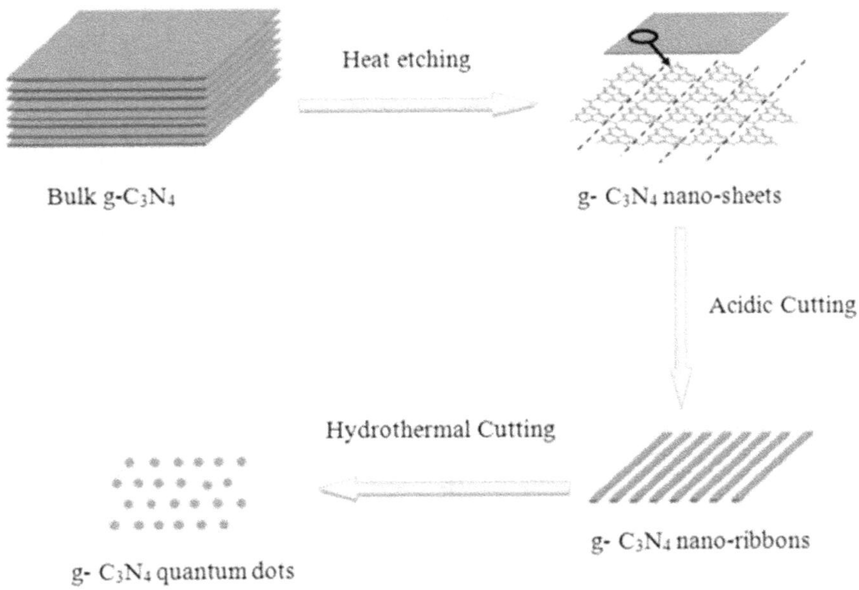

It is considered the best nano-hybrid photo-electrode due to its very low charge transfer (CT) resistance of electrode or insignificant electron-hole recombination rate. From the energy level diagram, it was noted that the conduction band edge is situated at a more negative end for the RGO nano-hybrid than the water reduction potential i.e. −0.197 V vs. Ag/AgCl. The valence band edge becomes more positive than the water oxidation potential i.e. 1.03 V vs. Ag/AgCl in the presence of 1(M) H_2SO_4 electrolyte solution which is required for the PEC water splitting process. According to the electronic arrangement of the band diagram for GO, it possesses a greater transportation role and faster electron acceptor ability.

Importance In PEC Water Splitting Process

Graphitic Carbon Nitride (g-CN)

The demand for polymeric CN is increasing drastically because of its rare properties, facile synthesis from earth-abundant, cheap starting materials and various applications starting from photo-catalysis. g-CN grabbed great attention as a semiconductor photo-

Figure 3. Thermal decomposition of melamine and the formation of graphitic carbon nitride

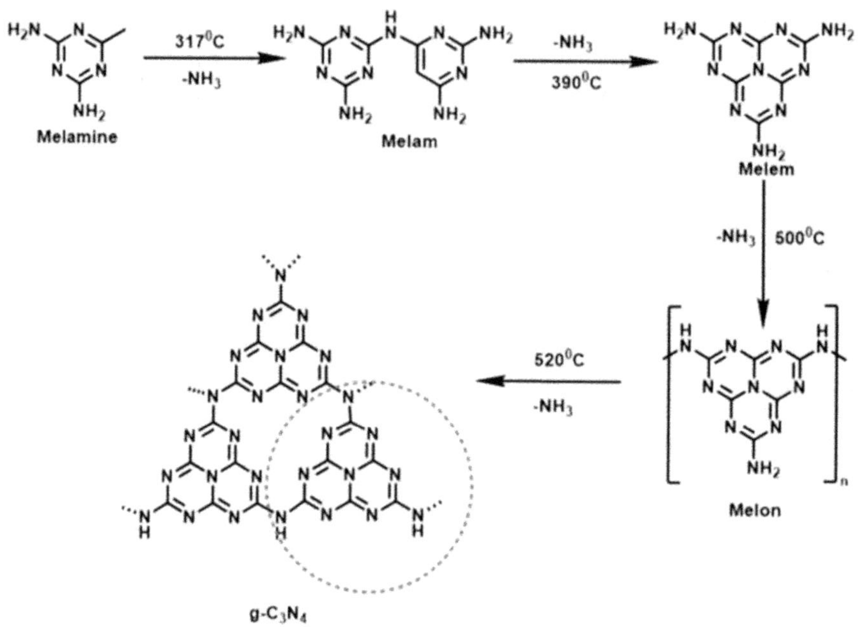

catalyst due to its polymeric nature in hydrogen evolution reaction (HER) particularly under visible light irradiation. This is due to greater stability, easy synthesis process with functionalization, favourable band gap energy with appropriate electronic band structure for absorption of visible light. The structure of g-CN composed of generally two units,tri-s-triazine (C_6N_7) and s-triazine(C_3N_3) rings, as presented in the Fig. 3.It has been reported that tri-s-triazine achieves more stability under ambient conditions. Among two basic units of g-CN, density functional theory calculations also confirmed that tri-s-triazine has the greater thermodynamic stability.

Being an analogue to graphene, g-CN is now-a-days mostly studied and popular for PEC water splitting process. Graphene has a two dimensional structure of sp^2-hybridized carbons whereas g-CN has graphitic planes with p-conjugation to produce sp^2-hybridized C as well as N atoms. Fina et al. explained 3D structure of g-CN with powder X-ray diffraction and neutron diffraction analysis. They also revealed that g-CN possesses a 3D arrangement with non-alignment layers of tri-s-triazine. The repulsive forces of p-electrons are eluded in adjacent layers and as a result they are non-aligned. The average reactivity of carbon nitride at the interface, all together with the aromatic π-conjugated foundation and intra-layer hydrogen bonds, provides enough possibilities to control its molecular structure and properties to meet certain task-specific applications. The water-splitting process is extremely endothermic in nature. However, hydrogen evolution from water requires a significant amount of input energy which makes it incompatible to exist energy generation. Therefore, an effective photo-catalyst for the water splitting process must have a potential difference of 1.23 Vat 0 pH. It is observed that the band-gap of pristine g-CN is ~2.7eV; the position of the conduction band is about −1.1 eV and the valence band is situated at +1.6 eV. As reported through DFT analysis, p_z orbitals of N atom is responsible for valence band formation and oxidation reaction takes place at that site whereas p_z orbitals of C atom participated in the conduction band formation and acted as a reduction reaction site. g-CN generally acts as a photo-catalyst i.e. can separate electrons and holes on photo-excitation and favours the photo-reduction reactions due to its sufficiently negative potential of the conduction band position. It is very praiseworthy in many fields because the basic property of nitrogen facilitates the π-bonding with improved stability. The unusual electronic behaviour of CN leads to the formation of valence band by the lone pair of N atoms and a conduction band comprising of p_z orbital of the C atom. In this polymeric structure, the availability of earth abundant C and N atom makes it very versatile to endow applications with excellent properties.

Photocatalytic (PC) Activity of The Composites Having Carbon Based Semiconductor

The carbon based semiconductors with internal π-conjugation must have appropriate HOMO as well as LUMO energy levels so that under illumination, there must be the excitation of charge carrier transfer. With the eventual passage of time, the aim of many researchers was concentrated at a point where they have been using methods to develop various bismuth based semiconductors photocatalysts modified with carbon-related components. At present, the two most popularly used carbon based materials for this purpose are graphene and carbon nitride (CN - which has 2D feature of graphene).

Effect of Incorporation of C-Based Nanomaterials on Bismuth Oxide for PC activity

Bi_2O_3(BO) is generally used as a well-known photocatalyst under the irradiation of visible light. The bandgap of this semiconductor belongs within a wide region from 2.1 eV to 2.8 eV. When modified with the incorporation of noble materials, say, using graphene, the photocatalytic activity can be increased significantly. Although the semiconductor possesses suitable bandgap energy, there are certain limitations in the overall photocatalytic ability. It has a tendency to quickly recombine the photo-generated charge carriers and also susceptible towards photo-corrosion. As the conduction band (CB) of Bi_2O_3 and the valence band (VB) of graphene are located at close proximity, they can be coupled to enhance the photocatalytic ability. In this way, the electrons accumulated in the CB of graphene shows significant reduction property whereas the holes that assembles in the VB of Bi_2O_3 shows tremendous oxidation property. The photocatalytic activities that involve the Z-scheme are more efficient when compared with a single component based on the redox and the updated photocatalytic performance for the case of conventional photocatalysts. S. Sharma et al. prepared a composite of α-BO/C-dots via facial sonochemical synthesis. The composite degraded 86% of Indigo Carmine (IC) dye in 120 min compared to 57% using pure α-BO. Similarly it degrades 79% of the antibiotic drug levofloxacin when visible light is irradiated for 120 min. The photocatalytic behaviour of dye in the absence of any scavenger was ~86%, which is ~ 58% for IPA. The photocatalytic decomposition of IC dye is a pseudo-first-order kinetic reaction and the rate constant (*k*) can be calculated from the slope of the Langmuir- Hinshelwood plot which turned out to be 0.01734 min^{-1}. In these processes $\dot{O}H$ and e^- are identified as dominating as well primary reactive species. Y. Li et al synthesized g-CN/BO composites by a mixing-calcination technique where g-CN/BO composites exhibited higher photocatalytic (PC) activity for degradation of Rhodamine B (Rh B) than that of pure form of either BO or g-CN. This photocatalytic decomposition of the dye follows first order kinetics. The pristine BO and g-CN photo-catalyst degrade Rh B by 39% and 42%, respectively whereas the composite decomposes 83% of the dye in 3 h under visible light radiation. Pristine BO and g-CN show very low photocatalytic activity with reaction rate constant (k) values of 0.140 h^{-1} and 0.175 h^{-1}, respectively whereas the g-CN/BO heterojunction with g-CN contents of 18.3%, 53.2%, 71.1% and 87.7% shows higher values of 0.418 h^{-1}, 0.556 h^{-1}, 0.243 h^{-1} and 0.210 h^{-1}, respectively, as presented in Fig.4. The presence of g-CN/BO composite was confirmed by the XRD pattern which exhibits the diffraction peak of monoclinic BO and g-CN.

B. T. Huy prepared heterojunction ZnO-BO/g-CN through the co-precipitation method for photocatalytic degradation of Indigo carmine dye. The different photocatalysts, namely, ZnO-BO, pure g-CN and ZnO-BO/g-CN heterojunction degraded 17.6%, 26.7% and 93% of the dye in 180 min. Under visible light irradiation, around 25.2, 38.7, 52.0, 68.8 and 45.3% of IC dye was decomposed by ZnO-BO, pure g-CN, ZnO-BO/g-CN (1%), ZnO-BO/g-CN (2%) and ZnO-BO/g-CN (3%), respectively in the presence of H_2O_2 in 180 min. The kinetics followed for this dye degradation was first order in nature with k values appearing as, for ZnO-BO/g-CN (2%): 0.0037 min^{-1}which was significantly higher than that of ZnO-BO: k = 0.0004 min^{-1}, ZnO-BO/g-CN (1%): k = 0.0022 min^{-1} and ZnO-BO/g-CN (3%): k = 0.0019 min^{-1}, respectively. Under visible light irradiation and in the presence of

Figure 4. Comparative study of the rate constant for different g-C_3N_4/Bi_2O_3 composites

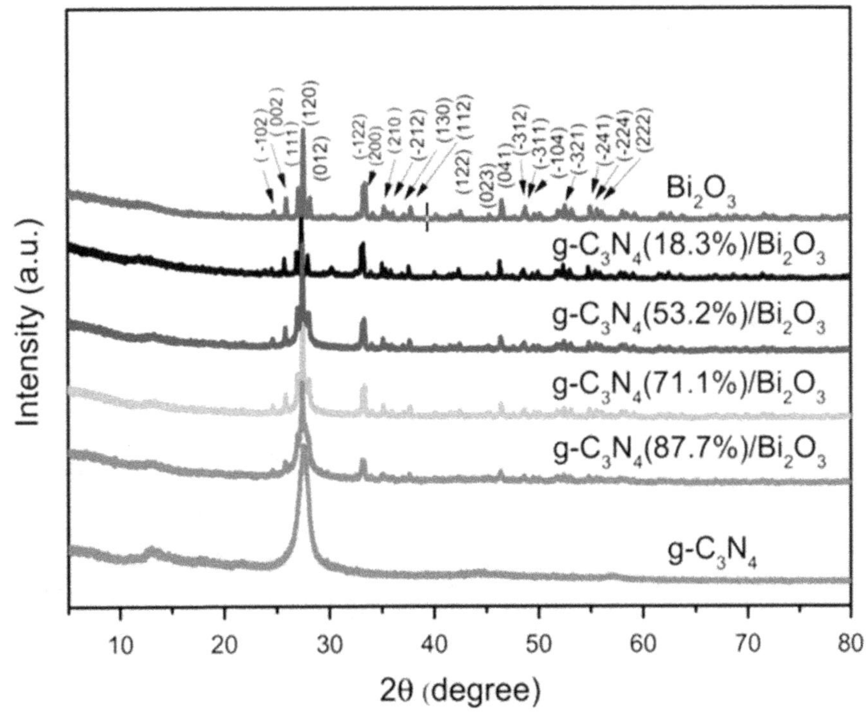

H_2O_2, the dye degradation reaction rate constant values were found to be 0.001 min^{-1}, 0.0036 min^{-1}, 0.0056 min^{-1}, 0.0032 min^{-1} for ZnO-BO, ZnO-BO/g-CN (1%), ZnO-BO/g-CN (2%) and ZnO-BO/g-CN(3%), respectively. L. Zhang et al. synthesized a heterojunction, for Z-scheme, of β-Bi_2O_3/g-C_3N_4 composite using $Bi_2O_2CO_3$ and g-CN nano-sheet precursors using hydrothermal and calcination methods. Pure g-C_3N_4 and β-Bi_2O_3 showed low activity of photocatalysis and degraded 87% and 67% of Rh B dye after 80 minutes. The composites of β-Bi_2O_3/ g-C_3N_4 achieved the highest photocatalytic activity of 98% after continuous exposure of light for 80 minutes and it follows first order kinetics. From XRD analysis it was confirmed that all peaks of the composite are well assigned with the main detectable diffraction peaks of pure BO and g-CN. The presence of g-CN was further confirmed by FTIR analysis. as presented in Fig. 5.

Y. Hong et al. synthesized β-Bi_2O_3decorated g-C_3N_4nano-composites (core/shell) and explored photocatalytic performance through degradation of tetracycline (TC) dye. The pure g-C_3N_4 and the pure β-Bi_2O_3 exhibited a performance of 15.4% and 56.6%, respectively whereas the 5% CN@BO sample demonstrated the highest PC

Figure 5. Analysis of FTIR patterns of different β-Bi_2O_3/g-C_3N_4 composite samples

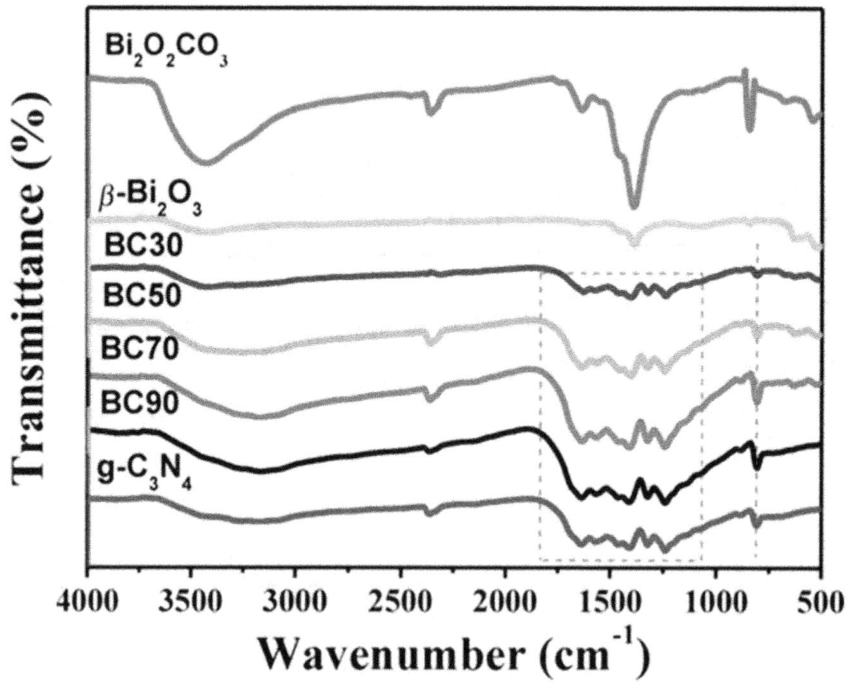

activity by degrading ~80% of the TC dye, as presented in Fig. 6. The photodegradation reaction involved pseudo first order kinetics.

R. He et al. developed g-C_3N_4 - Bi_2O_3composite system through in-situ depositing quantum-sized metal oxide particles on CN at a suitable condition and the process has been depicted in the schematic diagram as Fig. 7. The Bi_2O_3/g-C_3N_4composite, even with lower Bi_2O_3 content (~ 3% by weight) exhibited excellent photocatalytic activity compared to the pure Bi_2O_3matrix as measured throughdecomposition of phenol under visible light irradiation.

G. Fan et al. synthesized in-situ nanoparticles of Bi_2O_3 by coupling with g-C_3N_4 through thermal polymerization. The samples were prepared using various amounts of Bi_2O_3 and designated as0.1-Bi_2O_3/g-C_3N_4, 0.5-Bi_2O_3/g-C_3N_4, 1.0-Bi_2O_3/g-C_3N_4, and 1.5-Bi_2O_3/g-C_3N_4, where methylene blue (MB) was used as the dye. The photocatalytic decomposition efficiency of MB dye for the optimized composite(0.5-Bi_2O_3/g-C_3N_4) was reported as 100%without any quencher and in the presence of different quenchers it changes to ~50% (IPA), ~97% (BQ), ~89% (Na_2EDTA) and ~98%($K_2Cr_2O_7$), respectivelyafter 90 minutes of visible light illumination. The

Figure 6. The photocatalytic performance of as-fabricated samples for degradation of TC under visible-light irradiation (λ> 420 nm)

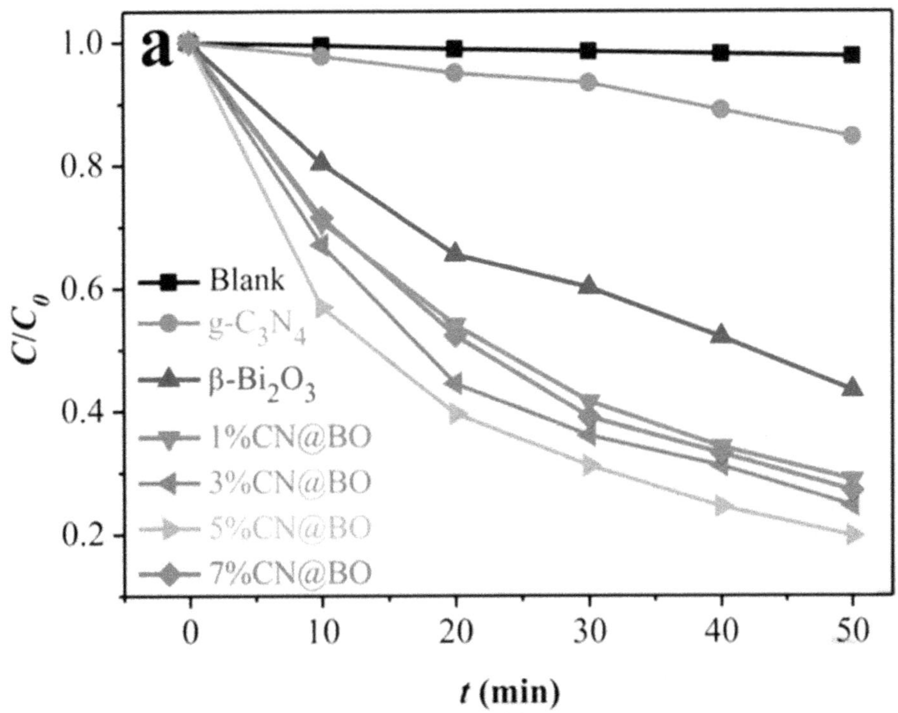

photodegradation efficiency for the pure Bi_2O_3 and pure g-C_3N_4 was noted as only ~25% and ~66% in similar experimental conditions. J. Zhang et al. synthesised heterojunctions of Bi_2O_3 and g-C_3N_4 via ball milling technique. The photodegradation efficiency of the g-C_3N_4 was 35% and 11% for both MB and Rh B, respectively after constant illumination of light for 60 minutes. Bi_2O_3 (1 wt.%)/g-C_3N_4 demonstrated the highest photocatalytic (PC) activity among all other composites whereas for the other composites, a descending trend for the PC performance was recorded as the proportion of Bi_2O_3 changes (in wt.%) like 3, 5, 0.5 and 0 and the overall reaction followed first order kinetics. For MB dye, the rate constant values of Bi_2O_3 (1 wt. %)/C_3N_4 and pure g-C_3N_4 were 0.0253 min^{-1} and 0.0074 min^{-1}, respectively while that for Rh B, the values were 0.0101 min^{-1} and 0.002min^{-1}, respectively. The rate constant for the sample Bi_2O_3 (1 wt. %)/C_3N_4 for both MB and Rh B dyes was around 3 times and 5 times higher compared to pristine g-C_3N_4. Y. Cui et al. used MOFs to synthesize ternary composites of MIL-53(Fe)/α-Bi_2O_3/g-C_3N_4 (MBC) employing hydrothermal technique followed by calcinations susing varying amounts of 10%,

Figure 7. The formation scheme of Bi_2O_3/g-C_3N_4, including intermediate products collected after centrifugation of the faint yellow and black suspensions

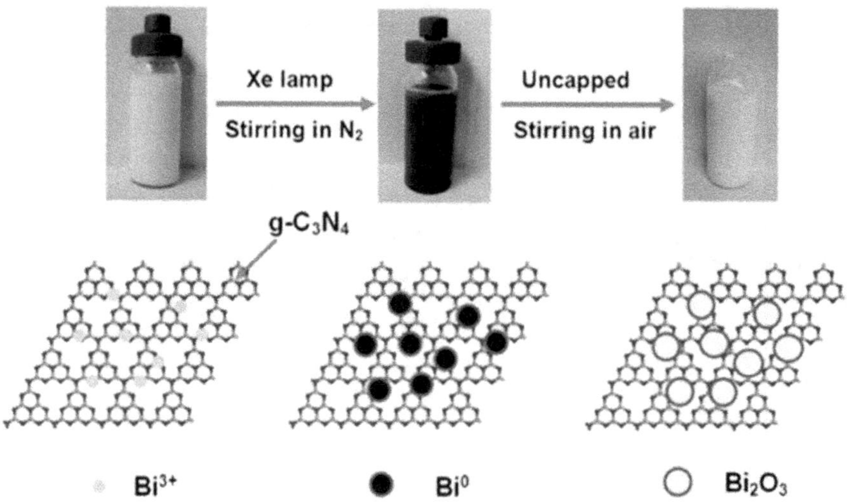

25%, 32% and 40%, on the surface of the binary heterojunction of α-Bi_2O_3 and g-C_3N_4 (α-Bi_2O_3/g-C_3N_4). Amido black 10B was used for the test under irradiation and the binary composite was capable of degrading 97% of the dye within 120 min exposure to light. Among those ternary composites, the 32%-MBC sample degraded a maximum of 100% of dye and achieved the highest degradation kinetics within 45 minutes. The dye degradation reaction followed pseudo-first order kinetics with rate constant value of 0.08434 min^{-1}. For the MIL-53(Fe), pure g-C_3N_4, and pure α-Bi_2O_3 and for the binary composite of α-Bi_2O_3/g-C_3N_4, the rate constant (k) values are of 0.00436, 0.00751, 0.01269 and 0.02661(in min^{-1}), respectively as presented in Fig. 8. The 32%-MBC sample exhibited 20, 11, 6 and 3 times greater rates than MIL-53(Fe), pristine g-C_3N_4 and α-Bi_2O_3 and the binary composite, respectively.

S. Xue et al. synthesized photocatalysts that are composites of nitrogen doped Bi_2O_3 and g-C_3N_4 through ultrasonic dispersion with Rhodamine B (Rh B) dye. Among the three binary composites, the sample containing 66% of g-C_3N_4 demonstrated the highest photocatalytic efficiency of ~95% after continuous exposure of light for 15 minutes whereas by the same time the efficiency of N-doped Bi_2O_3 was only75%. The rate constants (in min^{-1}) for the photodegradation of the RhB dye were 0.0331 and 0.1063 for N-doped Bi_2O_3 and N-doped g-C_3N_4/Bi_2O_3 respectively. The modification of the crystallites through the incorporation of dopants was confirmed by XRD

Figure 8. Successive photocatalytic degradation rate constant determination over as-prepared photocatalysts

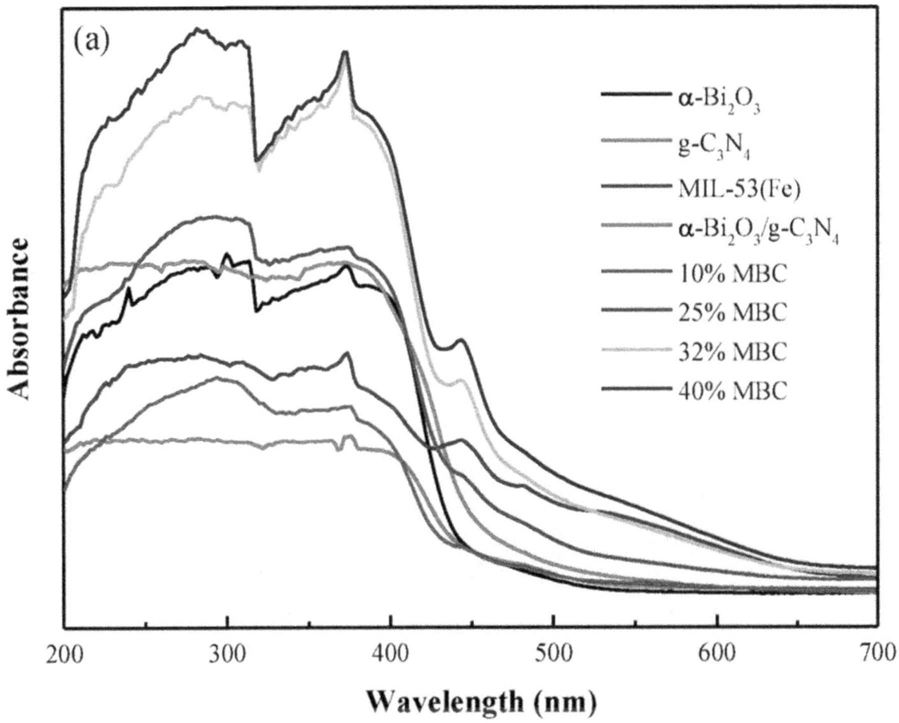

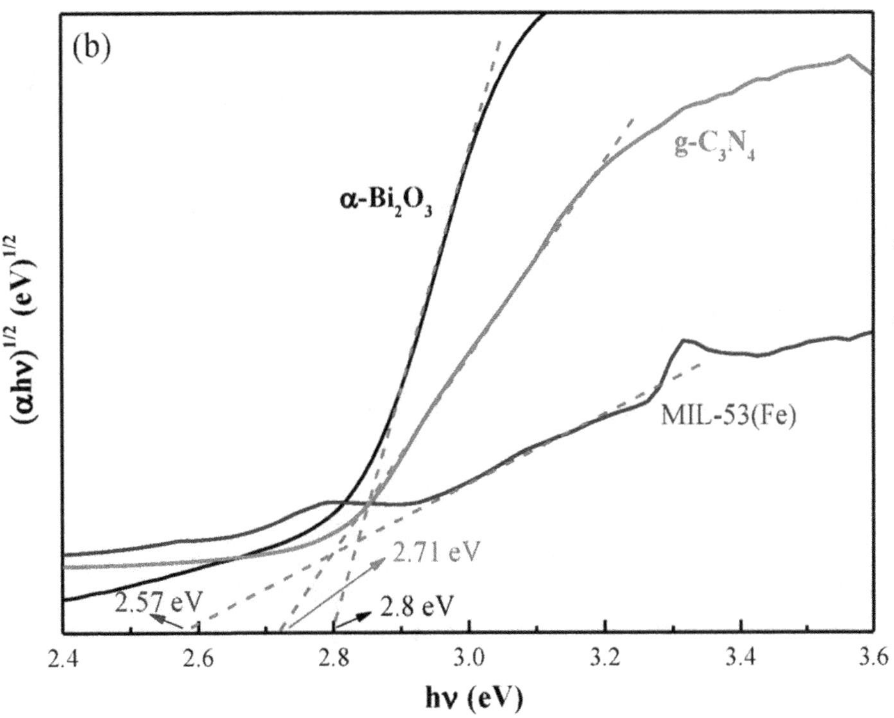

analysis as the diffraction peaks are wider in case of N-doped BO compared to pure one.H. Zhao et al. fabricated Bi_2O_3/g-C_3N_4/$Ag_6Si_2O_7$ ternary composite photocatalysts with mass ratio 1:25.5via blending calcinationmethod. The samples were labelled as Bi_2O_3/g-C_3N_4/$Ag_6Si_2O_7$(a:b) and this a:b ratio were 1:2, 1:5,1:20, 1:25, 1:26, 1:30, 1:40 to check the decomposition of MB dye. Among all, the ternary composites, Bi_2O_3/g-C_3N_4/$Ag_6Si_2O_7$ (1:25.5) sample achieved optimum photocatalytic ability (~94%) and the highest photodegradation rate (0.2278 min^{-1}), the rate for Bi_2O_3/g-C_3N_4/$Ag_6Si_2O_7$(1:25.5) sample was ~28, ~1.5, and ~2 foldsgreater thanpure samples ofbinary composites of Bi_2O_3/g-C_3N_4, Bi_2O_3/$Ag_6Si_2O_7$ and Bi_2O_3/g-C_3N_4/$Ag_6Si_2O_7$, respectively. The other ternary composites also exhibited greater photocatalytic rate with a rate constant of 0.083 min^{-1}which was 3.5 times higher than that using pure $Ag_6Si_2O_7$ ($k = 0.023$ min^{-1}) and ~2.5 times compared to the rate of Bi_2O_3/g-C_3N_4 ($k = 0.036$$min^{-1}$). The morphology of the binary and ternary composite was determined through HR-TEM analysis illustrated in Fig. 9. It is confirmed that the smooth surface of BO particles were randomly anchored at the surface of g-CN which indicates the efficacious fabrication of the composite by simple co-calcination. It is reported in the literature that, as presented in Fig. 9, the lattice spacing of 0.264 nm was assigned to (−1 1 5) plane of the $Ag_6Si_2O_7$ and that of 0.327 nm could be ascribed to (1 2 0) plane of Bi_2O_3 suggesting co-existence of $Ag_6Si_2O_7$, g-C_3N_4 and Bi_2O_3 phases in the composite.

Figure 9. Relative surface morphology analysis of the composite sample

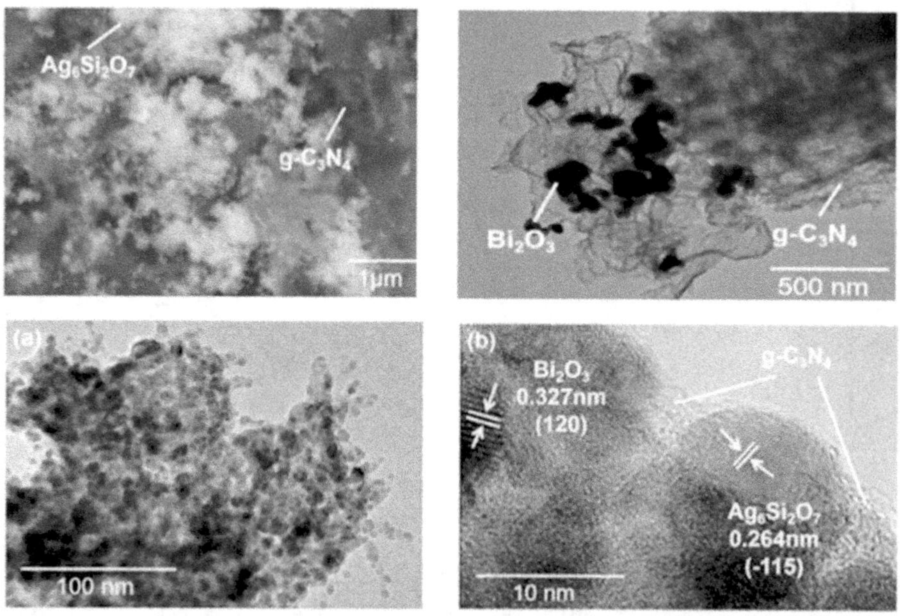

M. Alhaddad et al. fabricated Bi_2O_3/g-C_3N_4 nano-composites through sonication method. Various nano-composite samples of Bi_2O_3/g-C_3N_4 were prepared by tuning the molar amount, as denoted by x, of Bi_2O_3 in each case. The samples were named as xBi_2O_3/g-C_3N_4 and x represents 0.5%, 1%, 3% and 5%. The photocatalytic efficiency for the samples was determined by analyzing the amount of H_2 produced from aqueous glycerol solution under visible light irradiation. Pure Bi_2O_3 and g-C_3N_4can produce hydrogen of 510 μmol g^{-1} and 400 μmol g^{-1}, respectively whereas the nano-composite samples containing 3% Bi_2O_3 showed maximum capacity of 8542 μmol g^{-1}, i.e. 20 folds greater than that of pure g-C_3N_4. The increment in the PC activity for the binary composites was noted as the Bi_2O_3loading reached upto 3 wt.%.

Photoelectrochemical (PEC) Performances of The Carbon Based Composites Semiconductors

Role of incorporation of C-based nano-materials on bismuth oxide for PEC performance

Shaddad et al. synthesized Bi_2O_3 films via electro-deposition technique followed by modification through Moas dopant and developing oxygen vacancies (OVs) on the surface. The samples prepared were pure Bi_2O_3 and Mo-doped Bi_2O_3 and their corresponding oxygen vacancies. The XPS data shows the peaks for Na 2s, Bi 4f, Mo $3d_{5/2}$ (~232 eV), O 1s (~530.6 eV) regions. The photocurrent densities for pure Bi_2O_3, Mo-doped Bi_2O_3 and their corresponding OVs are 0.04, 0.49, 0.16, 1.06 (in mA cm^{-2}) whereas the photocurrent conversion efficiencies are 0.01%, 0.1%, 0.05% and 0.26%, respectively. From the Nyquist plots, the ohmic (R_s) and the charge transfer (R_{ct}) resistance values for the pure Bi_2O_3, Bi_2O_3 with oxygen vacancy, Mo-doped Bi_2O_3 and its oxygen vacancy are 0.04497,0.0236, 0.04155, 0.02796 (in kΩ) and 95966, 32749, 7760, 6426 (in kΩ), respectively. The MS plots of the different samples indicate n-type semiconductivity and the flat band potentials appears as 0.40, –0.034, 0.37, –0.123 (in V) for pure Bi_2O_3, Mo-doped Bi_2O_3 and OVs, respectively. The modification done by annealing in a vacuum and doping with Na_2Mo_4 showed 25 times better results than the other samples. Using the suitable co-precipitation technique, Reddy et al. fabricated Bi_2O_3 nanostructures doped with Fe_3O_4. The XRD data of this work inferred that both Fe_3O_4 and Bi_2O_3 has cubic crystalline phase. From the XPS study, two peaks of Fe $2p_{3/2}$ and $2p_{1/2}$, two peaks of Bi $4f_{7/2}$ and $4f_{5/2}$, and two peaks for O 1s were obtained. The photocurrent density was the highest for Bi_2O_3 when doped with Fe (0.36 mA cm^{-2}). The electrochemical impedance analysis indicates that the modified sample shows minimum charge transfer resistance (46.04 Ω) and that for the path conducting the ion is 26.11 Ωsuggesting maximum capability and the highest PEC water splitting efficiency using prepared $Bi_2O_3 \cdot Fe_3O_4$ nano-structure.

Yang et al. prepared heterojunctionsof Bi_2O_3 and $BiFeO_3$(BFO) films by the pulsed laser deposition method. From the XRD results, it was evident that the samples of BFO are in the rhombohedral crystalline phase, the Bi_2O_3 samples are majorly amorphous and for the heterojunctions peaks show their tetragonal phase. The XPS spectrum of the Bi_2O_3 film shows two peaks of Bi $4f_{5/2}$ and $4f_{7/2}$. The heterojunction of Bi_2O_3/BFO shows maximum photocurrent density (~84.07 μAcm^{-2}) and which is two times higher than that of pure BFO. Similarly, for IPCE measurements, the heterojunction showed better results (~2.6% at λ = 365nm) than pure BFO and Bi_2O_3 (~1% λ = 365-475 nm). The charge transfer resistance for the heterojunction is found to be much less (~ 10 Ω) when compared with the film of BFO (~ 18 Ω). Thus, the developed heterojunction has produced more promising results than the pure sample of Bi_2O_3 and BFO.

Sahnesarayi et al. fabricated Bi_2O_3 photocatalyst through the chemical bath deposition (CBD) method. The XRD analysis data confirmed that the prepared photocatalysts were polycrystalline in nature and the existence of a tetragonal phase was found in all samples. Different samples were obtained by varying the temperature of the bath and denoted as BT1 (at 45°C), BT2A2 (at 60°C), and BT3 (at 75°C). The BT2A2 sample exhibited the highest photocurrent density of ~470 $\mu A\ cm^{-2}$ at a potential of 1.2V RHE with the minimum resistance (1124 Ω) associated with the charge transfer process. Another variation was achieved by adding complexing agents to the chemical bath and was named as BT2A2, BA3, and BA4. The optimized BA3 sample exhibited the maximum photocurrent density of ~750 $\mu A/cm^2$ and the lowest charge transfer resistance. In the Nyquist plot, the radius of the semicircle represents the charge transfer resistance (R_{ct}) and varies in the order BT2A2 > BA4 > BA3. From the Mott-Schottky analysis, it was confirmed that the materials have the n-type semiconductivity, whereas the photoluminescence spectrum peak intensity of the photocatalyst follows the order BA3 < BA4 < BT2A2, indicating a similar trend in the recombination process between the photoactivated electron and hole pairs. As a result the BA3 semiconductor behaves as a most effective photocatalyst compared to others.

H. Wang et al.developed Z-scheme based sensors through synthesizing heterojunctions of β-Bi_2O_3 and g-C_3N_4 (CNN) by a suitable one pot method and explored their applications for the detection of chlorpyrifos. The binary composites were prepared by incorporating 1 ml, 2 ml and 3 ml g-C_3N_4 in β-Bi_2O_3. From the XRD data, it was revealed that β-Bi_2O_3 had a tetragonal crystalline phase, whereas the CNN had 2 characteristic peaks corresponding to the planes (100) and (002). From the XPS analysis, it is reported that,as presented in Fig. 10,the peak positions for all the elements of pure β-Bi_2O_3 (Bi $4f_{5/2}$ and Bi $4f_{7/2}$), g-C_3N_4 (C = C and for aromatic

Figure 10. Comparative study of XPS analysis of pure and composite samples

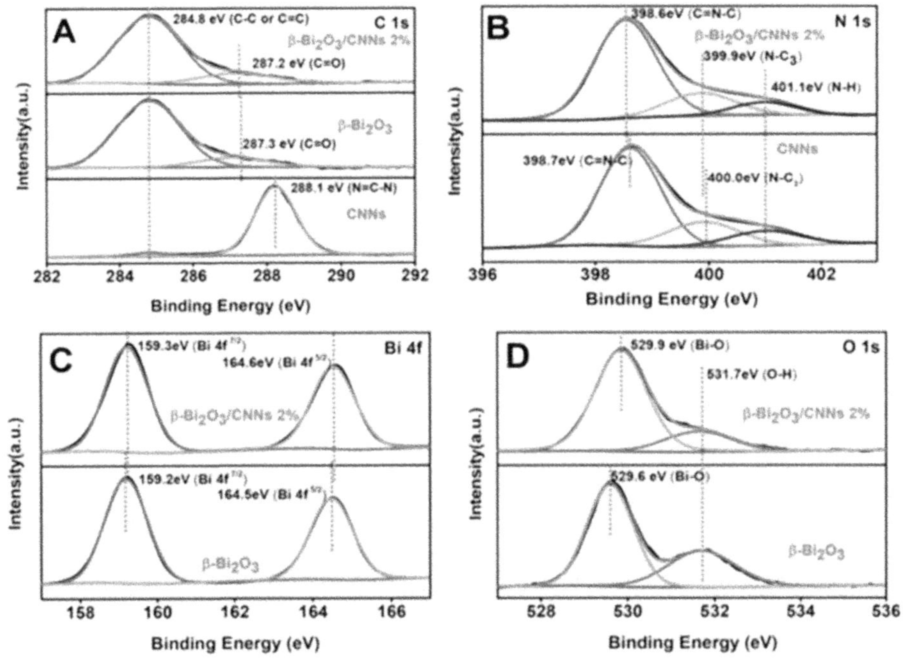

rings containing C and N) and their binary composite. The photocatalyst having 2% CNN showed the highest photocurrent among all prepared binary composites, which is ~12.56 and ~2.78 times higher than pure CNN and β-Bi_2O_3, respectively. The binary composite (2% CNN) has the smallest resistance for charge transfer (276.5 Ω), as obtained from the Nyquist plots, compared to CNN (~1413 Ω) and β-Bi_2O_3 (~1183 Ω), indicating rapid charge transfer across the electrode-electrolyte interface. It is reported that the CNN exhibited n-type semiconductivity whereas the β-Bi_2O_3 showed p-type in nature, confirmed by the Mott-Schottky analysis.

Cui et al· also reported the PEC performance of used MOFs to synthesize ternary composites of MIL-53(Fe), α-Bi_2O_3 and g-C_3N_4 (MBC) using the method of precipitation of crystals of MIL-53(Fe), with varying amounts 10%, 25%, 32% and 40%, on the surface of the binary heterojunction of α-Bi_2O_3 and g-C_3N_4 (α-Bi_2O_3/g-C_3N_4). The 32% MBC provides maximum photocurrent density (1.6 μAm^{-2}) compared to pure samples and heterojunctions, as presented in Fig. 12. The Mott-Schottky plots exhibit the n-type semiconductivity of the prepared photocatalysts and the flat band potential of MIL-53(Fe) was found to be –0.61 V (vs. SCE at pH 7). Due to a lower slope value, the 32% MBC sample exhibited much higher charge carrier density than pristine MIL-53(Fe) and the binary composite of α-Bi_2O_3/g-C_3N_4. From the XRD

Figure 11. Study of Transient photocurrent responses of β-Bi_2O_3/CNNs (2%)

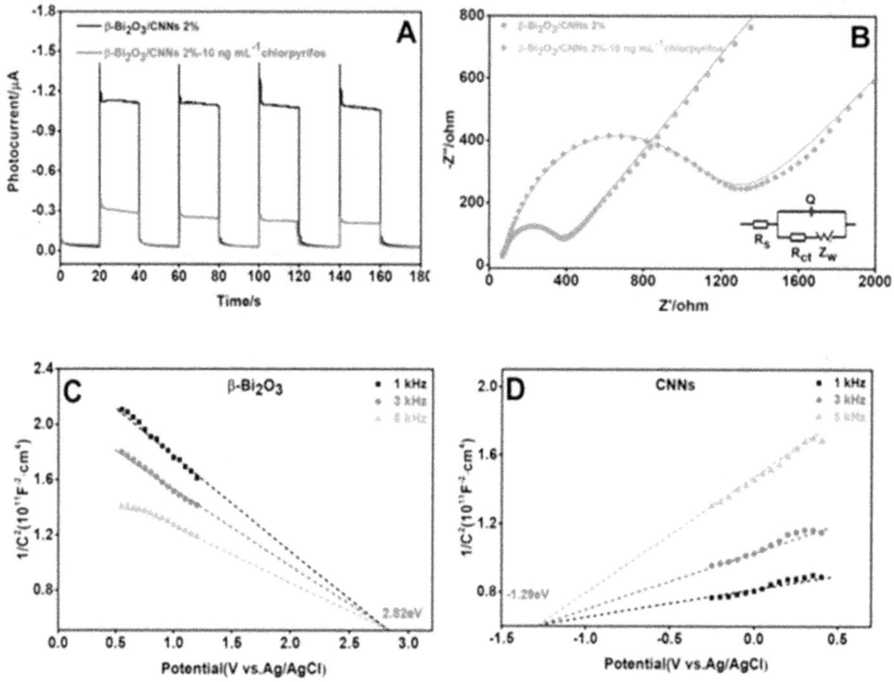

analysis, two characteristic peaks (at 13.09° and 27.47°) for pristine g-C_3N_4 were obtained whereas the monoclinic phase was confirmed for α-Bi_2O_3. The elemental analysis, as revealed from XPS spectra, represents suitable peaks of 1s orbital for C, N and O. Along with these, peaks were obtained for Bi $4f_{5/2}$ and Bi $4f_{7/2}$ and for Fe $2p_{1/2}$ and Fe $2p_{3/2}$. The BET specific surface area of the composite was slightly decreased compared to the pure one, reported in the literature, as presented in Fig. 13. This is due to the covering of MOF porosity through α-BO/g-CN. Generally, the greater specific surface areas can provide more active reaction sites, which promote the transfer of photo-generated carriers in the photocatalysts and also facilitates the adsorption of pollutants molecules over its surface.

Y. Lifabricated nano-composites of Bi_2O_3, modified with g-C_3N_4 which exhibited improved photocatalytic and photoelectrochemical performance. They varied the weigh % of g-C_3N_4 in the entire composite and termed the samples as g-CN (weight %)/BO (where weight %= 18.3%, 53.2%, 71.2%, 87.7%). The composite with 53.2% of g-C_3N_4 exhibited the highest photocurrent when compared not only with pure Bi_2O_3 (~7 folds higher) and pure g-C_3N_4 (~ 2 folds higher) but also among all other synthesized composites. Both g-CN and monoclinic BO-related diffraction peaks

Figure 12. Analysis of Transient photocurrent of g-C_3N_4, α-Bi_2O_3, MIL-53(Fe), α-Bi_2O_3/ g-C_3N_4, and 32%-MBC

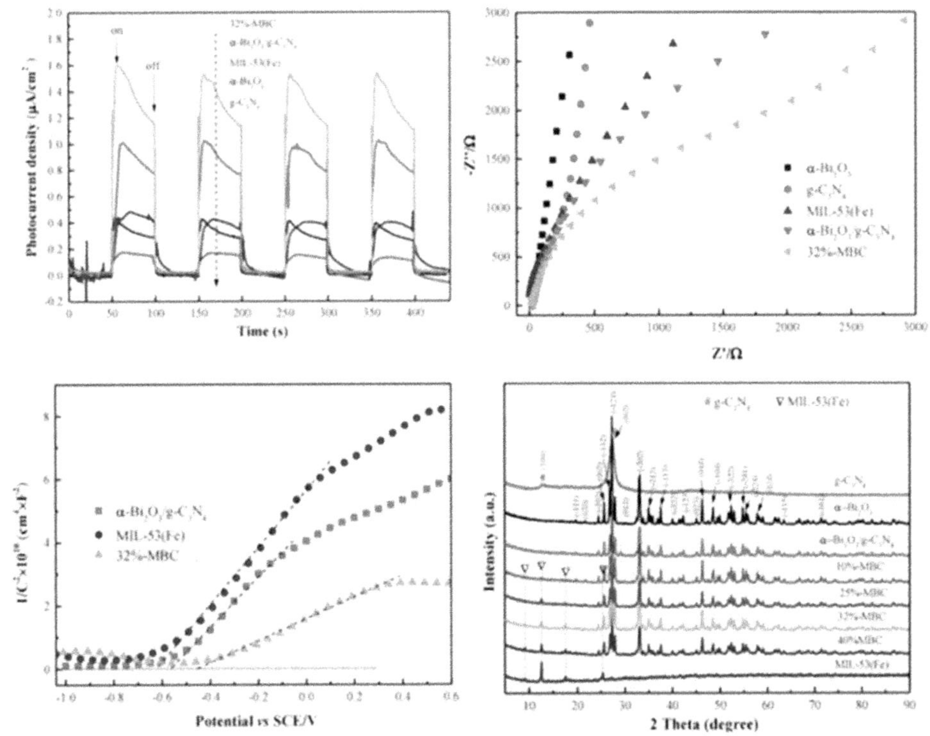

have been noted in the XRD patterns of the composite material. Additionally, as the composition of g-CN in the composite increases, the peak intensity of g-CN also increases gradually.The binding energy for the binary composite appears to have a slight positive shift at 530.7eV for O1s due to the adsorption of oxygen on the surface of the semiconductor composite, as confirmed by XPS analysis. In comparison with N1s binding energy of g-CN, the binary composite with the same energy indicates that the sp^2-hybridized nitrogen (C=N–C)is assigned to the composite material to confirm the presence of sp^2-bonded graphitic carbon nitride. Again Alhaddad et al. fabricated Bi_2O_3/g-C_3N_4 nano-composites through the sonication method to demonstrate as efficient photocatalysts for PEC water splitting application that generally favors the evolution of hydrogen in the presence of visible light. The composites have been developed by varying the content of Bi_2O_3 as 0.5%, 1%, 3% and 5%. The photocurrent density was observed as maximum for the composite loaded with 3% Bi_2O_3. This nano-composite proved to be the most proficient among all the other nano-composites and also when compared with pure g-C_3N_4 and Bi_2O_3.

Figure 13. BET surface analysis of MIL-53(Fe), α-Bi_2O_3/g-C_3N_4 and X-MBC composites (X=10%, 25%, 32% and 40%)

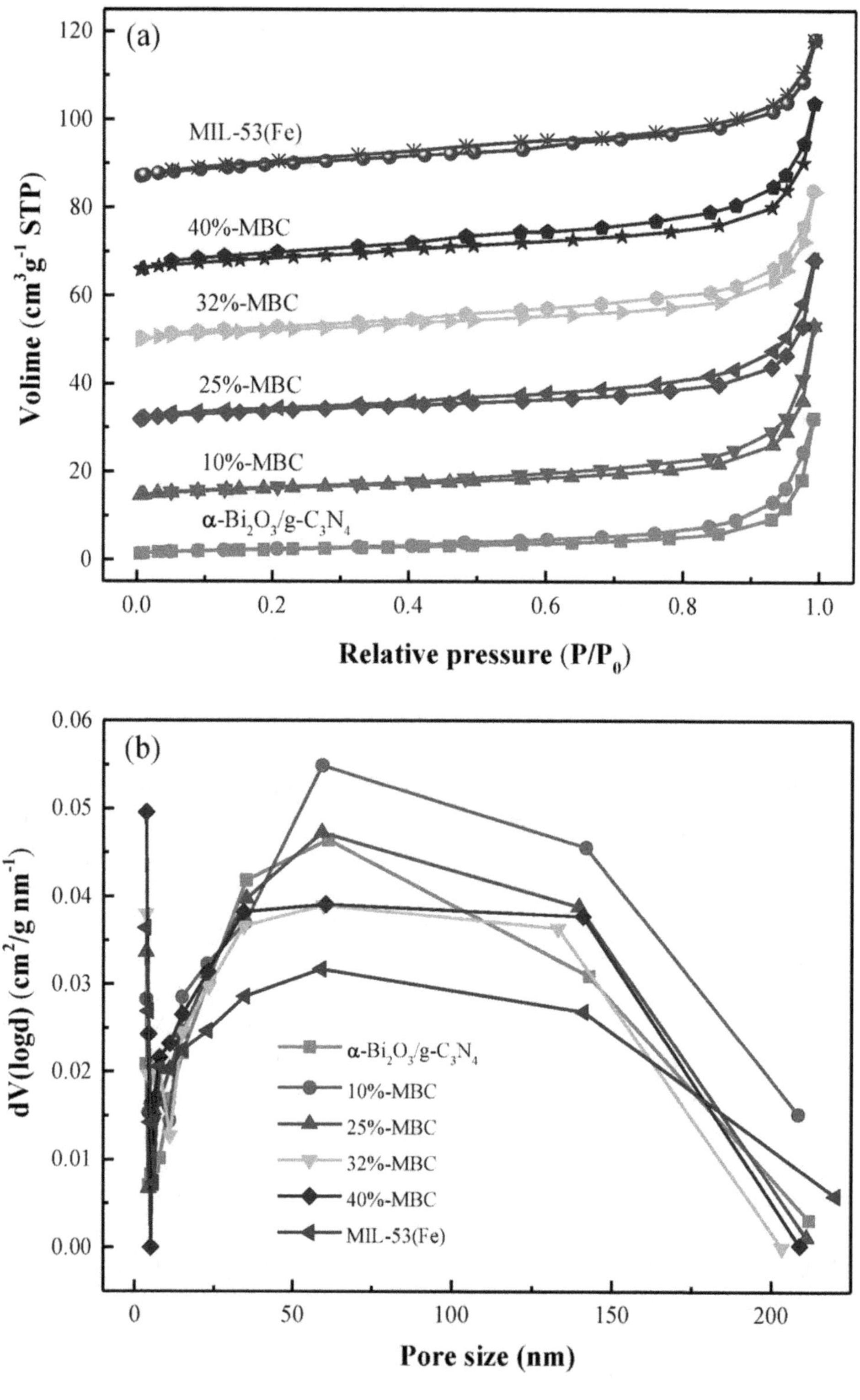

The XRD analysis demonstrated the presence of pure Bi_2O_3 that has low crystalline properties with smaller dimensions. Also the composites indicate the presence of both Bi_2O_3 and g-C_3N_4 according to the various proportions of loaded Bi_2O_3.X. Zhang developed heterojunctions comprising Bi_2O_3 and P-C_3N_4 in order to remove levoflaxin from water. In this work, the author noted that the composite of Bi_2O_3 and P-C_3N_4 showed outstanding results while their Nyquist plots were compared to that of pure Bi_2O_3 and P-C_3N_4 as the radius of the charge transfer resistance and solution resistance appeared to have a much smaller value than that of the pure samples. The XPS results concluded that the binding energy of$4f_{7/2}$of Bi has been shifted slightly (from 158.7 to 159.2 eV) as compared to pristine Bi_2O_3.Y. Hong [l]and his co-workers prepared stable nano-composites by loading different weight ratio of g-C_3N_4onto Bi_2O_3 as 1%, 3%, 5% and 7%. The photocatalyst having 5% g-C_3N_4 loaded onto Bi_2O_3 showed the maximum photoelectrochemical performance among all other composites as well as that of pure Bi_2O_3 or g-C_3N_4. The Nyquist plot of the 5% g-C_3N_4 loaded Bi_2O_3 had a much smaller arc radius that proved the sample was showing highest capacity of doing the experiment. Also the 5% g-C_3N_4 loaded Bi_2O_3 sample exhibited the highest photocurrent among all the other samples.

SUMMARY AND OUTLOOK

Bi-based photocatalysts have enormous potential in the remediation of polluted environments and to transform visible light into a usable form of energy, i.e. chemical or electrical energy. Recent developments of C-based nano-material composites with Bi_2O_3 and $BiVO_4$ in the fields of photocatalytic (PC) and photoelectrochemical (PEC) applications have been selectively summarized and discussed in this report. Due to their unique physiochemical characteristics these composites grabbed the attention of researchers toward visible-light driven photocatalytic activities. The composites have some photocatalytic (PC) ecological applications due to their proper arrangement in crystal, typical electrical capabilities and chemical characteristics. Solar light may generate robust and plentiful reactive species, which has a significant potential for nano-material photocatalysis, particularly nano-photocatalysts. To produce reactive species, a photocatalyst must be able to absorb visible light efficiently. Bismuth (Bi) and C-based nano-materials like g-C_3N_4, rGO, CNT, quantum dot etc. are well-investigated visible-light-active photocatalysts among different semiconductors and are focused as they have excellent chemical stability and bulk availability. The composite efficiently reduced the electron and hole pair recombination to promote interfacial charge transfer. As a result the reactive species generation for photocatalytic process is more facile and photoelectrochemical performance is also enhanced.

However, there are so many significant challenges related to the position of the conduction band, improvement of efficiency in PC performance, broad bandwidth and increase in charge carrier recombination rates that can be discussed from the point of view of practical utilization. Although the composite can be conveniently explored for environmental remediation, further efforts are still necessary to achieve significant advancements and to develop new strategies for the modification of the more efficient C-based nano-materials composite with Bi_2O_3 and $BiVO_4$ semiconductors.

ACKNOWLEDGMENT

The author SG sincerely acknowledges the Department of Science & Technology, Government of India, for financial support (vide reference no. SR/WOS-A/CS-10/2018, dated 02.01.2019) under DST-Women Scientist Scheme - A (DST-WOS-A) to carry out this work. Financial support through the sponsored project grant of the Govt. of India from DST (Sanction Order No. DST/INT/SL/P-33/2021, dt. 08-10-2021) to the Department of Chemistry, IIEST, Shibpur is gratefully acknowledged. The present work was also financially supported by CSIR (File no. 01(2876)/17/EMR-II, 02.05.2017) to the Department of Chemistry, IIESTS.

REFERENCES

Alhaddad, M., Navarro, R. M., Hussein, M. A., & Mohamed, R. M. (2020). Bi_2O_3/g-C_3N_4 nanocomposites as proficient photocatalysts for hydrogen generation from aqueous glycerol solutions beneath visible light. *Ceramics International*, *46*(16), 24873–24881. doi:10.1016/j.ceramint.2020.06.271

Aragaw, B. A. (2020). Reduced graphene oxide-intercalated graphene oxide nano-hybrid for enhanced photoelectrochemical water reduction. *Journal of Nanostructure in Chemistry*, *10*(1), 9–18. doi:10.100740097-019-00324-x

Che, G. L., Lakshmi, B. B., Martin, C. R., & Fisher, E. R. (1999). Metal-nanocluster-filled carbon nanotubes: Catalytic properties and possible applications in electrochemical energy storage and production. *Langmuir*, *15*(3), 750–758. doi:10.1021/la980663i

Chen, X., Liu, Q., Wu, Q., Du, P., Zhu, J., Dai, S., & Yang, S. (2016). Incorporating graphitic carbon nitride (g-C_3N_4) quantum dots into bulk heterojunction polymer solar cells leads to efficiency enhancement. *Advanced Functional Materials*, *26*(11), 1719–1728. doi:10.1002/adfm.201505321

Cui, Y., Nengzi, L., Gou, J., Huang, Y., Li, B., & Cheng, X. (2020). Fabrication of dual Z-scheme MIL-53(Fe)/α-Bi_2O_3/g-C_3N_4 ternary composite with enhanced visible light photocatalytic performance. *Separation and Purification Technology, 232*, 115959. doi:10.1016/j.seppur.2019.115959

Deng, Y., Tang, L., Zeng, G., Zhu, Z., Yan, M., Zhou, Y., Wang, J., Liu, Y., & Wang, J. (2017). Insight into highly efficient simultaneous photocatalytic removal of Cr(VI) and 2,4-diclorophenol under visible light irradiation by phosphorus doped porous ultrathin g-C_3N_4 nano-sheets from aqueous media: Performance and reaction mechanism. *Applied Catalysis B: Environmental, 203*, 343–354. doi:10.1016/j.apcatb.2016.10.046

Dubale, A. A., Su, W. N., Tamirat, A. G., Pan, C. J., Aragaw, B. A., Chen, H. M., Chen, C. H., & Hwang, B. J. (2014). The synergetic effect of graphene on Cu_2O nanowire arrays as a highly efficient hydrogen evolution photocathode in water splitting. *Journal of Materials Chemistry. A, Materials for Energy and Sustainability, 2*(43), 18383–18397. doi:10.1039/C4TA03464C

Fan, G., Ma, Z., Li, X., & Deng, L. (2021). Coupling of Bi_2O_3 nanoparticles with g-C_3N_4 for enhanced photocatalytic degradation of methylene blue. *Ceramics International, 47*(4), 5758–5766. doi:10.1016/j.ceramint.2020.10.162

Fina, F., Callear, S. K., Carins, G. M., & Irvine, J. T. S. (2015). Structural investigation of graphitic carbon nitride via XRD and neutron diffraction. *Chemistry of Materials, 27*(7), 2612–2618. doi:10.1021/acs.chemmater.5b00411

Fujishima, A., & Honda, K. (1972). Electrochemical photolysis of water at a semiconductor electrode. *Nature, 238*(5358), 37–38. doi:10.1038/238037a0 PMID:12635268

Gaillard, N., Chang, Y. C., De Angelis, A., Higgins, S., & Braun, A. (2013). A nanocompositephotoelectrode made of 2.2 eV band gap copper tungstate (CuWO4) and multi-wall carbon nanotubes for solar-assisted water splitting. *International Journal of Hydrogen Energy, 38*(8), 3166–3176. doi:10.1016/j.ijhydene.2012.12.104

Ghosh, S., Hajra, P., Sariket, D., Ray, D., Baduri, S., & Bhattacharya, C. (2022). *Modifications of $BiVO_4$ semiconductors for oxidation of water and detoxification of organic wastes: photoelectrochemical applications of semiconductors*. IGI Global. doi:10.4018/978-1-6684-4553-2.ch001

Ghosh, S., Laha, D., Hajra, P., Sariket, D., Ray, D., Baduri, S., Sahoo, H. S., & Bhattacharya, C. (2023). Development of transition metal incorporated bismuth-based oxide semiconductors as potential candidates for solar assisted wáter splitting applications. *ChemElectroChem, 10*(6), e202201062. doi:10.1002/celc.202201062

Gong, K. P., Du, F., Xia, Z. H., Durstock, M., & Dai, L. M. (2009). Nitrogen-doped carbon nanotube arrays with high electrocatalytic activity for oxygen reduction. *Science*, *323*(5915), 760–764. doi:10.1126cience.1168049 PMID:19197058

Gooding, J. J. (2005). Nanostructuring electrodes with carbon nanotubes: A review on electrochemistry and applications for sensing. *Electrochimica Acta*, *50*(15), 3049–3060. doi:10.1016/j.electacta.2004.08.052

Hajra, P., Kundu, S., Maity, A., & Bhattacharya, C. (2019). Facile photoelectrochemical water oxidation on Co^{2+}-adsorbed $BiVO_4$ thin films synthesized from aqueous solutions. *Chemical Engineering Journal*, *374*, 1221–1230. doi:10.1016/j.cej.2019.06.014

He, R., Zhou, J., Fu, H., Zhang, S., & Jiang, C. (2018). Room-temperature in situ fabrication of Bi_2O_3/g-C_3N_4 direct Z-scheme photocatalyst with enhanced photocatalytic activity. *Applied Surface Science*, *430*, 273–282. doi:10.1016/j.apsusc.2017.07.191

Hong, Y., Li, C., Yin, B., Li, D., Zhang, Z., Mao, B., Fan, W., Gu, W., & Shi, W. (2018). Promoting visible-light-induced photocatalytic degradation of tetracycline by an efficient and stable beta-Bi_2O_3@g-C_3N_4 core/shell nanocomposite. *Chemical Engineering Journal*, *338*, 137–146. doi:10.1016/j.cej.2017.12.108

Huy, B. T., Paeng, D. S., Thao, C. T. B., Phuong, N. T. K., & Lee, Y. I. (2020). ZnO-Bi2O3/graphitic carbon nitride photocatalytic system with H_2O_2-assisted enhanced degradation of Indigo carmine under visible light. *Arabian Journal of Chemistry*, *13*(2), 3780–3800. doi:10.1016/j.arabjc.2019.01.003

Kim, Y. K., & Park, H. (2011). Light-harvesting multi-walled carbon nanotubes and CdS hybrids: Application to photocatalytic hydrogen production from water. *Energy & Environmental Science*, *4*(3), 685–694. doi:10.1039/C0EE00330A

Kroke, E., Schwarz, M., Bordon, E. H., Kroll, P., Noll, B., & Norman, A. D. (2002). Tri-s-triazine derivatives. Part I. From trichloro-tri-s-triazine to graphitic C_3N_4 structures. *New Journal of Chemistry*, *26*(5), 508–512. doi:10.1039/b111062b

Li, Y., Wu, S., Huang, L., Xu, H., Zhang, R., Qu, M., Gao, Q., & Li, H. (2015). g-C3N4 modified Bi2O3 composites with enhanced visible-light photocatalytic activity. *Journal of Physics and Chemistry of Solids*, *76*, 112–119. doi:10.1016/j.jpcs.2014.08.012

Liang, Q., Li, Z., Yu, X., Huang, Z. H., Kang, F., & Yang, Q. H. (2015). Macroscopic 3D porous graphitic carbon nitride monolith for enhanced photocatalytic hydrogen evolution. *Advanced Materials*, *27*(31), 4634–4639. doi:10.1002/adma.201502057 PMID:26135383

Lin, X., Xu, D., Zhao, R., Xi, Y., Zhao, L., Song, M., Zhai, H., Che, G., & Chang, L. (2017). Highly efficient photocatalytic activity of g-C3N4 quantum dots (CNQDs)/Ag/Bi2MoO6 nanoheterostructure under visible light. *Separation and Purification Technology*, *178*, 163–168. doi:10.1016/j.seppur.2017.01.020

Liu, Z. L., Lin, X. H., Lee, J. Y., Zhang, W., Han, M., & Gan, L. M. (2002). Preparation and characterization of platinum-based electrocatalysts on multi-walled carbon nanotubes for proton exchange membrane fuel cells. *Langmuir*, *18*(10), 4054–4060. doi:10.1021/la0116903

Luo, B., Song, R., Geng, J., Jing, D., & Zhang, Y. (2018). Facile preparation with high yield of a 3D porous graphitic carbon nitride for dramatically enhanced photocatalytic H_2 evolution under visible light. *Applied Catalysis B: Environmental*, *238*, 294–301. doi:10.1016/j.apcatb.2018.07.039

Ngo, Y. L. T., Choi, W. M., Chung, J. S., & Hur, S. H. (2019). Highly biocompatible phenylboronic acid-functionalized graphitic carbon nitride quantum dots for the selective glucose sensor. *Sensors and Actuators. B, Chemical*, *282*, 36–44. doi:10.1016/j.snb.2018.11.031

Oh, T., Kim, M., Choi, J., & Kim, J. (2018). Design of graphitic carbon nitride nanowires with captured mesoporous carbon spheres for EDLC electrode materials. *Ionics*, *24*(12), 3957–3965. doi:10.100711581-018-2544-0

Rahman, M. Z., Davey, K., & Mullins, C. B. (2018). Tuning the intrinsic properties of carbon nitride for high quantum yield photocatalytic hydrogen production. *Advancement of Science*, *5*(10), 1800820. doi:10.1002/advs.201800820 PMID:30356987

Reddy, I. N., Jang, W. Y., Manjunath, V., & Shim, J. (2021). Fe_3O_4-Bi_2O_3 nanostructures for efficient energy generation application-water oxidation under visible light irradiation. *Journal of Electroanalytical Chemistry (Lausanne, Switzerland)*, *895*, 115484. doi:10.1016/j.jelechem.2021.115484

Rono, N., Kibet, J. K., Martincigh, B. S., & Nyamori, V. O. (2020). A review of the current status of graphitic carbon nitride. *Critical Reviews in Solid State and Material Sciences*, *46*(3), 189–217. doi:10.1080/10408436.2019.1709414

Sahnesarayi, M. K., Rastegari, S., & Sarpoolaky, H. (2022). Enhanced photoelectrochemical water splitting performance of vertically aligned Bi_2O_3 nanosheet arrays derived from chemical bath deposition method by controlling chemical bath temperature and complexing agent concentration. *Surfaces and Interfaces*, *30*, 101819. doi:10.1016/j.surfin.2022.101819

Shaddad, M. N., Arunachalam, P., Hezam, M. A. L., Saeedan, N. M., Gimenez, S., Bisquert, J., & Al Mayouf, A. M. (2021). Unprecedented solar water splitting of dendritic nanostructured Bi_2O_3 films by combined oxygen vacancy formation and Na_2MoO_4 doping. *International Journal of Hydrogen Energy*, *46*(46), 23702–23714. doi:10.1016/j.ijhydene.2021.04.184

Sharma, S., Mehta, S. K., Ibhadon, A. O., & Kansal, S. K. (2019). Fabrication of novel carbon quantum dots modified bismuth oxide (α-BO/C-dots): Material properties and catalytic applications. *Journal of Colloid and Interface Science*, *533*, 227–237. doi:10.1016/j.jcis.2018.08.056 PMID:30165300

Sun, S. P., Gu, S., Sun, J. H., Xia, F. F., & Chen, G. H. (2018). First principles investigation of the electronic properties of graphitic carbon nitride with different building block and sheet staggered arrangement. *Journal of Alloys and Compounds*, *735*, 131–139. doi:10.1016/j.jallcom.2017.11.061

Tang, Z., Zhang, X., Duan, L., Wu, A., & Lu, W. (2019). Three-Dimensional Carbon Nitride Nanowire Scaffold for Flexible Supercapacitors. *Nanoscale Research Letters*, *14*(1), 98. doi:10.118611671-019-2932-z PMID:30874966

Wang, H., Liang, D., Xu, Y., Liang, X., Qiu, X., & Lin, Z. (2021). A highly efficient photoelectrochemical sensor for detection of chlorpyrifos based on 2D/2D β-Bi_2O_3/g-C_3N_4 heterojunctions. *Environmental Science. Nano*, *8*(3), 773–783. doi:10.1039/D0EN01243B

Wang, W., Yu, J. C., Shen, Z., Chan, D. K. L., & Gu, T. (2014). g-C_3N_4 quantum dots: Direct synthesis, upconversion properties and photocatalytic application. *Chemical Communications (Cambridge)*, *50*(70), 10148–10150. doi:10.1039/C4CC02543A PMID:25052166

Wang, X., Maeda, K., Chen, X., Takanabe, K., Domen, K., Hou, Y., Fu, X., & Antonietti, M. (2009). Polymer semiconductors for artificial photosynthesis: Hydrogen evolution by mesoporousgraphitic carbon nitride with visible light. *Journal of the American Chemical Society*, *131*(5), 1680–1681. doi:10.1021/ja809307s PMID:19191697

Woan, K., Pyrgiotakis, G., & Sigmund, W. (2009). Photocatalytic Carbon-Nanotube-TiO_2 Composites. *Advanced Materials*, *21*(21), 2233–2239. doi:10.1002/adma.200802738

Wu, M., Zhang, J., Liu, C., Gong, Y., Wang, R., He, B., & Wang, H. (2018). Rational design and fabrication of noble-metal-free NixPco-catalyst embedded 3D N-TiO_2/g-C_3N_4 heterojunctions with enhanced photocatalytic hydrogen evolution. *ChemCatChem*, *10*(14), 3069–3077. doi:10.1002/cctc.201800197

Xie, M., Wei, W., Jiang, Z., Xu, Y., & Xie, J. (2016). Carbon nitride nanowires/nanofibers: A novel template-free synthesis from a cyanuric chloride-melamine precursor towards enhanced adsorption and visible-light photocatalytic performance. *Ceramics International*, *42*(3), 4158–4170. doi:10.1016/j.ceramint.2015.11.089

Xue, S., Hou, X., Xie, W., Wei, X., & He, D. (2015). Dramatic improvement of photocatalytic activity for N-doped Bi2O3/g-C3N4 composites. *Materials Letters*, *161*, 640–643. doi:10.1016/j.matlet.2015.09.067

Yan, X., Pu, R., Xie, R., Zhang, B., Shi, Y., Liu, W., Ma, G., & Yang, N. (2021). Design and fabrication of $Bi_2O_3/BiFeO_3$ heterojunction film with improved photoelectrochemical performance. *Applied Surface Science*, *552*, 149442. doi:10.1016/j.apsusc.2021.149442

Zhang, J., Hu, Y., Jiang, X., Chen, S., Meng, S., & Fu, X. (2014). Design of a direct Z-scheme photocatalyst: Preparation and characterization of $Bi_2O_3/g\text{-}C_3N_4$ with high visible light activity. *Journal of Hazardous Materials*, *280*, 713–722. doi:10.1016/j.jhazmat.2014.08.055 PMID:25232654

Zhang, L., Wang, G., Xiong, Z., Tang, H., & Jiang, C. (2018). Fabrication of flower-like direct Z-scheme $\beta\text{-}Bi_2O_3/g\text{-}C_3N_4$ photocatalyst with enhanced visible light photoactivity for Rhodamine B degradation. *Applied Surface Science*, *436*, 162–171. doi:10.1016/j.apsusc.2017.11.280

Zhang, X., Zhang, Y., Jia, X., Zhang, N., Xia, R., Zhang, X., Wang, Z., & Yu, M. (2021). In situ fabrication of a novel S-scheme heterojunctionphotocatalyts Bi2O3/P-C3N4 to enhance levofloxacin removal from water. *Separation and Purification Technology*, *268*, 118691. doi:10.1016/j.seppur.2021.118691

Zhao, H., Zhang, J., & Zheng, H. (2020). Facile preparation of dual Z-scheme $Bi_2O_3/g\text{-}C_3N_4/Ag_6Si_2O_7$ photocatalyst for highly efficient visible-light photocatalytic degradation of organic pollutants. *Applied Surface Science*, *527*, 146901. doi:10.1016/j.apsusc.2020.146901

Zhu, Z. Z., Wang, Z., & Li, H. L. (2009). Self-assembly of palladium nanoparticles on functional multi-walled carbon nanotubes for formaldehyde oxidation. *Journal of Power Sources*, *186*(2), 339–343. doi:10.1016/j.jpowsour.2008.10.006

Chapter 4
Titanium Oxide for Photodegradation of Organic Pollutants:
Synthesis, Limitations, and Future Prospects

Wilfrida N. Nyairo
Maseno Universiy, Kenya

Victor Odhiambo Shikuku
https://orcid.org/0000-0002-2295-293X
Kaimosi Friends University, Kenya

ABSTRACT

Water scarcity has been exacerbated by environmental pollution caused by increasing global industrialization and urbanization. This has led to continued research for more ecologically safe and cleaner technologies for eliminating pollutants. In the recent past, nano-engineered metal oxides such as titanium oxide (TiO_2) have shown promising photocatalytic effect of completely mineralizing organic pollutants to form harmless byproducts. TiO_2 is the most frequently used photocatalyst, however its effectiveness is limited to light absorption within the ultra violet region due to its wide band-gap. A redshift by TiO_2 has been achieved through tuning of the anatase-rutile phase transitions, doping, surface modifications and creation of multi component heterojunctions. This chapter discusses the synthesis of TiO_2–based photocatalysts and photocatalytic activity enhancement of TiO_2 through doping and formation of hybrids. Their application in the photo-degradation of dyes and other organic pollutants as well as their limitations and future prospects are also addressed.

DOI: 10.4018/978-1-6684-8743-3.ch004

INTRODUCTION

Availability of clean drinking water in the world today is increasingly becoming a challenge due to the pollution of water resources by effluents emanating from industries and run-offs from agricultural fields. Among the pollutants released to the aquatic systems are heavy metals, dyes, pesticides among other organic pollutants which pause potential harm to both humans and aquatic life. Several techniques have been employed in the removal of these pollutants from water including reverse osmosis, coagulation, precipitation, electrochemical processes, membrane filtration, adsorption and irradiation or advanced oxidation processes (Ngeno et al., 2022; Tome et al., 2023). However, collectively these technologies have diverse drawbacks such as need for expensive equipment, trained personnel and production of secondary pollutants that require further treatment. Moreover, non-biodegradable organic compounds are difficult to remove by the conventional water treatment techniques such as coagulation and precipitation (Poole, 2004).

Amongst these water treatment techniques, advanced oxidation process (AOP) has demonstrated to be the most effective alternative route for removal of these organic pollutants from aqueous medium (Shikuku & Nyairo, 2019). AOP is a chemical process that involves the use of highly reactive hydroxyl (•OH) radicals to completely destroy a wide range of organic pollutants to mineralization with no selectivity. Various oxidation processes used to produce the hydroxyl species by use of UV light include Fenton's reaction, H_2O_2/UV, O_3/UV, H_2O_2/O_3/UV, TiO_2/UV among others (Xiang, Yu, & Wong, 2011). Fenton oxidation utilises H_2O_2 and Fe^{2+} ion source to produce •OH radicals to decompose organic molecules. Fenton's approach is considered to be fast and easy to apply. However, it has the disadvantages of high maintenance costs and the need for iron sequestration from the effluent system after use (Bello, Raman, & Asghar, 2019). On the other hand, coupling of TiO_2, a heterogeneous catalyst, with UV for the production of •OH offers the advantage of integrating it into the water treatment. Furthermore, TiO_2/UV photocatalysis can be carried out at room temperature and it makes use of atmospheric oxygen as the oxidant for complete mineralization of organic pollutants. As a result, TiO_2/UV has proved to be the most effective AOP and in the recent past, therefore, the photocatalytic process using TiO_2 has received immense attention (Lee & Park, 2013). The merits and demerits of various oxidation processes and photocatalysis are enumerated in Table 1.

The technology of applying semi-conducting solids such as TiO_2, ZnO, ZnS/CuS, Fe_2O_3, $ZnWO_4$, WO_3 among others, for their catalytic activity has been studied widely. In environmental remediation, these semi-conductors have been used as ultra violet or visible light driven photocatalysts that offer ecologically safe and cleaner degradation of organic pollutants by use of advanced oxidation process (Devi &

Table 1. The merits and demerits of various oxidation processes and photocatalysis.

AOP	Advantages	Disadvantages	Ref.
Fenton reaction	Wide coverage. Simple and easy to use. No off gas treatment required.	Requirement for iron recovery from the system. Relatively high maintenance costs involved. Works only at low pH.	(Bello, Raman, & Asghar, 2019)
H_2O_2/UV	No off gas treatment required. More effective than UV or H_2O_2 used independently.	Gives turbidity which inhibits UV light penetration. Contamination caused by UV lamp failure.	(Poole, 2004)
O_3/UV	Produces more •OH than H_2O_2/UV.	Relatively costly. Contamination due to failure of UV lamp.	(Lim, Shi, von Gunten, & McCurry, 2022)
Photocatalysis	Use of solar irradiation. Low operational and installation costs. Low energy requirement compared to other AOPs. Unmanned operation is possible.	It is pH sensitive. Reduction of catalytic activity with time.	(Kwon, Fan, Cooper, & Yang, 2008)

Kavitha, 2013). TiO_2 catalyst under UV light or solar irradiation has gained a lot of attention for degradation of organic pollutants due to its relatively low cost, chemical stability, biocompatibility and photocatalytic activity. The effectiveness of TiO_2 photocatalyst is hindered by its inability to work within the full light range, fast recombination of the photo induced charges (h^+/e^-) and migration of the photogenerated electrons (Zu, et al., 2021). The modification of TiO_2 through doping, surface modifications, use of heterojunctions and manipulation of the anatase-rutile phases has been employed in order to improve its photocatalytic performance (Low, Yu, Janroniec, Wageh, & Al-Ghamdi, 2017). The integration of nanotechnology to form nano-engineered catalysts (usually 1-100nm) is largely correlated to higher performance ranges of the photocatalysts for water treatments. Studies show that the nano-sized TiO_2 catalysts are more effective photocatalysts due to their higher surface area to volume ratio compared to the micro-sized catalysts (Kwon, Fan, Cooper, & Yang, 2008; Xu, Ravi Anusuyadevi, Aymonier, Luque, & Marre, 2019). However, the long-life exposure effects of these nanomaterials to human health is a debatable subject (Baranowska-Wojak, Szwajgeir, Oleszczuk, & Winiarsk-Mieczan, 2019). Another challenge associated with the nano-scale phenomenon is the need for nanoparticle recovery from the water treatment systems.

This chapter gives highlights on the techniques of the synthesis and modification of TiO_2-based photocatalysts, the treatment of wastewater containing organic pollutants using the micro-scale and nano-engineered TiO_2-based photocatalysts, the benefits of using engineered TiO_2-based photocalysts over conventional ones. The principle and mechanisms involved in catalytic photodegradation of organic pollutants and the influencing conditions involved as well as the limitations of these photocatalysts.

TIO_2 AS A PHOTOCATALYST AND THE PHOTOCATALYTIC MECHANISM

Photocatalysis is a solar photo reaction that occurs when light strikes a photo sensitive chemical compound (photocatalyst) such as TiO_2. The semiconductor has stable valence electrons occupying the valence band (VB) and an empty higher energy conduction band (CB). When illuminated, the TiO_2 photocatalyst absorbs photons with energy equal or greater than its band gap energy (3.0-3.2 eV). These supra band gap photons cause excitation of electrons from VB to the CB of the semiconductor. The photo induced charge carriers, positive holes (h^+) and electrons (e^- then initiate reduction and oxidation reactions that generate highly reactive oxidative species at VB and the CB, respectively. The organic molecules are then destroyed by the highly reactive •OH radicals through a series of reactions on the surface of the semiconductor (Millis & Le Hunte, 1997; Fugishima, Zhang, & Tyrk, 2008).

Water treatment by advanced oxidative technology through photocatalysis using TiO_2 is, therefore, achieved by the photo induced charge separation on the surface of the semiconductor and generation of •OH radicals which cause mineralization of the recalcitrant pollutants. When the TiO_2 integrated in water treatment system is irradiated with UV light, the photo induced positive holes on the VB of TiO_2 react with adsorbed water molecules to form •OH. On the other hand, the photo induced electrons on the CB, react with dissolved oxygen in the air to form superoxide radical anions ($\bullet O_2^-$). The •OH plays the central role in the photocatalytic oxidative process, the general reaction mechanism of the photo induced charge carriers (h^+/e^-) is as suggested in the equations below (Xiang, Yu, & Wong, 2011);

The photo excitation that generates h^+/e^- pairs: $TiO_2 + hv \rightarrow e^- + h^+$ (1)

Ionization of water: $h^+ + H_2O \rightarrow OH^- + H^+$ (2)

Protonation of hydroxyls: $h^+ + OH^- \rightarrow \cdot OH + H^+$ (3)

Reduction of oxygen to form oxide radicals: $O_2 + e^- \rightarrow \cdot O_2^-$ (4)

Protonation of superoxide: $\cdot O_2^- + \cdot OH \rightarrow \cdot OOH$ (5)

Photodegradation: $RH + \cdot OH \rightarrow \cdot R' + H_2O$ (6)

Scavenging of e$^-$: $\cdot OOH + e^- \rightarrow HO_2^-$ (7)

Formation of H_2O_2: $HO_2^- + H^+ \rightarrow H_2O_2$ (8)

UNDOPED AND DOPED TIO$_2$

The polymorphs of TiO_2 that show high photocatalytic activity are anatase and rutile, and they have reported band gap energy values of 3.2 eV and 3.0 eV, respectively. This implies that their photocatalytic activity of TiO_2 operates at UV irradiation of a wavelength < 400 nm. Hence, the optimization and application of this transition phase has been widely investigated with the most commonly utilized TiO_2 being the commercial mixed phase of anatase and rutile known as Degussa P25 (Dahl, Liu, & Yin, 2014). The mixed phase takes advantage of the synergistic effects of the redshift through excitation of the rutile phase and decreased recombination rate of the photo-induced charges from the anatase TiO_2 (Hurum, Gray, Rajh, & Thurnaer, 2004).

The application of TiO_2 photocatalyst for water treatment is a widely utilized technique because it is fast and photo efficient. However, the undoped TiO_2 photocatalyst, Degussa P25 can only utilize 5% of sunlight due to its wide band gap. The band gap of TiO_2 photocatalysts has been altered by doping with other elements, coupling with other materials and tuning both the structure and properties of TiO_2 in order to obtain a broad spectrum in the visible light range. Doping aims at promoting the separation of charge carriers by introducing a new impurity energy level above the valence band of the semiconductor (Khlyustrova, et al., 2020).

SYNTHESIS AND MODIFICATION OF TIO_2-BASED PHOTOCATALYSTS

The structural properties and functional performance of TiO_2 is dependent on the method of synthesis, the doping method and type of dopant (Javanbakht & Mohammadian, 2021). Numerous approaches have been employed in enhancing the effectiveness of TiO_2, and the most established is the use of nano-engineered particles which have shown to have better photocatalytic activities compared to bulk powder (Kwon, Fan, Cooper, & Yang, 2008). The nanoparticles have higher surface area to volume ratio which provides better charge carrier separation and trapping of ions on the surface of TiO_2 and therefore have improved performance in the photodegradation process in water treatment systems. Nanotechnology has not only allowed for the manipulation of size of TiO_2 particles with the aim of optimizing their photocatalytic performance but also modification of their shape. As a result, different morphologies of TiO_2 particles such as microspheres, fibres, rods, nanotubes and nanosheets have been studied. Compared to the unmodified TiO_2 (Degussa P25), the TiO_2 with modified morphology have shown improved adsorption capabilities (Zhao & Zhu, 2017).

Methods of Synthesis

Numerous fabrication methodologies for TiO_2 particles have been reported and they include the wet chemistry methods such as hydrothermal (Wang G., 2007) or solvothermal (Wang, Tang, Leiw, & Lim, 2011), sol-gel (Mohammed, 2020), microwave assisted (Muthuvel, Said, Jothibas, Gurushankar, & Mohana, 2021), sonochemical (Potdar, Huang, Praveen, Manickam, & Sonawane, 2022) and ionic liquid assisted synthesis (Diaz-Sanchez, et al., 2021). The sol-gel process involves dissolving the precursor materials in a solvent while managing the pH until the gel forms. The gel is then separated and dried, pulverized and calcinated in order to yield the TiO_2 nanoparticles as shown in Fig 1. The process is a simple low temperature method that offers the merits of short duration, completely uniform mixture and low calcination temperature. The product formed after calcination is highly crystalline and possesses high specific surface area (Zhang, Zhang, Chen, & Oh, 2012). Hydrothermal and solvothermal approaches are similar except that the former utilises water as a solvent. In a typical a process, the precursor material and the solvent are placed in a sealed autoclave operating at a specified temperature and pressure. The merits of the processes include the ability to control variables such as pH, temperature and additives (Wang, Tang, Leiw, & Lim, 2011). In the sonochemical approach the reactants undergo a chemical reaction through the application of powerful ultrasound radiation using a sonocator probe (Potdar, Huang,

Praveen, Manickam, & Sonawane, 2022). Microwave assisted technique employs microwave irradiation for the reacting molecules to interact. It is associated with high reaction rate, simplicity and efficiency (Muthuvel, Said, Jothibas, Gurushankar, & Mohana, 2021).

Despite being the most established fabrication methods for TiO_2 particles, the wet chemistry process involves many steps followed by calcination of the precipitates formed at optimum temperatures to obtain the nanoparticles. There are alternative approaches which offer shorter routes such as electro-deposition, chemical vapour deposition (Li, Ismat Shah, Huang, Jung, & Ni, 2002), flame spray pyrolysis (Ismail, Hedhili, Anjum, Singaravelu, & Chung, 2021) among others.

Figure 1. Typical procedure of doped TiO2 preparation by sol-gel method (Zaleska, 2008)

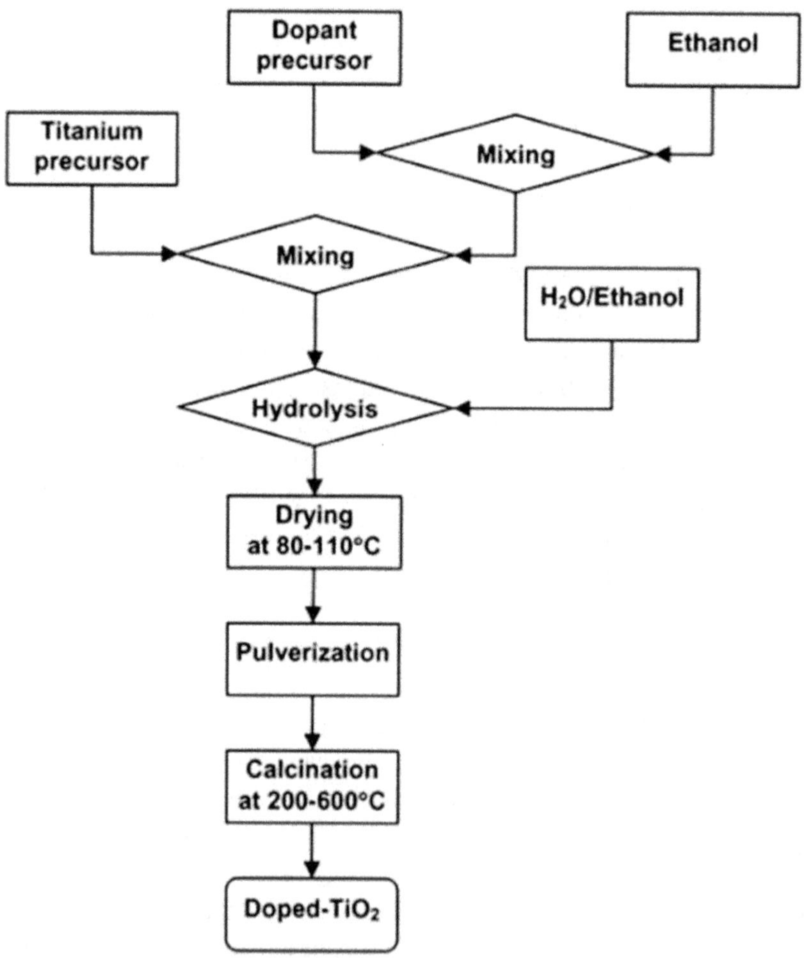

Doping With Metals and Non-Metals

Doping of TiO_2 surfaces through the introduction of metallic nanoparticles such as Ag (Abbad, et al., 2020), Fe (Ismail, Hedhili, Anjum, Singaravelu, & Chung, 2021), Au (Mohammed, 2020), Pt (Potdar, Huang, Praveen, Manickam, & Sonawane, 2022), Pd (Hariharan, et al., 2019), Al, Cu, Mo and W (Khlyustrova, et al., 2020) or co-doping Pt/Pd (Zhang, Zhang, Chen, & Oh, 2012) has been seen to enhance its photocatalytic activity. The metal alters the photo induced electrons and holes recombination behaviour, by trapping the photo induced electrons hence causing the positive holes of the TiO_2 surface to become stable. As a result, the lifetime of the photo induced holes is increased thus more radicals are produced and this improves the redox reaction. Besides incorporating metals, substituting of the O atoms on TiO_2 lattice with heteroatoms such as N (Ansari, Khan, Ansari, & Cho, 2013), S (Cravanzola, Cesano, Geziano, & Scarano, 2017), B (Liu, et al., 2012), F (Diaz-Sanchez, et al., 2021) among others has been reported to improve the performance of TiO_2. The presence of these heteroatoms enables the TiO_2 photocatalysts to work in the full light range.

The nature of dopant influences the characteristics and performance of TiO_2 in photocatalytic activity. For instance, a study conducted by Khlyustrova et al. (2020) compared the activity of TiO_2 doped with different elements for the photodegradation of Rhodamine B. The catalyst was synthesized through sol-gel method, and a low-temperature underwater plasma doping method was used with Al, Cu, Mo and W acting as electrodes. It was found that the introduction of Al and Cu increased its photocatalytic activity to 70% while doping with Mo and W increased TiO_2 activity to 90% in 60 min visible light irradiation.

Co-doped TiO_2 nanomaterials have also been seen to possess better photocatalytic capabilities than those TiO_2 doped with the corresponding single elements. As observed by Zhang and workers (2012), Pt/Pd doped TiO_2 synthesized *via* sol-gel process showed significantly higher photodegradation activity for methyl orange compared to the individual Pt and Pd doped TiO_2. This phenomenon was attributed to simultaneous effect of Pt and Pd in cascading electron transfer from TiO_2 to Pt and Pd hence reducing electron-hole recombination rates (Zhang, Zhang, Chen, & Oh, 2012)

Studies have also shown that the precursor material for TiO_2 and the concentration of TiO_2 influence its activity. A study investigated the photodegradation of 4-cholorophenol by N-doped TiO_2 prepared using different TiO_2 precursors but the same N precursor material-urea. The catalysts were synthesized through a solution based method using titanium tetraisopropoxide and titanium tetrachloride and urea, and it was observed that the titanium tetrachloride and thiourea combination showed superior photocatalytic activity (Sakthivel, Janczarek, & Kisch, 2004)

Hybrids of TiO_2 Nanocomposites

Hybrids of TiO_2 nanocomposites have also been reported to promote charge carrier separation and also act as charge carrier sinks. In addition, the hybrids play the synergistic effect of co-catalysts in promoting specific reactions (Prasannalakshmi & Shanmugam, 2017). Anchoring TiO_2 with graphene oxide (GO), carbon nanotubes (CNTs), graphitic carbon nitride (g-C_3N_4) semiconductors, metal-organic frameworks (MOFs) prevents recombination of charge carries and increases the lifetime of the photo induced electrons. On the other hand, when applied directly in water treatment nanoscale TiO_2 tend to aggregate during advanced oxidation process but these anchors offer useful interface between the photocatalysts and the aqueous organic pollutants (Zhang, et al., 2022).

Recently nanocarbon/TiO_2 composites have increasingly gained attention for photocatalytic activities (Thambiliyagodage, 2022). The carbon nanomaterials, particularly GO, CNTs and g-C_3N_4 have the potential of enhancing the photocatalytic performance of TiO_2 due to their inherent properties such as large specific surface area, excellent active sites, and inhibition of e^-/h^+ recombination and red shift catalysis through manipulation of the band gap (Mohammed, 2020). In the evaluation of CNTs/TiO_2 nanocomposites prepared by sol–gel methods for the degradation of Rhodamine B dye under simulated sunlight, the findings showed that the CNT enhanced the composite photocatalytic activity to approximately twice that of prisitine TiO_2 (Gao, Chen, & Li Puma, 2009). Similarly, studies have shown that GO enhances the photocatalytic activity of TiO_2 (Nguyen & Juang, 2019; Hasan, Li, Tian, & Qin, 2020). According to Nguyen & Juang (2019), the evaluation of the composite r-GO/TNT prepared by hydrothermal process showed that the incorporation reduced GO (r-GO) onto titanium nanotubes (TNT) improved TNT's photocatalytic activity in the photodegradation of methylene blue dye. This was ascribed to the absorption intensity of visible light and separation of photoiduced carrier charges brought about by the presence of the r-GO.

Coupling of TiO_2 with other semiconductors is another strategy applied in order to achieve effective transfer of carrier charges in photocatalysis. Studies have shown that TiO_2 combined with other semiconductors such as ZnO, ZnS, CuS, Fe_2O_3, $ZnWO_4$, WO_3, CdS results to improved photocatalytic performance of TiO_2. Meng and group prepared WO_3/TiO_2 composites of different weight ratios and observed that regardless of their composition the photocatalytic activities of the WO_3/TiO_2 composites were much better that those of pure WO_3 or TiO_2 applied individually (Meng, et al., 2018).

Metal organic frameworks (MOFs) are a class of porous crystalline materials with enormous internal surface area comprised of metal ions or clusters harmonized by organic ligands. The novel enhanced crystal hybrid substances have been investigated and reported for photocatalytic activity because of their chemical and structural tunability (Liu, et al., 2022). A highly efficient MOF derived TiO_2/carbon nitride (TCN-x) was fabricated *via* pyrolysis of a mixture of NH_2-MIL 125/melamine to investigate the photodecomposition of tetracycline (TC) and methylene blue dye (MB). The study reported that compared to pristine TiO_2 and carbon nitride, the MOF supported catalyst, TCN-x had better photocatalytic activity of 91.40% in 120 min for TC and 86.59% in 60 min for MB. This excellent performance was ascribed to improved separation and migration of the photo induced charge carriers due to the MOF (Liu, et al., 2022).

Most of the aforementioned methodologies for production and modification of TiO_2 photocatalysts are associated with good results but they have inherent drawbacks. These methods are energy intensive, expensive and they require advanced tools, they also use toxic chemicals and thus not eco-friendly. To address these challenges, a lot of attention has been focused on sustainable fabrication of nanoparticles through green synthesis of TiO_2 photocatalysts (Zeebaree, et al., 2022). This approach utilizes bio-reduction of metal salts using microorganisms or plant extracts from leaves, flowers, stems, roots or fruits (Nabi, et al., 2018). For instance, metal oxide particles synthesized biologically using lemon leaves extract were used to investigate the photodegradation of methylene blue. The powder calcinated at 150-450 °C, yielded anatase TiO_2 nanopaticles of 6-14 nm which showed photodegradation of 94% towards methylene blue dye within 180 min at pH 11 (Vishali, Manikandan, John, & Murali, 2021).

PHOTODEGRADATION OF ORGANIC POLLUTANTS USING TIO_2 PHOTOCATALYSTS

Photocatalysis is an effective and eco-friendly technique for removing organic and recalcitrant compounds from wastewater. This technique can effectively remove contaminants such as dyes, pesticides, pharmaceuticals, and other organic compounds from water and wastewater. Due to its biocompatibility, affordability and photostability TiO_2 has demonstrated to be the most effective catalyst for the photodegradation of organic pollutants. This section discusses the application of TiO_2 photocatalysts in the removal of dyes such as have such as methylene blue (MB), Rhodamine B (RhB), methyl orange (MO) among others, and other organic compounds such as phenols and pharmaceutical compounds, the mechanism and factors influencing their photocatalytic activity.

Photocatalytic Degradation of Organic Dye Compounds by TiO_2

Textile wastewater is a major environmental concern and several efforts have been made towards the removal of the dyes present in such waters. Dyes are bio-persistent and highly soluble in water and therefore, conventional water treatment techniques have become increasingly ineffective in eliminating these dyes from water. TiO_2 photocatalysis has shown to be a promising approach in the mineralization of these organic dyes (Diaz-Sanchez, et al., 2021). Table 2 shows a summary of some of the TiO_2 photocatalysts applied in the photodegradation of dyes and the operation parameters.

Several factors influence the photocatalytic degradation of organic pollutants in the presence of a photocatalyst. The operation parameters include: the catalytic preparation method, initial concentration of the pollutant, presence of dopant or anchor, catalytic loading, pH, temperature, light intensity, and irradiation time. For instance, F-doped TiO_2 catalyst with high surface area that obtained by incorporating an ionic liquid as a structure directing agent during synthesis was evaluated for mineralization of selected organic pollutants. The study observed that the multifunctional photocatalyst prepared through a combination of hydrothermal and sol-gel process presented a high catalytic efficiency in the photodegradation of MB, ciprofloxacin and naproxen. The findings reported a photocatalytic degradation of up 96% for MB, 90% for ciprofloxacin and 79% for naproxen under UV light (300 W) in 10 min (Diaz-Sanchez, et al., 2021). A ternary compound of Cr_2S_3-GO/TiO_2 prepared by combining GO and TiO_2 *via* sol-gel exhibited photodegradation of 98.3, 96.6, and 86.3% for MB, Rh B and MO, respectively under visible light for 120 min using 0.3 mol TiO_2 loading. The efficiency of the hybrid material was attributed to the synergistic effects of the photocatalyst Cr_2S_3 and increased defects on TiO_2 caused by the presence of GO (Hasan, Li, Tian, & Qin, 2020). Mohammed (2020) investigated the efficiencies of TiO_2, TiO_2/Au and TiO_2/Au/SWCNTs catalysts for the photodegaradation of MB dye under solar simulated light at room temperature. The study observed that only the ternary nanohybrid absorbed radiation in the visible light range. At the same time, the ternary composite had the highest photocatalytic activity of 77% after 30 min of visible light illumination, this was attributed to the narrow band due to the presence of SWCNTs. A proposed CNTs mechanisms in enhancing photocatalytic activity is that CNTs create Schottky barrier at TiO_2/CNTs interface, thus causing photo induced electrons to migrate to CNTs surface because it has a lower Fermi level as that of a metal (Mohammed, 2020). This implies that the visible light driven photocatalytic activity of TiO_2 was not influenced by Au alone but SWCNTs played the role of efficiently transferring electrons to the CNTs surface. The scavenging test also indicated that the SWCNTs supporting TiO_2/Au nanoparticles acted as photosensitizers that transferred electrons to the TiO_2 surface

Table 2. TiO2 photocatalysts applied in the photodegradation of dyes and the operation parameters

Pollutant	Photocatalyst	Method	Light source	Irrad. time (min)	Effect %	Ref.
MB	F-doped TiO_2	Hydrothermal and sol-gel	UV (300 W)	10	96	(Diaz-Sanchez, et al., 2021)
MB	Cr_2S_3-GO/TiO_2	Sol gel	Visible	120	98.3	(Hasan, Li, Tian, & Qin, 2020)
RhB	Cr_2S_3-GO/TiO_2	Sol gel	Visible	120	96.6	
MO	Cr_2S_3-GO/TiO_2	Sol gel	Visible	120	86.3	
MB	TiO_2/Au/SWCNTs	Sol gel	Solar simulated	30	77	(Mohammed, 2020)
MB	r-GO/TNT	Hydrothermal	Visible	180	78	(Nguyen & Juang, 2019)
MB	r-GO/TNT	Hydrothermal	UV (100 W)	60	77.4	
MB	TCN-x		Visible	120	91.4	(Liu, et al., 2022)

that were seized by oxygen molecules at the CB to form reactive oxidative species (ROS) that generated the hydroxyl radicals which are used to degrade the dye.

Photocatalytic Degradation of Pesticides by TiO_2

Pesticides are widely used in controlling pests and weeds in the agricultural industry, however the unutilised chemical substances pose adverse impacts on the ecosystem if they do not go through efficient water treatment systems. Photocatalytic degradation of these compounds has proved to be one of the most efficient pathway for removing these pesticides from the environment. This section discusses the application of TiO_2-based photocatalysts in the mineralization of pesticides.

A study on the photocatalytic performance of a catalyst prepared by the photo deposition of Fe_2O_3 onto TiO_2 examined the degradation of 2, 4-dichlorophenoxyacetic acid (2, 4-D) under UV-light (Lee, Lintang, & Yuliati, 2017). The results indicated that the catalyst with the optimum loading of Fe/Ti ratio at 0.5 mol % had a removal efficiency of 88% within 60 min. The authors attributed the remarkable performance to the improved charge transfer due do the formation of the heterojunction between Fe_2O_3 onto TiO_2 which also inhibited the e^-/h^+ recombinations, the mechanism is illustrated in Fig. 2. A visible-light activated In,S-TiO2@rGO nanocomposite prepared *via* ultrasonic assisted solvothermal method was evaluated for degradation of atrazine (ATZ) in water. It was observed that co-doping and compositing of TiO_2 with rGO enhanced its visible light photocatalytic activity. The nanocomposite attained 95.5% mineralization of 20 mg/L at 3.0 mol% In, 1.0 mol% S and 5 wt% rGO within 20 min. Furthermore, the TiO_2 hybrid could retain its photocatalytic activity during four consecutive recycles in the removal (Khavar, Moussavi, Mahjoub, M, & Abdolmaleki, 2018).

Abdelhaleem & Chu (2019) investigated the photocatalytic activity of a hybrid process for carbofuran (CBF) degradation. The peroxymonosulfate (PMS) activation under visible LED (Vis LED) by Fe (III) impregnated N-doped TiO_2 photocatalyst (FeNT/PMS) achieved a degradation efficiency of about 90% of CBF. Additionally, the hybrid still showed an efficient CBF degradation of more than 70% even in the presence of NO_3^-, SO_4^{2-}, HCO_3^- and Cl^- ions (Abdelhaleem & Chu, 2019). Sol-gel synthesized carbon doped TiO_2 coupled with graphitic carbon nitride (C- TiO_2/g-C_3N_4) was investigated for the degradation of diazinon under visible radiation. The degradation efficiency of diazinon with the photocatalyst C- TiO_2 (30 wt%)/g-C_3N_4 was 98.34% under the optimum conditions of 15 mg/L initial concentration of diazinon, 0.1 g/L catalyst dosage, pH 5.5 and contact time of 30 min (Amiri & Anbia, 2023).

The degradation of 2, 4, 6-tri-chlorophenol- a phenolic derivative which is also a pesticide was conducted by Phattepur and group (2023). The Zn-doped black TiO_2 nanoparticles synthesized through sol gel method effectively degraded 2, 4, 6- tri-chlorophenol under visible light. The degradation efficiency of 95% for the pesticide by Zn (2 mol%)-TiO_2 was attributed to narrowing band gap to 2.73 eV (Phattepur,

Figure 2. Illustration of the mechanism for major charge transfer pathways on Fe2O3 (0.5)/TiO2 for 2, 4-dichlorophenoxyacetic acid degradation
(Lee, Lintang, & Yuliati, 2017)

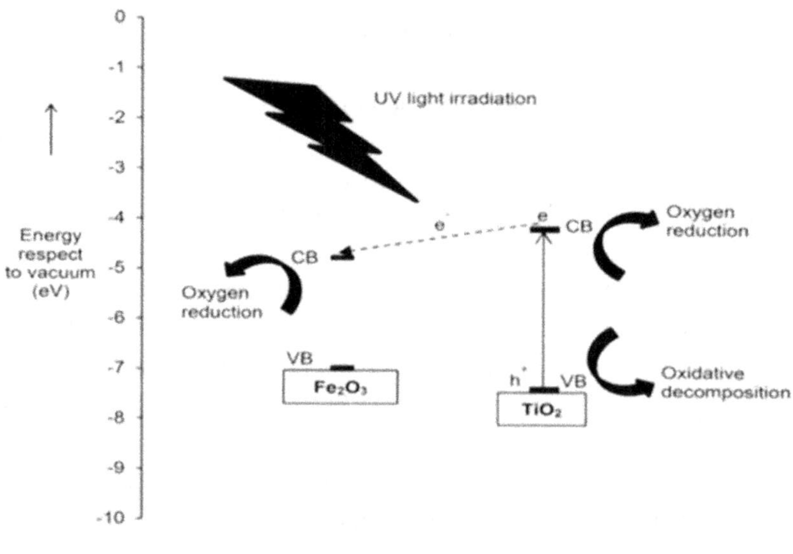

Table 3. TiO2 photocatalysts applied in the photodegradation of dyes and the operation parameters

Pollutant	Photocatalyst	Method	Light source	Irrad. Time (min)	Efficiency %	Ref.
2, 4-D	Fe_2O_3(0.5)/ TiO_2	Photo-deposition	UV	60	88	(Lee, Lintang, & Yuliati, 2017)
Atrazine	In,S-TiO2@ rGO	Ultrasonic assisted solvothermal	Visible	20	95.5	(Khavar, Moussavi, Mahjoub, M, & Abdolmaleki, 2018)
Carbofuran	FeNT/PMS		Vis-LED		90	(Abdelhaleem & Chu, 2019)
Diazinon	C- TiO_2/g-C_3N_4	Sol gel	Visible	30	98.3	(Amiri & Anbia, 2023)
2,4,6-Tri chlorophenol	Zn (2 mol%)- TiO_2	Sol gel	Visible		95	(Phattepur, Nagaraju, Shruthi, Sandhya, & Banupriya, 2023)

Nagaraju, Shruthi, Sandhya, & Banupriya, 2023). Table 3 shows a summary of selected TiO_2 photocatalysts applied in the photodegradation of pesticides and the operation parameters

Photocatalytic Degradation of Other Organic Pollutants by TiO_2

This section will discuss the photocatalytic degradation of other organic compounds such as phenols and pharmaceuticals among others. This class of contaminants are generally considered to be less commonly monitored yet they have potential adverse ecological or human health effects (Bolong, Ismail, Salim, & Matsuura, 2009). Therefore, the removal of these harmful chemicals from wastewater is a global environmental concern that should be addressed (Rizzi, et al., 2021).

Phenols are organic compounds with a chemical structure comprising of hydroxyl group(s) attached to one or more of its aromatic rings. Their presence in the environment can be either through natural or anthropogenic activities. The natural sources of phenols are mainly decomposition of organic matter present in water and metabolism waste from human bodies, animals and microorganisms. The occurrence of phenols due to anthropogenic activities include run offs from agricultural fields, industries, domestic and municipal wastewater (Said, Ismail, Karim, Abdullah, & Hafeez, 2021). Pharmaceuticals (hormones, antibiotics, etc.) are mainly used as medicine for preventing and management of diseases in both animals and humans. Once consumed, these organic compounds undergo metabolic reactions in the body but not all of the compounds are completely degraded hence they get excreted (Debska, Kot-Wasik, & Namiesnik, 2004). Residues of these compounds may eventually enter aquatic systems through surface run offs and wastewater containing human and animal waste from inadequate sanitary systems and animal farms.

The efficiency and feasibility of the photodegradation of these compounds using TiO_2-based photocatalysts has been widely reported and Table 4 shows a summary of the degradation of selected phenolic and pharmaceutical compounds using TiO_2 photocatalysts. Wang et al. (2011) incorporated Fe^{3+} dopant and C in form of carbonates to TiO_2 to form the composite Fe/C- TiO_2 to investigate its catalytic activity. The solvothermally synthesized composite evaluated for the photodegradation of Bisphenol A (BPA) and clofibric acid (CA) under different light sources exhibited better photocatalytic activity compared to P25. It was observed that the lattice defects induced by Fe dopant and carbonates species were responsible for the visible light photocatalytic activity and retardation of recombination of photogenerated positive holes and electrons (Wang, Tang, Leiw, & Lim, 2011). A co-doped catalyst of TiO_2 containing cobalt and carbon (Co-C-TiO_2) synthesized using co-precipitation technique was investigated for the degradation of phenol. The degradation of phenol by visible light irradiation indicated that doping and

Table 4. TiO2 photocatalysts applied in the photodegradation of phenols, pharmaceuticals and other organic compounds and their operation parameters

Pollutant	Photocatalyst	Method	Light source	Irrad. Time	Effect %	Ref
Phenol	Co- TiO_2	Co-precipitation	Visible		98.75	(Venkatesan, et al., 2022)
Ciprofloxacin	F-doped TiO_2	Hydrothermal and sol-gel	UV (300 W)	10	90	(Diaz-Sanchez, et al., 2021)
Naproxen	F-doped TiO_2	Hydrothermal and sol-gel	UV (300 W)	10	79	
Metrodinazole	TiO_2/Fe^{3+}		UV-C	120	97	(Malakootian, Olama, Malakootian, & Nasisri, 2019)
BPA	Fe/C- TiO_2	Solvothermal	White LED	180	97	(Wang, Tang, Leiw, & Lim, 2011)
Clofibric acid	Fe/C- TiO_2	Solvothermal	White LED	120	95	
Tetracycline	TCN-x		Visible	120	91.4	(Liu, et al., 2022)

co-doping enhanced the photocatalytic activity of TiO_2. The study reported a maximum degradation efficiency of phenol of 98.75 and 15% by the Co-TiO_2 and pristine TiO_2, respectively. The improved photocatalytic activity was attributed to the absorption in the visible light due to Co- and C-doping of TiO_2 and the reduced recombination rate of the photo induced electron/hole charges. Additionally, the Co-C-TiO_2 nanoparticles retained their degradation efficiency at 95% after several consecutive reuse cycles (Venkatesan, et al., 2022) .

A study on the photocatalytic degradation of metrodinazole using TiO_2/Fe^{3+} in UV-C radiation indicated that the optimal conditions for catalyst using synthetic solutions were pH 11, metrodinazole concentration of 500 mg/L and contact time of 120 min. The reported efficiency of TiO_2/Fe^{3+} in the removal for metrodinazole was 97% in synthetic solutions and 69.85% in wastewater containing the pharmaceuticals (Malakootian, Olama, Malakootian, & Nasisri, 2019).

LIMITATIONS AND FUTURE PROSPECTS

One of the main drawbacks of TiO_2 is its wide band gap and low conductivity which limits its applications without appropriate structural modifications. Most of the established methods for fabricating TiO_2 photocatalysts have worked well in achieving good results in the modification of TiO_2 nanoparticles which possess remarkable photocatalytic activity. However, the production of these photocatalysts is energy-intensive, time consuming and less eco-friendly. Green and inexpensive methods of structural modification that are sustainable, especially for emerging economies, are areas of future research. Post-recovery of TiO_2 photocatalyst after the water treatment remains unresolved. Future studies should focus on stabilization and separability of magnetic TiO_2 at pilot-scale. Additionally, studies that maximize both the photocatalytic and adsorptive properties of TiO_2 are needed since TiO_2 generally has low adsorption capacity. Enhancement of its adsorption capacity will make TiO_2 a multifunctional water purifying material.

Since photocatalytic degradation rarely leads to complete mineralization of the pollutant, studies focussing on integration of TiO_2 with other treatment strategies, such as electrochemical degradation, biological treatment etc., are needed. The by-products of photocatalytic degradation cause secondary pollution and may be more toxic than the parent molecules. Finally, for sustainable use of TiO_2-based photocatalysts, life cycle assessment (LCA) are required for complete assessment of the photocatalysts from cradle to grave.

CONCLUSION

Modified TiO_2 nanoparticles under UV or solar irradiation have become prominent and suitable photocatalysts for the photodegradation of organic pollutants. Besides the progress made in production of TiO_2 with suitable pore size distribution and morphologies, an array of modification approaches have been employed for TiO_2 fabrication including metal doping, non-metal doping, co-doping of either metal-metal or non–metal and non–metal. Other strategies applied are the coupling of TiO_2 with carbon nanomaterials like GO and CNTs, semiconductors, or metal-organic frameworks to form of hybrid composites. The aim of these modifications is to achieve better charge carrier separation and shift of photocatalytic activity in the visible light region. The modified TiO_2 photocatalysts performed better than the pristine TiO_2. Large scale and sustainable utilization of the modified TiO_2 photocatalysts is possible in the future when the production of these catalysts becomes relatively cheap and the catalysts can effectively be used in the natural environment with a response in the full light range. Life cycle assessment (LCA) studies on TiO_2 photocatalysts are needed before implementation.

REFERENCES

Abbad, S., Guergouri, K., Gazaout, S., Djebabra, S., Zertal, A., Barille, R., & Zaabat, M. (2020). Effect of silver doping on the photocatalytic activity of TiO2 nanapowders synthesized by sol-gel route. *Journal of Environmental Chemical Engineering*, *8*(3), 103718. doi:10.1016/j.jece.2020.103718

Abdelhaleem, A., & Chu, M. (2019). Insights into peroxymonosulfate activation for carbofuran degradation under visible LED via a double-component photocatalyst of Fe (III) impregnated N-doped TiO2. *Chemosphere*, *237*, 124487. doi:10.1016/j.chemosphere.2019.124487 PMID:31398605

Amiri, S., & Anbia, M. (2023). Enhanced degradation of diazinon in aqueous solution using C-TiO2/g-C3N4 nanocomposite under visible light: Synthesis, characterization, kinetics, and mechanism studies. *Materials Research Bulletin*, *165*, 112289. doi:10.1016/j.materresbull.2023.112289

Ansari, S., Khan, M., Ansari, M., & Cho, M. (2013). Nitrogen-doped titanium dioxide (N-doped TiO2) for visible light photocatalysis. *New Journal of Chemistry*, *00*, 1–3.

Baranowska-Wojak, E., Szwajgeir, D., Oleszczuk, P., & Winiarsk-Mieczan, A. (2019). Effects of Titanium Dioxide Nanoparticles Exposure to Human Health-A Review. *Biological Trace Element Research*, *193*(1), 118–129. doi:10.100712011-019-01706-6 PMID:30982201

Bello, M., Raman, A., & Asghar, A. (2019). A review on approaches for addressing the limitations of Fenton oxidation for recacitrant wastewater. *Process Safety and Environmental Protection*, *126*, 119–140. doi:10.1016/j.psep.2019.03.028

Cravanzola, S., Cesano, F., Geziano, F., & Scarano, D. (2017). Sulfur-doped TiO2: Structure and surface properties. *Catalysis*, *7*(7), 214.

Dahl, M., Liu, Y., & Yin, X. (2014). Composite Titanium Dioxide Nanomaterials. *Chemical Reviews*, *114*(19), 9853–9889. doi:10.1021/cr400634p PMID:25011918

Debska, J., Kot-Wasik, A., & Namiesnik, J. (2004). Fate and analysis of pharmaceutical residues in aquatic environment. *critical. Critical Reviews in Analytical Chemistry*, *34*(1), 51–67. doi:10.1080/10408340490273753

Devi, L., & Kavitha, R. (2013). A review of non metal ion doped titania for the catalytic degradation of organic pollutants under UV/solar light: Role of photogenerated charge carrier dynamics in enhancing the activity. *Applied Catalysis B: Environmental*, *140-141*, 559–587. doi:10.1016/j.apcatb.2013.04.035

Diaz-Sanchez, M., Renones, P., Mena-Palomo, I., Lopez-Coleazo, E., Fresno, F., Oropeza, F., & Gomez-Ruiz, S. (2021). Ionic liquid assisted synthesis of F-doped titanium dioxide nanomaterials with high surface area for multifunctional catalytic and photocatalytic applications. *Applied Catalysis A, General*, *613*, 118029. doi:10.1016/j.apcata.2021.118029

Fugishima, A., Zhang, X., & Tyrk, D. (2008). TiO2 Photocatalysis and related surface phenomena. *Surface Science Reports*, *63*(12), 515–582. doi:10.1016/j.surfrep.2008.10.001

Gao, B., Chen, G. Z., & Li Puma, G. (2009). Carbon nanotubes/titanium dioxide (CNTs/TiO2) nanocomposites prepared by conventional and novel surfactant wrapping sol–gel methods exhibiting enhanced photocatalytic activity. *Applied Catalysis B: Environmental*, *89*(3-4), 503–509. doi:10.1016/j.apcatb.2009.01.009

Hariharan, D., Christy, A. J., Pitchaiya, S., Sagadevan, S., Thangamaniyandi, P., Devan, U., & Nehru, L. (2019). Green hydrothermal synthesis of gold and palladium doped titanium dioxide nanoparticles for multifunctional performance. *Journal of Materials Science Materials in Electronics*, *30*(13), 12812–12819. doi:10.100710854-019-01647-9

Hasan, J., Li, H., Tian, G., & Qin, C. (2020). Fabrication of Cr2S3-GO-TiO2 composite with high visible-light-driven photocatalytic activity on degradation of organic dyes. *Chemical Physics*, *539*, 110950. doi:10.1016/j.chemphys.2020.110950

Hurum, D., Gray, K., Rajh, T., & Thurnaer, M. (2004). Photoinitiated Reactions of 2,4,6 TCP on Degussa P25 Formulation TiO2: Wavelength Sensitive Decomposition. *The Journal of Physical Chemistry B*, *108*(42), 16483–16487. doi:10.1021/jp047097m

Ismail, M., Hedhili, M., Anjum, D., Singaravelu, V., & Chung, S. (2021). Synthesis and characterization of iron doped TiO2 nanoparticles using ferrocene from flame spray pyrolysis. *Catalysts*, *11*(4), 438–454. doi:10.3390/catal11040438

Javanbakht, V., & Mohammadian, M. (2021). Photo assisted advanced oxidation processes form efficient removal of anionic and cationic dyes using Bentonite/TiO2 nanophocatalyst immobilized with silver nanoparticles. *Journal of Molecular Structure*, *1239*, 130496. doi:10.1016/j.molstruc.2021.130496

Khavar, A., Moussavi, G., Mahjoub, A. M. S., & Abdolmaleki, P. (2018). Synthesis and visible-light photocatalytic activity of In,S-Ti)2@rGO nanocomposite for degradation and detoxification of pesticide atrazine in water. *Chemical Engineering Journal*, *345*, 300–311. doi:10.1016/j.cej.2018.03.095

Khlyustrova, A., Sirotkin, N., Kusova, T., Kraev, A., Titov, V., & Agafonov, A. (2020). Doped TiO2: The effect of doping elements on photocatalytic activity. *Materials Advances*, *1*(5), 1193–1201. doi:10.1039/D0MA00171F

Kwon, S., Fan, M., Cooper, A., & Yang, H. (2008). Photocatalytic application of micro- and nano TiO2 in environmental engineering. *Critical Reviews in Environmental Science and Technology*, *38*(3), 197–227. doi:10.1080/10643380701628933

Lee, S., Lintang, H., & Yuliati, L. (2017). HIghly photocatalytic activity of Fe2O3/TiO2 nanocomposites prepared by photodeposition for degradation of 2,4-dicholorophenoxyacetic acid. *Beilstein Journal of Nanotechnology*, *8*, 915–926. doi:10.3762/bjnano.8.93 PMID:28546886

Lee, S.-Y., & Park, S. (2013). TiO2 photocatalysts for water treatment applications. *Journal of Industrial and Engineering Chemistry*, *19*(6), 1761–1769. doi:10.1016/j.jiec.2013.07.012

Li, W., Ismat Shah, S., Huang, C.-P., Jung, O., & Ni, C. (2002). Metallorganic chemical vapour deposition and characterization of TiO2 nanoparticles. *Materials Science and Engineering B*, *96*(3), 247–253. doi:10.1016/S0921-5107(02)00352-5

Lim, S., Shi, J., von Gunten, U., & McCurry, D. (2022). Ozonation of organic compounds in water and wastewater: A critical review. *Water Research*, *213*, 118053. doi:10.1016/j.watres.2022.118053 PMID:35196612

Liu, G., Pan, J., Yin, L., Irvine, J., Li, F., Tan, J., & Cheng, H.-M. (2012). Heteroatom-Modulated Switching of photocatalytic Hydrogen and Oxygen Evolution Preferences of Anatase Another approach that has been applied in the Heteroatom-Modulated Switching of photocatalytic Hydrogen and Oxygen Evolution Preferences of Anatase TiO2. *Advanced Functional Materials*, *22*, 3233–3238. doi:10.1002/adfm.201200414

Liu, S., Zou, Q., Ma, Y., Chi, D., Chen, R., Fang, H., Hu, W., Zhang, K., & Chen, L.-F. (2022). Metal-organic frameworks derived TiO2/carbon nitride heterojunction photocatalyst with efficient catalytic performance under visible light. *Inorganica Chimica Acta*, *536*, 120918. doi:10.1016/j.ica.2022.120918

Low, J., Yu, J., Janroniec, M., Wageh, S., & Al-Ghamdi, A. (2017). Heterojunction photocatalysts. *Advanced Materials*, *29*(20), 1601694. doi:10.1002/adma.201601694 PMID:28220969

Malakootian, M., Olama, N., Malakootian, M., & Nasisri, A. (2019). Photocatalytic degradation of metrodinazole from aquatic systems by TiO2- doped Fe3+ nanophotocatalyst. *International Journal of Environmental Science and Technology*, *16*(8), 4275–4284. doi:10.100713762-018-1836-2

Meng, S., Sun, W., Zhang, S., Zheng, X., Fu, X., & Chen, S. (2018). Insight in the transfer mechanism of photogenerated carriers for WO3/TiO2 heterojunction photocatalysts is it transfer of Band-Band or Z-scheme? Why? *Journal of Physical Chemistry C*, *122*, 26326-26336.

Millis, A., & Le Hunte, S. (1997). An overview of semiconductor photocatalysis. *Journal of Photochemistry and Photobiology A Chemistry*, *108*(1), 1–35. doi:10.1016/S1010-6030(97)00118-4

Mohammed, M. K. (2020). Sol-gel synthesis of Au-doped TiO2 supported SWCNT nanohybrid with visible-light driven photocatalytic for high degradation performance toward methylene blue dye. *Optik (Stuttgart)*, *223*, 165607. doi:10.1016/j.ijleo.2020.165607

Muthuvel, A., Said, N., Jothibas, M., Gurushankar, K., & Mohana, V. (2021). Microwave assisted green synthesis of nanoscaled: Photocatalyst, antibacterial and antioxidant properties. *Journal of Materials Science Materials in Electronics*, *32*(18), 23522–23539. doi:10.100710854-021-06840-3

Nabi, G., Khalid, N., Tahir, M., Rafique, M., Rizwan, M., Hussain, S., ... Majid, A. (2018). A Review on Novel Eco-friendly Green Approach to Synthesis TiO2 nanoparticles using different extracts. *Journal of Inorganic and Organometallic Polymers and Materials*, *28*(4), 15521564. doi:10.100710904-018-0812-0

Nguyen, C., & Juang, R.-S. (2019). Efficient removal of methylene blue dye by a hybrid adsorption–photocatalysis process using reduced graphene oxide/titanate nanotube composites for water reuse. *Journal of Industrial and Engineering Chemistry*, *76*, 296–309. doi:10.1016/j.jiec.2019.03.054

Phattepur, H., Nagaraju, G., Shruthi, R., Sandhya, R., & Banupriya, C. (2023). Synthesis of Zn doped black TiO2 nanoparticles for degradation of 2, 4, 6 tri-chloro-phenol under visible light. *Journal of the Indian Chemical Society*, *100*, 100881. doi:10.1016/j.jics.2023.100881

Poole, A. (2004). Treatment of biorefactory organic compounds in wool scour effluents by hydroxyl radical oxidation. *Water Research*, *38*(14-15), 3458–3464. doi:10.1016/j.watres.2004.06.001 PMID:15276763

Potdar, S., Huang, C.-M., Praveen, B., Manickam, S., & Sonawane, S. (2022). Highly photoactive titanium dioxide supported platinum catalyst: Synthesis using cleaner ultrasound approach. *Catalysts*, *12*(1), 78. doi:10.3390/catal12010078

Prasannalakshmi, P., & Shanmugam, N. (2017). Fabrication of TiO2/ZnO nanocomposites for solar energy driven photocatalysis. *Materials Science in Semiconductor Processing*, *61*, 114–124. doi:10.1016/j.mssp.2017.01.008

Rizzi, V., Gubitosa, T., Signovile, R., Fini, P., Cecone, P., Matencio, A., & Cosma, P. (2021). Cyclodextrin nanosponges as adsorbent material to remove hazardous pollutants from water. *Chemical Engineering Journal*, *411*, 128514. doi:10.1016/j.cej.2021.128514

Said, K., Ismail, A. F., Karim, Z., Abdullah, M., & Hafeez, A. (2021). A review of technologies for the phenolic compounds recovery and phenol removal from wastewater. *Process Safety and Environmental Protection*, *151*, 257–289. doi:10.1016/j.psep.2021.05.015

Sakthivel, S., Janczarek, M., & Kisch, H. (2004). Visible light activity and photoelectrochemical properties of nitrogen-doped TiO2. *The Journal of Physical Chemistry B*, *108*(50), 19384–19387. doi:10.1021/jp046857q

Thambiliyagodage, C. (2022). Efficient photocatalysis of carbon coupled TiO2 to degrade pollutants in wastewater – A review. *Environmental Nanotechnology, Monitoring & Management*, *18*, 100737. doi:10.1016/j.enmm.2022.100737

Venkatesan, A., Al-onazi, W. A., Elshikh, M., Pham, T. H., Suganya, S., Boobas, S., & Priyadharsan, A. (2022). Study of synergistic effect of cobalt and carbon codoped with TiO2 photocatalyst for visible light induced degradation of phenol. *Chemosphere*, *305*, 135333. doi:10.1016/j.chemosphere.2022.135333 PMID:35709834

Vishali, D., Manikandan, B., John, R., & Murali, K. (2021). Degradation of Methylene Blue Dye by TiO2 Nanoparticles BIologically Synthesized Using Leaf Extract of Citrus Aurantifolia. *Transactions on Electrical and Electronic Materials*, *22*(5), 622–629. doi:10.100742341-020-00270-4

Wang, G. (2007). Hydrothermal synthesis and photocatalytic activity of nanocrystalline TiO2 powders in ethanol-water mixed solutions. *Journal of Molecular Catalysis A Chemical*, *274*(1-2), 185–191. doi:10.1016/j.molcata.2007.05.009

Wang, X., Tang, Y., Leiw, M.-Y., & Lim, T.-T. (2011). Solvothermal synthesis of Fe-C codoped TiO2 nanoparticles for visible light photocatalytic removal of emerging organic contaminants in water. *Applied Catalysis A, General*, *409-410*, 257–266. doi:10.1016/j.apcata.2011.10.011

Xiang, Q., Yu, J., & Wong, P. (2011). Quantitative characterization of hydroxyl radicals produced by various photocatalysts. *Journal of Colloid and Interface Science*, *357*(1), 163–167. doi:10.1016/j.jcis.2011.01.093 PMID:21349533

Xu, C., Ravi Anusuyadevi, P., Aymonier, C., Luque, R., & Marre, S. (2019). Nanostructured materials for photocatalysis. *Chemical Society Reviews*, *48*(14), 3868–3902. doi:10.1039/C9CS00102F PMID:31173018

Zaleska, A. (2008). Doped TiO2: A review. *Recent Patents on Engineering*, *2*(3), 157–164. doi:10.2174/187221208786306289

Zeebaree, A., Zeebaree, S., Rashid, R., Zebari, O., Albarwry, A., Ali, A., & Zebari, A. Y. (2022). Sustainable engineering of plant-synthesized TiO2 nanocatalysts: Diagnosis, properties and their photocatalytic performance in removing of methylene blue dye from effluent. A review. *Current Research in Green and Sustainable Chemistry*, *5*, 100312. doi:10.1016/j.crgsc.2022.100312

Zhang, F., Zhang, K., Chen, T., & Oh, W. (2012). Photocatalytic degradation of methyl orange on platinum and palladium co-doped TiO2 nanoparticles. *Synthesis and Reactivity in Inorganic, Metal-Organic, and Nano-Metal Chemistry*, *42*(5), 685–691. doi:10.1080/15533174.2011.615040

Zhang, Y., Hawboldt, K., Zhang, L., Lu, J., Chang, L., & Dwyer, A. (2022). Carbonaceous nanomaterial TiO2 heterojunctions for visible light driven photocatalytic degradation of aqueous organic pollutants. *Applied Catalysis A, General*, *630*, 118460. doi:10.1016/j.apcata.2021.118460

Zhao, C., Zhu, D. C., Cheng, X., & Cao, S. (2017). Highly ordered Ag-TiO2 nanocomposited arrays with high visible photocatalytic activity. *Frontiers of Materials Science*, *11*(3), 241–249. doi:10.100711706-017-0386-8

Zu, M., Zhou, X., Zhang, S., Qian, S., Li, D.-S., Liu, X., & Zhang, S. (2021). Sustainable engineering of TiO2-based advanced oxidation technologies: From photocatalysis to application devices. *Journal of Materials Science and Technology*, *78*, 202–222. doi:10.1016/j.jmst.2020.10.061

Chapter 5
A Review on Multifunctional Silver Nanoparticles Produced With Green Routes for Photocatalysis, Sensing, and Imaging

Gizem Karabulut
https://orcid.org/0000-0003-0930-5380
Istanbul University-Cerrahpasa, Turkey

Nuray Beköz Üllen
Istanbul University-Cerrahpasa, Turkey

ABSTRACT

Metallic nanoparticles (NPs) have great potential for biological and chemical sensor applications because the surface of the NPs has a sensitive spectral response, and the light signal can be easily monitored by their absorption or scattering. In particular, silver nanoparticles (Ag NPs) are used in various applications due to their advanced physicochemical properties such as their excellent antibacterial, catalytic, and conductivity at the nanoscale. Plant-mediated synthesis methods have come to the fore with the green chemistry approach in the synthesis of NPs. The aim of this chapter is to provide a detailed resource by considering the studies in the literature on the use of Ag NPs produced by green routes mainly in the fields of photocatalysis, sensing, and imaging. Additionally, this chapter aims to determine the importance and potential of plant-mediated synthesized Ag NPs in these applications will be determined and to be a guide for industrial and academic personnel working in this field.

DOI: 10.4018/978-1-6684-8743-3.ch005

INTRODUCTION

Metallic nanoparticles (NPs) have found usage in various fields by showing improved physicochemical and biological properties as a modified nano-engineering materials at the nano-scale compared to their macro-scale structures (Das & Ansari, 2009; Iliger et al., 2021; Yongfei Wang et al., 2019). Metallic NPs recently come to the fore in nanotechnology applications due to reasons such as having less toxicity at the nanoscale compared to the atomic level, free electrons in the conduction bands improve the optical properties of metallic NPs and thus surface plasmon resonance (SPR) can be observed in the UV-Visible regions, allowing photothermal treatment due to their ability to convert light/radio frequencies into heat (Vaid et al., 2020). NPs synthesis methods were classified as bottom-up and top-down synthesis. In this classification, the methods were classified according to their initial size. While the methods that perform synthesis from atomic size to nano size are divided as bottom-up, methods that perform synthesis by reduction from bulk/macro size to nano size are divided as top-down. Another classification of methods is physical, chemical, and biological (Chavali & Nikolova, 2019; Jamkhande et al., 2019; Nasrollahzadeh et al., 2021; Rana et al., 2020; Vaid et al., 2020). A general classification of methods was shown in Figure 1.

Figure 1. Classification of the NPs synthesis methods

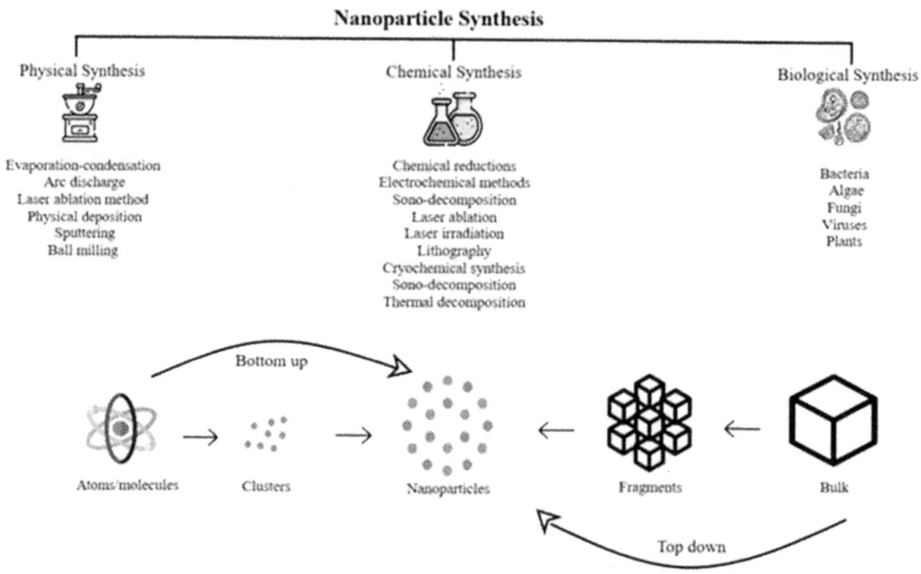

In physical synthesis, the synthesis is carried out with the help of a physical power obtained with the help of equipment. In general, it has disadvantages such as high equipment cost, high energy consumption, and solvent pollution (Nasrollahzadeh et al., 2021). In chemical synthesis methods, the driving force of synthesis is a chemical reaction. Chemical synthesis methods are not preferred due to the use of chemical products and harmful byproducts (Nasrollahzadeh et al., 2021). Nowadays, the world is facing problems such as global warming, pollution, and the depletion of natural resources, which is why the use of more environmentally friendly methods in production processes is coming to the fore in many fields. Biological synthesis have created an alternative to other methods with features such as being more environmentally friendly, being more cost-effective than other methods, and being able to synthesize relatively faster (Ashique et al., 2022; Iravani, 2014; Nasrollahzadeh et al., 2021; Tortella et al., 2022). Recently, biological synthesis methods using eukaryotic and prokaryotic organisms have come to the fore as part of the green chemistry approach to the synthesis of NPs. Biological synthesis can be performed using bacteria, fungi, viruses, and plants. Bacteria, as prokaryotic organism have generally developed resistance to extreme conditions due to their exposure to metal-rich environments and are therefore suitable for synthesis (Habibullah et al., 2021; Iravani, 2014). The mechanism in bacteria-mediated synthesis is the synthesis of metallic NPs through various reactions with metal ions in or around the cell by enzymes or secondary metabolites (Al-Sheddi et al., 2018; Tortella et al., 2022). Bacteria-mediated synthesis of metallic NPs is more environmentally friendly, cost-effective, biodegradable, and less toxic than chemical and physical methods. However, it has disadvantages in terms of health and safety compared to other biological methods (Ashique et al., 2022; Habibullah et al., 2021). Eukaryotic microorganisms are preferred due to their high intracellular metal uptake capacity, ability to synthesize NPs with different chemical compositions, ability to produce large amounts of enzymes per unit biomass, and ease of use under laboratory conditions (Dilshad et al., 2020; Jamkhande et al., 2019). In fungal-mediated synthesis, metal salt solutions are used for the synthesis of metallic NPs. Enzymes and proteins secreted by fungi act as reducing agents (Jamkhande et al., 2019). Since fungi secrete more proteins than bacteria, their production capacity is higher. One of the advantages is that the NPs are easy to scale and their shape can be controlled (Habibullah et al., 2021). However, the synthesis method is risky due to safety factors (Jamkhande et al., 2019). Today, the plant-mediated synthesis of metal NPs with a green approach is one of the most widely used methodologies (Habibullah et al., 2021). Synthesis of metallic NPs from plant sources involves purification of the bio-reducing agent in the plant and controlled mixing with the precursor metal solution. Then, reactions take place at room temperature leading to the formation of the NPs (Vaid et al., 2020). First, the plant to be used is collected and the extract is prepared.

Figure 2. Plant-mediated synthesis mechanism of the metallic NPs

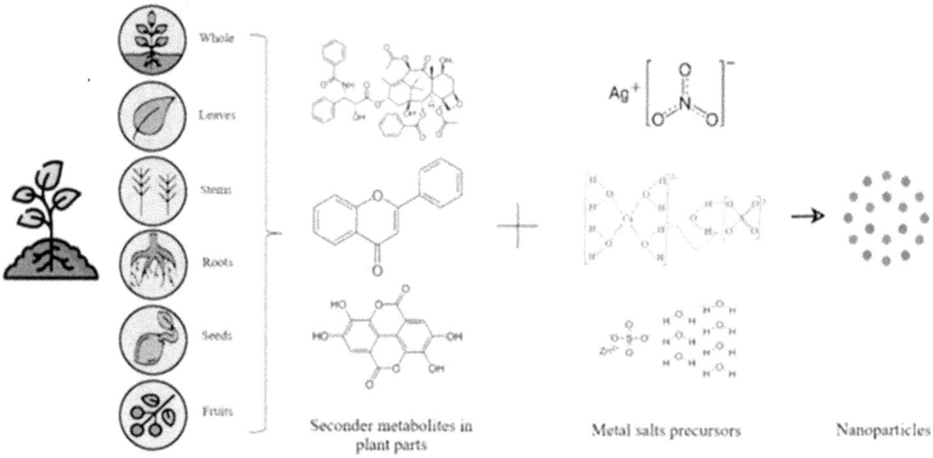

For the plant-mediated synthesis process, the extract of the whole plant or specific parts such as leaves, stems, roots, and seeds are used. The primary and secondary metabolites in the plant content are constantly involved in the metabolic reactions that take place in the plant. These metabolites are used as reducing agents in the plant-mediated synthesis of metallic NPs (Jamkhande et al., 2019). Plant extracts contain secondary metabolites such as phenolic acid, flavonoid, and terpenoid, which are selective for metal ions and provide for the formation of NPs. These secondary metabolites in the prepared extract reduce the metal ions when mixed with the prepared metal precursor solution, providing synthesis and stabilization of the metallic NPs (Jamkhande et al., 2019; Sankar Sana et al., 2022). A schematic diagram of the plant-mediated synthesis of the metallic NPs is given in Figure 2. Plant-mediated synthesis is generally preferred because the source is easy to find, is relatively suitable for high production volumes, and produces environmentally friendly by-products and waste products. Because of these advantages, it has become an alternative to other synthesis methods (Al-Sheddi et al., 2018).

In the literature, there are many studies on the green synthesis and use in various fields of metallic NPs such as copper (Ilgar et al., 2022), gold (Philip et al., 2011), platinum (Konishi et al., 2007), palladium (Bi & Ahmad, 2022), and silver (Arboleda et al., 2018; Karabulut et al., 2023; Karakuş et al., 2021; Zhu et al., 2020). Among of them, silver nanoparticles (Ag NPs) come to the fore in nano-engineered materials due to their electrical conductivity, chemical stability, catalytic properties, and bioactivity (Dilshad et al., 2020; Karabulut et al., 2023). Because of these properties, Ag NPs have found a wide range of applications. Ag

NPs are widely known for their applications in various fields such as photonics, lithography, microelectronics, photocatalysis, and medicine (Alwhibi et al., 2020; Hossain et al., 2022). In this chapter, the role of silver NPs synthesized by green routes in photocatalysis, sensing, and imaging are discussed in detail. A detailed resource has been created to understand the importance of the green method and to determine the role of NPs synthesized in this way in various applications.

PHOTOCATALYSIS

Photocatalysis is a reaction with a light source. The light incident on the surface of the material acting as a photocatalyst is absorbed by the material and converted into chemical energy. When the band gap of the photocatalyst is equal to or less than the energy of the incident light, the electrons in the valence band of the photocatalyst will absorb the photon and reach the conduction band. The moving electrons' place remain vacant and its play the main role in photocatalysis. Oxidation and redox reactions take place at the vacant points and in the conduction band, resulting in the desired products. The photocatalyst breaks down the components accumulated on the surface of the material and ensures their elimination. Photocatalytic systems are used for purposes such as bacteria removal, water purification, air purification, self-purification, and odor elimination (Mokammel et al., 2022; Sankar Sana et al., 2022; Tahir et al., 2022).

Ag NPs can be activated by a visible light source due to its localized surface plasmon resonance (LSPR) property. It also acts as an electron scavenger that reduces the recombination capacity of e-h+ pairs, thus prolonging the lifetime of charge carriers and increasing photocatalytic efficiency (Aravind et al., 2021; Ha et al., 2021). In addition to these properties, green-mediated synthesized Ag NPs are coming to the forefront for the removal of undesirable substances by photocatalysis because they can be produced in a completely environmentally friendly manner, are inexpensive, and have the potential for many applications (Aravind et al., 2021; Marimuthu et al., 2020). General mechanism of the Ag NPs in photocatalysis is given in Figure 3. The mechanism of green synthesized Ag NPs is explained as follows: The formation of electron-hole pairs occurs in Ag NPs in solution irradiated by a light source. The OH group from the carbon derived from the vegetative structure interacts with the dye molecules and forms the OH free radicals. The larger surface area of the metallic NPs increases the excitation of electrons from the valence band to the conduction band to increase e^- transfer and form free radicals and superoxides. During the oxidation and reduction process takes place, free radicals and superoxides OH and O_2 are formed. The free radicals formed participate in the degradation process (Aravind et al., 2021).

Figure 3. General photocatalysis mechanism of Ag NPs

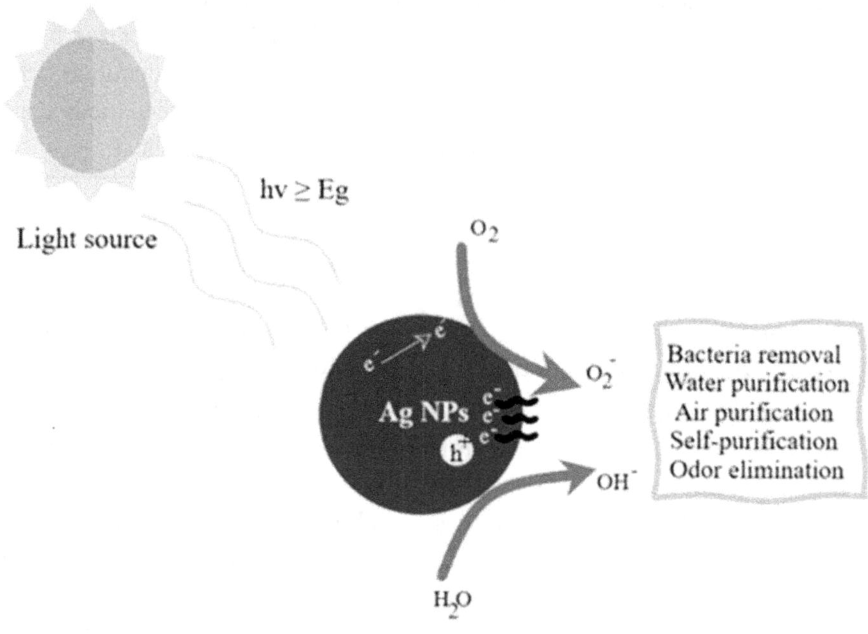

In a study by Aryan et al., (2021) plant-mediated synthesized Ag NPs were prepared using the *Kalanchoe pinnata*, which is also known as life plant, and the removal of dyes by photocatalytic degradation was investigated. In their study, the removal of the dye rhodamine B ($C_{28}H_{31}ClN_2O_3$) was successfully carried out. The degradation was studied under UV irradiation and in dark environment. It has been stated that the degradation in the presence of UV light was 83% in 45 min with a rate constant (k) of 0.042 min, which was three times higher than in the dark environment. As a result of the study, the authors were stated that plant-mediated synthesized Ag NPs are very useful for dye degradation due to their photocatalytic properties. In the study by Aravind et al. (2021) Ag NPs were synthesized in green way using jasmine flower extract and the degradation of methylene blue dye by photocatalysis was investigated. It was reported that the amount of degradation of green-mediated synthesized Ag NPs with a size of 22 nm increased with time and reached 71% after 2 hours. These Ag NPs are presented as a biodegradable and environmentally friendly solution for dye degradation. Sankareswari et al. (2022) synthesized Ag NPs with *Abrus precatorius* seed extract and then studied the photodegradation of ciprofloxacin (CIP) under visible light. In the examinations, it has been reported that green-mediated synthesized Ag NPs with *Abrus precatorius* seeds, which show a high catalyst effect, provide 90% CIP degradation at the end of 1.5 hours. In the

Table 1. Plant-mediated synthesized Ag NPs used as photocatalysts in various applications

Stabilizer	Application	Activity	References
Kalanchoe pinnata	Degradation of rhodamine B dye	83% after 45 min	(Aryan et al., 2021)
Kalanchoe daigremontiana	Removal of methylene blue	100% after 1 min	(Molina et al., 2019)
Cadiospermum halicacabum and *Moringa oleifera*	Degradation of methylene blue dye	92.7% after 48 h	(Balamurugan et al., 2022)
Tomato, orange, and grapefruit extracts	Environmental remediation of estrone and estrio	27.1, 52.1, and 24.8% after 180 min, respectively	(Weng et al., 2022)
Abrus precatorius	Photodegradation of CIP antibiotic	Above the 90% after 90 min	(Sankareswari et al., 2022)
Antidesma acidum	Methylene blue and congo red dyes degradation	81% and 90% after 18 min and 20 min, respectively	(Basumatary et al., 2023)
Trigonella foenum-graecum	Degradation of rhodamine B dye	~93% after 9 days	(Awad et al., 2021)
Cocos nucifera	Degradation of malachite green dye	75% after 6 h	(Sumi et al., 2016)
Coleus Vettiveroids	Methylene blue, malachite green, and crystal violet	70% after 3h	(Ajay et al., 2022)
Crinum asiaticum	Methylene blue	83.2% after 20 min	(Shukla et al., 2022)

stability and recyclability examinations, it was seen that it was 83% in the result of the 5th repetition, which shows the potential of green-mediated synthesized Ag NPs. Table 1 summarizes the use of plant-mediated synthesized Ag NPs in various pollutant removal applications with their photocatalysis function in recent years. It can be seen that studies on the removal of commercial dyes used in the textile industry from wastewater are widespread.

SENSING

Today, humanity is faced with many problems such as environmental problems and diseases. Early detection, prevention, and elimination of these problems have become very important for the continuity of the ecological system. Thank you to increasing technological developments, rapid, accurate, highly sensitive, and reusable detection of these situations is made possible by biosensors. Biosensors are innovative, effective, and independent analytical devices that react selectively and reversibly depending on the concentration or activity of the analyte to be detected in the sample (Malhotra et al., 2017; Tetyana et al., 2021; Vigneshvar et al., 2016). The biosensor is a compact device that contains a biological or biomimetic sensing element. It consists of three major components: Bioreceptor, analyte, and transducer (Purohit et al., 2020). The structure of a biosensor and its components is shown schematically in Figure 4. A target analyte is a structure that is detected in a sample. A bioreceptor is a structure that produces a measurable signal due to a physicochemical change that occurs as a result of interaction with an analyte through physical or chemical binding. These include enzymes, tissues, nucleic acids (DNA or RNA), antibodies, organelles, aptamers, etc (Fatima et al., 2022; Tetyana et al., 2021; Thomas et al., 2022). Following the interaction of analytes and bioreceptors, many physicochemical changes can occur, including pH change, mass change, electron or heat transfer, etc (Velasco-Garcia & Mottram, 2003). When at a certain temperature, sound, light, weight, or pressure is produced, the transducer converts it into an electrical signal that can be read or measured (Kumar et al., 2022). The unique sensing performance of nanostructure-based biosensors has enabled various sensing nanoplatforms to be used for enzymatic, non-enzymatic, optical, electrochemical, photoelectrochemical, thermal, acoustic, fluorescence, electrochemiluminescence, chemo resistive, colorimetric, and catalytic sensors. Nano-sized materials can have unique properties compared to bulk structures. They have also proven beneficial and revolutionized biosensor applications. Quickly analyze multiple samples, anytime and anywhere (Dede & Altay, 2018). The selectivity, detection, non-toxicity, biocompatibility, reversibility, rapid response, and sensitivity required of transducer materials can be achieved with nanomaterials (Bose et al., 2022; Karabulut et al., 2022; Thomas et al., 2022). For this reason, plant-mediated synthesized Ag NPs have been widely preferred for sensing applications in recent years due to their bioluminescent and electrochemical properties.

Santhosh et al. (2021) prepared an extract from the peels of *Allium cepa L* (red onion) and synthesized new plant-mediated Ag NPs. The authors tested these Ag NPs for the detection of heavy metals. The plant-mediated synthesized Ag NPs were tested in solutions of various heavy metal salts and a distinct color change was observed in $HgCl_2$ solution. This result paves the way for the detection of mercury in the liquid phase by simple observation and without the need for an instrument. The authors noted that green-produced Ag NPs could be used as a simple and cost-effective sensing partner in biosensor applications. In another study, Zamarchi and Vieira (2021) used the extract of *Araucaria angustifolia* (pine nut) to biologically synthesize Ag NPs through the plant and developed an electrochemical sensor for the detection of paracetamol. The chemical sensor developed with graphite and plant-mediated synthesized Ag NPs on the surface of a glassy carbon electrode

Figure 4. General structure and components of the biosensor

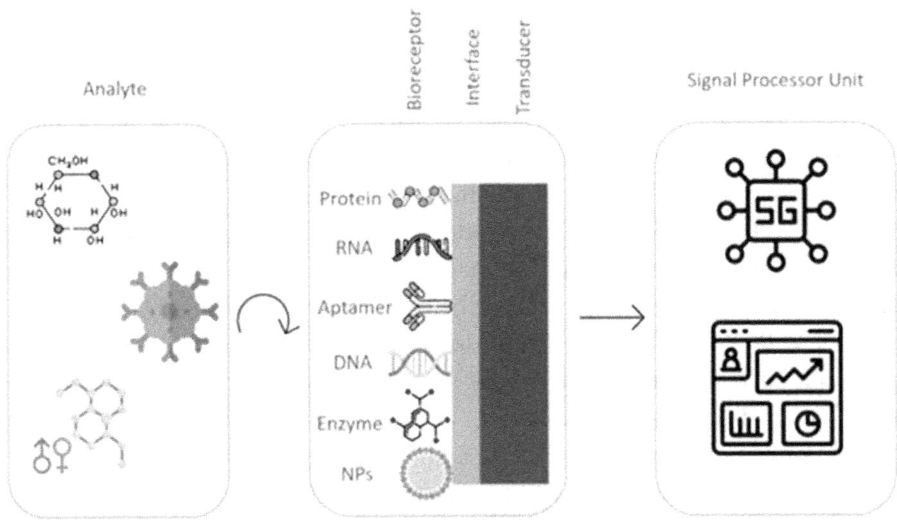

showed good repeatability, reproducibility, low detection limit (8.50×10-8 mol L-1), and sensitivity. The authors stated that these metallic NPs, which have surface properties such as greater surface area and conductivity, offer great advantages and can be developed and used industrially. Table 2 summarizes the studies on sensors modified with plant-mediated synthesized Ag NPs and their applications. In general, it was found that sensors modified with Ag NPs are widely preferred and beneficial in the detection of heavy metals.

Table 2. Plant-mediated synthesized Ag NPs used in sensing

Plant	Applications	References
Allium cepa L	Detection of Hg^{2+}	(Santhosh et al., 2021)
Araucaria angustifolia	Detection of paracetamol	(Zamarchi & Vieira, 2021)
Acacia chundra	Detection of Hg^{2+}	(Ramachandiran & Rajesh, 2022)
Matricaria recutita	Detection of Hg^{2+}	(Uddin et al., 2017)
Cissusquadrangularis	Detection of Cd^{2+}	(Pragathiswaran et al., 2021)
Hyacinth	Detection of Hg^{2+}	(Oluwafemi et al., 2019)
Mangifera indica	Detection of H_2O_2 and Hg^{2+}	(Hai et al., 2022)
Bambusa arundinacea	Detection of glucose	(Jayarambabu et al., 2023)
Allium sativum L.	Detection of cholesterol	(El-Naka et al., 2023)
Camellia sinensis	Detection of H_2O_2	(Tan et al., 2021)

IMAGING

Another area in which Ag NPs are used due to their advanced optical properties caused by the surface plasmon feature is in biological imaging applications (Majeed et al., 2022). Thanks to their small size, NPs exhibit optical scattering properties due to their enhanced surface plasmon property. Optical scattering properties depend on the size, shape, metal content, environment, and surface properties of the NPs (K. J. Lee et al., 2007; K. S. Lee & El-Sayed, 2006). This optical scattering property of metallic NPs makes them useful for imaging systems. Imaging with NPs is widely used in the diagnosis of cancer, one of the most common fatal diseases of mankind today. Here, the main mechanism is the tendency of NPs to accumulate in molecular form in cellular structures. Cancer cells are targeted with various functionalization of the NPs and imaging is sought with their accumulation in the cell and optical scattering (Çetin Uyanıkgil & Selcen Salmanoğlu, 2020; Dağlıoğlu, 2018; K. S. Lee & El-Sayed, 2006). In this field, there are advantages such as easy production and cost-effectiveness of NPs (Çetin Uyanıkgil & Selcen Salmanoğlu, 2020). It has already been said that the use of plant-mediated synthesized Ag NPs for biological applications is more advantageous than other synthesis techniques. Since a toxic material such as Ag NP is still absorbed into the body in applications such as cancer imaging, the fact that it is synthesized in a plant-based manner reduces its toxicity (Mohd Faheem et al., 2022; Yunjian Wang et al., 2021). There are many studies in the literature investigating the anticancer properties of Ag NPs synthesized by plant-based synthesis (Fan et al., 2022; Mohd Faheem et al., 2022; Ossoliński et al., 2023; Wen et al., 2020). The rationale in these studies is the same as the imaging mechanism. Ag NPs targeted to the cancer cell accumulate and destroy the cancer cell with a cytotoxic effect with the reactive oxygen species (ROS) mechanism (Mohd Faheem et al., 2022). For example, Wang et al. (2021) performed the plant-mediated synthesis of Ag NPs using *Zingiber officinale* leaves and studied them in pancreatic cancer cells. In another study, Faheem et al. (2022) synthesized Ag NPs using the whole parts of the *Bergenia ligulata* plant and investigated their breast cancer diagnostic and anticancer properties. It was found that the studies in the literature on plant-mediated synthesized Ag NPs were mainly focused on determining the anticancer property and investigating the cytotoxic properties. It was found in the literature that the studies dealing with imaging were few. This could be due to the fact that magnetic metallic NPs (e.g. iron) have proven properties in this field (Amani et al., 2019; Chavan et al., 2022; Li et al., 2021).

CONCLUSION

Nowadays, NPs are preferred in many fields due to their superior properties at the nanoscale. The synthesis methods of NPs significantly influence their applications. Plant-based synthesized NPs are used in electronics, textile, surface coating, and medicine industries because they are low cost, environmentally friendly, low toxic, and easy to synthesize. In particular, Ag NPs exhibit surface plasmon behavior due to their large surface area. Depending on this behavior, Ag NPs with their optical properties can be used in photocatalysis, sensing and imaging. In this review, Ag NPs and the importance of plant-mediated synthesis, the studies on photocatalysis, sensing, and imaging of Ag NPs synthesized in this manner were examined in detail and a resource was created for researchers working in this field. Plant-mediated synthesized Ag NPs were found to be widely used and suitable for commercial use, particularly in photocatalysis and sensing applications. In imaging applications, it has been found to be particularly useful in cancer diagnosis and treatment, but studies in this field are insufficient for Ag NPs. For this reason, it is recommended that researchers who will work in this area work to develop the properties of Ag NPs synthesized in this way and disseminate their use.

REFERENCES

Ajay, S., Panicker, J. S., Manjumol, K. A., & Subramanian, P. P. (2022). Photocatalytic activity of biogenic silver nanoparticles synthesized using Coleus Vettiveroids. *Inorganic Chemistry Communications*, *144*, 109926. doi:10.1016/j.inoche.2022.109926

Al-Sheddi, E. S., Farshori, N. N., Al-Oqail, M. M., Al-Massarani, S. M., Saquib, Q., Wahab, R., Musarrat, J., Al-Khedhairy, A. A., & Siddiqui, M. A. (2018). Anticancer potential of green synthesized silver nanoparticles using extract of nepeta deflersiana against human cervical cancer cells (HeLA). *Bioinorganic Chemistry and Applications, Nov*(1), 9390784. doi:10.1155/2018/9390784

Alwhibi, M. S., Soliman, D. A., al khaldy, H., Alonaizan, A., Abdulhaq Marraiki, N., El-Zaidy, M., & AlSubeie, M. S. (2020). Green biosynthesis of silver nanoparticle using Commiphora myrrh extract and evaluation of their antimicrobial activity and colon cancer cells viability. *Journal of King Saud University. Science*, *32*(8), 3372–3379. doi:10.1016/j.jksus.2020.09.024

Amani, A., Begdelo, J. M., Yaghoubi, H., & Motallebinia, S. (2019). Multifunctional magnetic nanoparticles for controlled release of anticancer drug, breast cancer cell targeting, MRI/fluorescence imaging, and anticancer drug delivery. *Journal of Drug Delivery Science and Technology*, *49*, 534–546. doi:10.1016/j.jddst.2018.12.034

Aravind, M., Ahmad, A., Ahmad, I., Amalanathan, M., Naseem, K., Mary, S. M. M., Parvathiraja, C., Hussain, S., Algarni, T. S., Pervaiz, M., & Zuber, M. (2021). Critical green routing synthesis of silver NPs using jasmine flower extract for biological activities and photocatalytical degradation of methylene blue. *Journal of Environmental Chemical Engineering*, *9*(1), 104877. doi:10.1016/j.jece.2020.104877

Arboleda, D. M., Santillán, J. M. J., Arce, V. B., Fernández van Raap, M. B., Muraca, D., Fernández, M. A., Torres Sanchez, R. M., Schinca, D. C., & Scaffardi, L. B. (2018). A simple and "green" technique to synthesize long-term stability colloidal Ag nanoparticles: Fs laser ablation in a biocompatible aqueous medium. *Materials Characterization*, *140*, 320–332. doi:10.1016/j.matchar.2018.04.021

Aryan, R., Ruby, & Mehata, M. S. (2021). Green synthesis of silver nanoparticles using Kalanchoe pinnata leaves (life plant) and their antibacterial and photocatalytic activities. *Chemical Physics Letters*, *778*, 138760. doi:10.1016/j.cplett.2021.138760

Ashique, S., Upadhyay, A., Hussain, A., Bag, S., Chaterjee, D., Rihan, M., Mishra, N., Bhatt, S., Puri, V., Sharma, A., Prasher, P., Singh, S. K., Chellappan, D. K., Gupta, G., & Dua, K. (2022). Green biogenic silver nanoparticles, therapeutic uses, recent advances, risk assessment, challenges, and future perspectives. *Journal of Drug Delivery Science and Technology*, *77*, 103876. doi:10.1016/j.jddst.2022.103876

Awad, M. A. A., Hendi, A., Ortashi, K. M., Alzahrani, B., Soliman, D., Alanazi, A., Alenazi, W., Taha, R. M., Ramadan, R., El-Tohamy, M., AlMasoud, N., & Alomar, T. S. (2021). Biogenic synthesis of silver nanoparticles using Trigonella foenum-graecum seed extract: Characterization, photocatalytic and antibacterial activities. *Sensors and Actuators. A, Physical*, *323*, 112670. doi:10.1016/j.sna.2021.112670

Balamurugan, V., Muruganandam, L., & Natarajan, R. (2022). Photocatalytic activity of silver nanoparticles synthesized using cadiospermum halicacabum and moringa oleifera leaf extract. *Materials Today: Proceedings*, *62*, 5365–5370. doi:10.1016/j.matpr.2022.03.540

Basumatary, S., Kumar, K. J., Daimari, J., Mondal, A., Kalita, S., Dey, K. S., & Deka, A. K. (2023). Biosynthesis of silver nanoparticles using Antidesma acidum leaf extract: Its application in textile organic dye degradation. *Environmental Nanotechnology, Monitoring & Management*, *19*, 100769. doi:10.1016/j.enmm.2022.100769

Bi, S., & Ahmad, N. (2022). Green synthesis of palladium nanoparticles and their biomedical applications. *Materials Today: Proceedings*, *62*, 3172–3177. doi:10.1016/j.matpr.2022.03.441

Bose, S., Maity, S., & Sarkar, A. (2022). Nano-materials as biosensor for heavy metal detection. *Food, Medical, and Environmental Applications of Nanomaterials*, 493–526. doi:10.1016/B978-0-12-822858-6.00018-2

Çetin Uyanıkgil, E. Ö., & Selcen Salmanoğlu, D. (2020). Metalik nanopartiküllerin hedeflendirilmesi. *Ege Tıp Dergisi*, *59*(1), 71–81. doi:10.19161/etd.698596

Chavali, M. S., & Nikolova, M. P. (2019). Metal oxide nanoparticles and their applications in nanotechnology. *SN Applied Sciences, 1*(6), 1–30. doi:10.1007/s42452-019-0592-3

Chavan, N., Dharmaraj, D., Sarap, S., & Surve, C. (2022). Magnetic nanoparticles – A new era in nanotechnology. *Journal of Drug Delivery Science and Technology*, *77*, 103899. doi:10.1016/j.jddst.2022.103899

Dağlıoğlu, C. (2018). A Comparative Study of Passive and Active Tumor Targeting Using Nanoparticles as Drug Delivery Systems. *Academic Platform - Journal of Engineering and Science, 6*(1), 1–7. doi:10.21541/apjes.349889

Das, I., & Ansari, S. A. (2009). Nanomaterials in science and technology. *Journal of Scientific and Industrial Research*, *68*(8), 657–667.

Dede, S., & Altay, F. (2018). Biosensors from the First Generation to Nano-biosensors. *International Advanced Researches and Engineering Journal*, *02*(02), 200–207.

Dilshad, E., Bibi, M., Sheikh, N. A., Tamrin, K. F., Mansoor, Q., Maqbool, Q., & Nawaz, M. (2020). Synthesis of functional silver nanoparticles and microparticles with modifiers and evaluation of their antimicrobial, anticancer, and antioxidant activity. *Journal of Functional Biomaterials*, *11*(4), 76. doi:10.3390/jfb11040076 PMID:33113975

El-Naka, M. A., El-Dissouky, A., Ali, G. Y., Ebrahim, S., & Shokry, A. (2023). Garlic capped silver nanoparticles for rapid detection of cholesterol. *Talanta*, *253*, 123908. doi:10.1016/j.talanta.2022.123908 PMID:36087411

Fan, N., Li, P., Wang, J., Gongsun, X., Xue, L., Bai, J., Morovvati, H., & Goorani, S. (2022). Novel formulation, characterization, cytotoxicity, antioxidant, and anti-lung cancer activities of silver nanoparticles green-formulated by plant extract. *Inorganic Chemistry Communications*, *143*, 109707. doi:10.1016/j.inoche.2022.109707

Fatima, T., Bansal, S., Husain, S., & Khanuja, M. (2022). Biosensors. *Electrochemical Sensors: From Working Electrodes to Functionalization and Miniaturized Devices*, 1–30. doi:10.1016/B978-0-12-823148-7.00001-5

Ha, L. P. P., Vinh, T. H. T., Thuy, N. T. B., Thi, C. M., & Van Viet, P. (2021). Visible-light-driven photocatalysis of anisotropic silver nanoparticles decorated on ZnO nanorods: Synthesis and characterizations. *Journal of Environmental Chemical Engineering*, *9*(2), 105103. doi:10.1016/j.jece.2021.105103

Habibullah, G., Viktorova, J., & Ruml, T. (2021). Current Strategies for Noble Metal Nanoparticle Synthesis. *Nanoscale Research Letters, 16*(1), 1–12. doi:10.1186/s11671-021-03480-8

Hai, N. D., Dat, N. M., Huong, L. M., Tai, L. T., Thinh, D. B., Nam, N. T. H., Dat, N. T., Phong, M. T., & Hieu, N. H. (2022). Phytosynthesis of silver nanoparticles using Mangifera indica leaves extract at room temperature: Formation mechanism, catalytic reduction, colorimetric sensing, and antimicrobial activity. *Colloids and Surfaces. B, Biointerfaces*, *220*, 112974. doi:10.1016/j.colsurfb.2022.112974

Hossain, N., Islam, M. A., & Chowdhury, M. A. (2022). Synthesis and characterization of plant extracted silver nanoparticles and advances in dental implant applications. *Heliyon*, *8*(12), e12313. doi:10.1016/j.heliyon.2022.e12313 PMID:36590472

Ilgar, M., Baytemir, G., Taşaltın, N., Güllülü, S., Yeşilyurt, İ. S., & Karakuş, S. (2022). Multifunctional maca extract coated CuO nanoparticles with antimicrobial and dopamine sensing activities: A dual electrochemical – Smartphone colorimetric detection system. *Journal of Photochemistry and Photobiology A Chemistry*, *431*, 114075. doi:10.1016/j.jphotochem.2022.114075

Iliger, K. S., Sofi, T. A., Bhat, N. A., Ahanger, F. A., Sekhar, J. C., Elhendi, A. Z., Al-Huqail, A. A., & Khan, F. (2021). Copper nanoparticles: Green synthesis and managing fruit rot disease of chilli caused by Colletotrichum capsici. *Saudi Journal of Biological Sciences*, *28*(2), 1477–1486. doi:10.1016/j.sjbs.2020.12.003 PMID:33613075

Iravani, S. (2014). Bacteria in Nanoparticle Synthesis: Current Status and Future Prospects. *International Scholarly Research Notices*, *2014*, 1–18. doi:10.1155/2014/359316 PMID:27355054

Jamkhande, P. G., Ghule, N. W., Bamer, A. H., & Kalaskar, M. G. (2019). Metal nanoparticles synthesis: An overview on methods of preparation, advantages and disadvantages, and applications. *Journal of Drug Delivery Science and Technology*, *53*, 101174. doi:10.1016/j.jddst.2019.101174

Jayarambabu, N., Saraswathi, K., Akshaykranth, A., Anitha, N., Venkatappa Rao, T., & Rakesh kumar, R. (2023). Bamboo-mediated silver nanoparticles functionalized with activated carbon and their application for non-enzymatic glucose sensing. *Inorganic Chemistry Communications*, *147*, 110249. doi:10.1016/j.inoche.2022.110249

Karabulut, G., Beköz Üllen, N., Akyüz, E., & Karakuş, S. (2023). Surface modification of 316L stainless steel with multifunctional locust gum/polyethylene glycol-silver nanoparticles using different coating methods. *Progress in Organic Coatings*, *174*, 107291. doi:10.1016/j.porgcoat.2022.107291

Karabulut, G., Üllen, N. B., & Karakuş, S. (2022). *Nanostructures in Biosensors: Development and Applications*. IntechOpen., doi:10.5772/intechopen.108508

Karakuş, S., Taşaltın, N., Taşaltın, C., & Üllen, N. B. (2021). Synthesis and Characterization of Konjac Gum/Polyethylene Glycol-Silver Nanoparticles and their Potential Application as a Colorimetric Sensor for Hydrogen Peroxide. *Journal of Inorganic and Organometallic Polymers and Materials*, *1*(9), 1–14. doi:10.100710904-021-01984-5

Konishi, Y., Ohno, K., Saitoh, N., Nomura, T., Nagamine, S., Hishida, H., Takahashi, Y., & Uruga, T. (2007). Bioreductive deposition of platinum nanoparticles on the bacterium Shewanella algae. *Journal of Biotechnology*, *128*(3), 648–653. doi:10.1016/j.jbiotec.2006.11.014 PMID:17182148

Kumar, R., Chauhan, G., & Martinez-Chapa, S. O. (2022). Metal oxides and their composites as flow-through biosensors for biomonitoring. *Metal Oxides for Biomedical and Biosensor Applications*, 291–319. doi:10.1016/B978-0-12-823033-6.00010-7

Lee, K. J., Nallathamby, P. D., Browning, L. M., Osgood, C. J., & Nancy Xu, X. H. (2007). In vivo imaging of transport and biocompatibility of single silver nanoparticles in early development of zebrafish embryos. *ACS Nano*, *1*(2), 133–143. doi:10.1021/nn700048y PMID:19122772

Lee, K. S., & El-Sayed, M. A. (2006). Gold and silver nanoparticles in sensing and imaging: Sensitivity of plasmon response to size, shape, and metal composition. *The Journal of Physical Chemistry B*, *110*(39), 19220–19225. doi:10.1021/jp062536y PMID:17004772

Li, X., Li, W., Wang, M., & Liao, Z. (2021). Magnetic nanoparticles for cancer theranostics: Advances and prospects. *Journal of Controlled Release*, *335*, 437–448. doi:10.1016/j.jconrel.2021.05.042 PMID:34081996

Majeed, M., Hakeem, K. R., & Rehman, R. U. (2022). Synergistic effect of plant extract coupled silver nanoparticles in various therapeutic applications- present insights and bottlenecks. *Chemosphere, 288*, 132527. doi:10.1016/j.chemosphere.2021.132527 PMID:34637861

Malhotra, S., Verma, A., Tyagi, N., & Kumar, V. (2017). *BIOSENSORS: PRINCIPLE, TYPES AND APPLICATIONS, 3*(2).

Marimuthu, S., Antonisamy, A. J., Malayandi, S., Rajendran, K., Tsai, P. C., Pugazhendhi, A., & Ponnusamy, V. K. (2020). Silver nanoparticles in dye effluent treatment: A review on synthesis, treatment methods, mechanisms, photocatalytic degradation, toxic effects and mitigation of toxicity. *Journal of Photochemistry and Photobiology. B, Biology, 205*, 111823. doi:10.1016/j.jphotobiol.2020.111823 PMID:32120184

Mohd Faheem, M., Bhagat, M., Sharma, P., & Anand, R. (2022). Induction of p53 mediated mitochondrial apoptosis and cell cycle arrest in human breast cancer cells by plant mediated synthesis of silver nanoparticles from Bergenia ligulata (Whole plant). *International Journal of Pharmaceutics, 619*, 121710. doi:10.1016/j.ijpharm.2022.121710 PMID:35367334

Mokammel, M. A., Islam, M. J., Hasanuzzaman, M., & Hashmi, M. S. J. (2022). Nanoscale Materials for Self-Cleaning and Antibacterial Applications. Encyclopedia of Smart Materials, 315–324. doi:10.1016/B978-0-12-815732-9.00015-2

Molina, G. A., Esparza, R., López-Miranda, J. L., Hernández-Martínez, A. R., España-Sánchez, B. L., Elizalde-Peña, E. A., & Estevez, M. (2019). Green synthesis of Ag nanoflowers using Kalanchoe Daigremontiana extract for enhanced photocatalytic and antibacterial activities. *Colloids and Surfaces. B, Biointerfaces, 180*, 141–149. doi:10.1016/j.colsurfb.2019.04.044 PMID:31039515

Nasrollahzadeh, M., Shafiei, N., & Nezafat, Z. (2021). Synthesis of biopolymer-based metal nanoparticles. *Biopolymer-Based Metal Nanoparticle Chemistry for Sustainable Applications*, 255–316. doi:10.1016/B978-0-12-822108-2.00018-1

Oluwafemi, O. S., Anyik, J. L., Zikalala, N. E., & Sakho, E. H. M. (2019). Biosynthesis of silver nanoparticles from water hyacinth plant leaves extract for colourimetric sensing of heavy metals. *Nano-Structures & Nano-Objects, 20*, 100387. doi:10.1016/j.nanoso.2019.100387

Ossoliński, K., Ruman, T., Ossoliński, T., Ossolińska, A., Arendowski, A., Kołodziej, A., Płaza-Altamer, A., & Nizioł, J. (2023). Monoisotopic silver nanoparticles-based mass spectrometry imaging of human bladder cancer tissue: Biomarker discovery. *Advances in Medical Sciences*, *68*(1), 38–45. doi:10.1016/j.advms.2022.12.002 PMID:36566601

Philip, D., Unni, C., Aromal, S. A., & Vidhu, V. K. (2011). Murraya Koenigii leaf-assisted rapid green synthesis of silver and gold nanoparticles. *Spectrochimica Acta. Part A: Molecular and Biomolecular Spectroscopy*, *78*(2), 899–904. doi:10.1016/j.saa.2010.12.060 PMID:21215687

Pragathiswaran, C., Violetmary, J., Faritha, A., Selvarani, K., & Nawas, P. M. A. (2021). Photocatalytic degradation, sensing of Cd2+ using silver nanoparticles synthesised from plant extract of cissus quadrangularis and their microbial activity. *Materials Today: Proceedings*, *45*, 3348–3356. doi:10.1016/j.matpr.2020.12.656

Purohit, B., Vernekar, P. R., Shetti, N. P., & Chandra, P. (2020). Biosensor nanoengineering: Design, operation, and implementation for biomolecular analysis. *Sensors International*, *1*, 100040. doi:10.1016/j.sintl.2020.100040

Ramachandiran, D., & Rajesh, K. (2022). Selective colorimetric and fluorimertic sensor of Hg (II) ion from silver nanoparticles using Acacia chundra leaves extract. *Materials Chemistry and Physics*, *287*, 126284. doi:10.1016/j.matchemphys.2022.126284

Rana, A., Yadav, K., & Jagadevan, S. (2020). A comprehensive review on green synthesis of nature-inspired metal nanoparticles: Mechanism, application and toxicity. *Journal of Cleaner Production*, *272*, 122880. doi:10.1016/j.jclepro.2020.122880

Sankar Sana, S., Haldhar, R., Parameswaranpillai, J., Chavali, M., & Kim, S. C. (2022). Silver nanoparticles-based composite for dye removal: A comprehensive review. *Cleaner Materials*, *6*, 100161. doi:10.1016/j.clema.2022.100161

Sankareswari, M., Amutha, C., Vasantha, V. S., Arunpandian, M., & Nagarajan, E. R. (2022). Biosynthesized silver nanoparticles using Rosary Pea seed Extract: Evaluation of Antibacterial, cytotoxic and photocatalytic activity. *Inorganic Chemistry Communications*, *145*, 109977. doi:10.1016/j.inoche.2022.109977

Santhosh, A., Theertha, V., Prakash, P., & Smitha Chandran, S. (2021). From waste to a value added product: Green synthesis of silver nanoparticles from onion peels together with its diverse applications. *Materials Today: Proceedings*, *46*, 4460–4463. doi:10.1016/j.matpr.2020.09.680

Shukla, S., Masih, A., Aryan, & Mehata, M. S. (2022). Catalytic activity of silver nanoparticles synthesized using Crinum asiaticum (Sudarshan) leaf extract. *Materials Today: Proceedings*, *56*, 3714–3720. doi:10.1016/j.matpr.2021.12.468

Sumi, M. B., Devadiga, A., Shetty, K. V., & Saidutta, M. B. (2016). Solar photocatalytically active, engineered silver nanoparticle synthesis using aqueous extract of mesocarp of Cocos nucifera (Red Spicata Dwarf), *12*(1), 14–32. doi:10.1080/17458080.2016.1251622

Tahir, M. B., Sohaib, M., Sagir, M., & Rafique, M. (2022). Role of Nanotechnology in Photocatalysis. Encyclopedia of Smart Materials, 578–589. doi:10.1016/B978-0-12-815732-9.00006-1

Tan, E., Kahyaoğlu, İ. M., & Karakuş, S. (2021). A sensitive and smartphone colorimetric assay for the detection of hydrogen peroxide based on antibacterial and antifungal matcha extract silver nanoparticles enriched with polyphenol. *Polymer Bulletin*, 1–27. doi:10.1007/S00289-021-03857-W/TABLES/5 PMID:34413556

Tetyana, P., Shumbula, P. M., & Njengele-Tetyana, Z. (2021). *Biosensors: Design, Development and Applications*. Nanopores. doi:10.5772/intechopen.97576

Thomas, S., Saji, K. J., & Jayaraj, M. K. (2022). An introduction to biosensors. *Nanomaterials for Sensing and Optoelectronic Applications*, 91–107. doi:10.1016/B978-0-12-824008-3.00013-8

Tortella, G., Rubilar, O., Diez, M. C., Cuozzo, S., Pieretti, J. C., & Seabra, A. B. (2022). Role of bacteria and actinobacteria in the biosynthesis of silver nanoparticles. *Green Synthesis of Silver Nanomaterials*, 493–512. doi:10.1016/B978-0-12-824508-8.00003-4

Uddin, I., Ahmad, K., Khan, A. A., & Kazmi, M. A. (2017). Synthesis of silver nanoparticles using Matricaria recutita (Babunah) plant extract and its study as mercury ions sensor. *Sensing and Bio-Sensing Research*, *16*, 62–67. doi:10.1016/j.sbsr.2017.11.005

Vaid, P., Raizada, P., Saini, A. K., & Saini, R. V. (2020). Biogenic silver, gold and copper nanoparticles - A sustainable green chemistry approach for cancer therapy. *Sustainable Chemistry and Pharmacy, 16*(August 2019). doi:10.1016/j.scp.2020.100247

Velasco-Garcia, M. N., & Mottram, T. (2003). Biosensor Technology addressing Agricultural Problems. *Biosystems Engineering*, *84*(1), 1–12. doi:10.1016/S1537-5110(02)00236-2

Vigneshvar, S., Sudhakumari, C. C., Senthilkumaran, B., & Prakash, H. (2016). Recent advances in biosensor technology for potential applications - an overview. *Frontiers in Bioengineering and Biotechnology*, *4*(FEB), 11. doi:10.3389/fbioe.2016.00011 PMID:26909346

Wang, Y., Chinnathambi, A., Nasif, O., & Alharbi, S. A. (2021). Green synthesis and chemical characterization of a novel anti-human pancreatic cancer supplement by silver nanoparticles containing Zingiber officinale leaf aqueous extract. *Arabian Journal of Chemistry*, *14*(4), 103081. doi:10.1016/j.arabjc.2021.103081

Wang, Y., Chou, S., & Zhang, Z. (2019). Nanomaterials Innovation. *Small*, *15*(32), 15–17. doi:10.1002mll.201902246 PMID:31397544

Wen, X., Wang, Q., Dai, T., Shao, J., Wu, X., Jiang, Z., Jacob, J. A., & Jiang, C. (2020). Identification of possible reductants in the aqueous leaf extract of mangrove plant Rhizophora apiculata for the fabrication and cytotoxicity of silver nanoparticles against human osteosarcoma MG-63 cells. *Materials Science and Engineering C*, *116*, 111252. doi:10.1016/j.msec.2020.111252 PMID:32806252

Weng, X., Yang, K., Owens, G., & Chen, Z. (2022). Biosynthesis of silver nanoparticles using three different fruit extracts: Characterization, formation mechanism and estrogen removal. *Journal of Environmental Management*, *316*, 115224. doi:10.1016/j.jenvman.2022.115224 PMID:35550961

Zamarchi, F., & Vieira, I. C. (2021). Determination of paracetamol using a sensor based on green synthesis of silver nanoparticles in plant extract. *Journal of Pharmaceutical and Biomedical Analysis*, *196*, 113912. doi:10.1016/j.jpba.2021.113912 PMID:33581590

Zhu, M., Li, X., Ge, L., Zi, Y., Qi, M., Li, Y., Li, D., & Mu, C. (2020). Green synthesis of κ-carrageenan@Ag submicron-particles with high aqueous stability, robust antibacterial activity and low cytotoxicity. *Materials Science and Engineering C*, *106*, 110185. doi:10.1016/j.msec.2019.110185 PMID:31753345

Chapter 6
Multifunctional Materials for the Sensing of Gases

Asma Rafiq
University of the Punjab, Lahore, Pakistan

Misbah Naz
University of Education, Lahore, Pakistan

Shehnila Altaf
University of Engineering and Technology, Lahore, Pakistan

Saira Riaz
University of the Punjab, Lahore, Pakistan

Shahzad Naseem
https://orcid.org/0000-0002-7814-7577
University of the Punjab, Lahore, Pakistan

ABSTRACT

Air contamination and a growing number of harmful gases releasing into the environment is a big issue of present era. Gases like CO, NH_3, CO_2, SO_2, and CH_4 can prompt lethal health threats. Gas sensors have attained noteworthy consideration of researchers in monitoring such kind of pollutants. Detection of gases by sensors typically depends upon the sensing properties, like sensitivity, selectivity, response, and recovery time. With the progress of the internet-of-things technology, the uses of gas sensors in the fields of smart homes, wearable gadgets, and smart mobile terminals have developed rapidly. In this chapter, various groups of the sensing materials are outlined thoroughly, comprising gas sensors based on multifunctional materials such as: metals, graphene, cellulose, MXenes, acoustic wave sensors, and catalytic sensors. This chapter also outlines the current accomplishments in the field of gas detection and signifies the current challenges and future outlooks in this field.

DOI: 10.4018/978-1-6684-8743-3.ch006

1. INTRODUCTION

The fast-growing population, industrial discharges, vehicle exhaust and open burning of trash are the primary cause of contaminants which frequently worsen the natural ecological environment. In order to keep the environment safe, there is a need of continuous monitoring of these pollutants. Different sorts of instruments are utilized to screen harmful gases, which are tedious, costly and seldom utilized for practical applications (Yamazoe & Miura, 1994). Gas sensors assist in classifying and analyzing different chemical compounds and are utilized for numerous fields like food production (J. Chen, Gu, Zhang, Mao, & Tian, 2019), detection of venoms in packed food parcels (Matindoust, Baghaei-Nejad, Abadi, Zou, & Zheng, 2016), monitoring of outdoor and indoor air quality to check the radon level (Blanco-Novoa, Fernández-Caramés, Fraga-Lamas, & Castedo, 2018), diagnosis of diseases like liver cancer and Alzheimer's detection, detection of gas leakage (Guo, Yang, Guo, & Liu, 2019) and threats detections in livestock (Ashari, Widodo, & Suryono, 2019).

Gas sensors generate electrical signals that form unique gas fingerprints in response to target. To identify these gas fingerprints easily, the choice of sensing element in sensors is a serious matter. One of the most commonly used strategies to select the sensing material is based on the original substrate material plated or mixed with other material suitable for reducing the noise (Kroutil et al., 2018). The criteria to assess the capacity or the performance of gas sensors is to check their selectivity, sensitivity, stability, and response time (Hunter et al., 2020)

Selectivity describes the ability of the gas sensor to distinguish a specific gas from a mixture of a variety of gases. The term sensitivity can be illustrated by the alteration in sensor's response per unit change of the input (Z. Chen, Chen, Song, Ye, & Fan, 2019). In contrast to harsh environmental conditions, under normal circumstances, gas sensors are exposed to numerous gases of varying concentrations (Feng et al., 2019). Here, the reliability of a gas sensor is assessed by its aptitude to be extremely selective for a particular gas, and in different applications to be sensitive towards trace amount of the gases. Instead, stability means the capacity of a sensor to repeat its response for some duration of time, while, response time refers to the time reserved by the sensor to extract a meaningful sensory signal response (Hunter et al., 2020). The response time is a serious constraint, particularly when dealing with poisonous gases, because immediate detection is mandatory to activate timely safety measures to avoid disastrous consequences. Gas sensors show various responses and functionalities for different applications. For example, in many cases, single sensor does not sufficiently fulfil the requirements of the application thus the need of smart sensing becomes intense.

Herein, we focus on the development of various materials used for gas sensing applications. This review outlines the relevant researches published in the last decade to provide a comprehensive study on multifunctional materials for gas sensing applications.

2. MULTIFUNCTIONAL MATERIALS FOR GAS SENSING APPLICATIONS

Gas sensor arrays (GSA) constituting different materials possessing diverse sensitivity and selectivity (Persaud & Dodd, 1982). However, less accuracy and drift like risks did not allow the GSA technology to flourish till next ten years. Recently, there is high demand to develop multifunctional nanomaterial-based gas sensors which operates at ambient temperature. The use of sensors is widespread, involving in the monitoring of pollution, the estimation of leaks, and harmful gases (Z. Wang, Zhu, Sun, Wang, & Yan, 2021). For monitoring leakage in fuel, filters, and in cars, they are frequently employed in automobiles (F. Yang et al., 2021). The medical and pharmaceutical sectors also extensively employ gas sensors for breathing analysis (X. Chen, Leishman, Bagnall, & Nasiri, 2021). It is significant to point out that gas sensors are widely used in a number of different sectors, but there is a rising need for sensors that are low-cost, extremely sensitive, and energy-efficient. Consequently, the researchers are focusing on creating trustworthy materials to continue producing cost-effective and effective sensors.

Based on their mobility, gas sensors can be separated into three sections: area monitoring, fixed, and portable sensor. In order to monitor the environment after working in sensitive environments, portable gas sensors are typically used (Bag & Lee, 2021; Hernández-Gordillo, Ruiz-Correa, Robledo-Valero, Hernández-Rosales, & Arriaga, 2021; Zhou et al., 2022). Mostly fixed ones are employed in homes, even for surveillance of outdoor areas of companies (Butterfield, 2021). Usually area monitoring sensors involves combination of fixed and mobile gas sensors (Gomes, Rodrigues, Rabêlo, Kumar, & Kozlov, 2019).

Gas sensors classified into different types based on the sensing method, the most common sensor types are electro-chemical, acoustic, optical, photo-ionization, and catalytic beads. The most researched gas sensors are electro-chemical ones because of their low power requirements, nearly perfect repeatability, accuracy, and resistance to contamination by other gases (Ramu et al., 2021). Optical sensors are employed because of their high specificity and quick response times caused by optical absorption (K. Chen, Yuan, & Zhao, 2021; Q. Ma, Kutilike, Kari, Abliz, & Yimit, 2021). Another fairly common type is the acoustic wave gas sensor, which offers the advantages of being a wireless sensor, being affordable, able to be installed in demanding environments like those with moving parts, and requiring little power (Midhuna, SD, & Sreenidhi, 2021). The most sought-after gas sensors include those that use photoionization. They are favored for their compact design, ease of handling, affordability, and response over a wide range (Farsalinos, Gillman, Hecht, Polosa, & Thornburg, 2016). Resistive sensors are still among the most researched ones due to their relative simplicity, affordability, works with different sensing materials, and relative ease of fabrication, (Comini, Faglia, Sberveglieri, Pan, & Wang, 2002; Kim et al., 2013).

3. GAS SENSORS: MATERIALS AND TECHNOLOGIES

3.1. Metal Based Gas Sensors

Metal oxides based gas sensing materials are popular among researchers because of their advantageous characteristics including sensitivity towards gases, economic,

Table 1. Comparison of numerous metal oxides, fabrication techniques, target gas, operating temperature, and their response

Materials	Synthesis path	Morphology	Gas	Response/ sensitivity	Temp./ Conc (ppm)	Response time (s)	Reference
Co-doped ZnO	Solvothermal method	Quasi-spherical	CH_4	7%	140 °C / 375	25.2	(Aghagoli & Ardyanian, 2018)
ZnO-rGO nanocomposite	Hydrothermal method	Nanorods	CH_4	14%	190 °C / 1500	30	(D. Zhang, Yin, & Xia, 2015)
Au/VO_x films	Sputtering	Thin film	CH_4	22%	Room temp. / 1000	–	(Liang, Liu, Li, & Li, 2016)
Nanoporous GO	Hummer process	Thin film	CO_2	37.40%	Room temp./ 60 sccm	25	(Shaban, Ali, & Rabia, 2019)
rGO-NiO(8)-In_2O_3(2)	Sonication + hydrothermal process	Nanoball	CO_2	49%	Room temp./ 50	6	(Amarnath & Gurunathan, 2021)
p-type Cr_2O_3 films	RF magnetron sputtering	Thin film	H_2	85%	400 °C / 1000	–	(Almaev et al., 2021)
rGO/Co_3O_4	Hydrothermal treatment	Nanosheet	NO_2	26.80%	Room temp.5	90	(B. Zhang et al., 2018)
α-Fe_2O_3 @ graphite	Low temp. hydrolysis	Nanorod	C_3H_6O	16.90%	260 °C / 50	5	(Yu, Cheng, Feng, Jia, & Song, 2019)
ZnO/MoS_2	Solvothermal	Nanosheet	NO_2	3050%	RT / 5	40	(B. Zhang et al., 2018)
rGO/Co_3O_4	Solvothermal	Nanosheet	NH_3	1.78%	RT / 20	351	(Srirattanapibul, Nakarungsee, Issro, Tang, & Thongmee, 2021)

transportability, easy handling, unique optical characteristics, large bandgaps, and dielectric constants, diverse electrical properties(Ayyala & Covington, 2021; Kaur, Singh, & Comini, 2022; D. Yang, Gopal, Lkhagvaa, & Choi, 2021). Most commonly, nanocomposites of metal oxides are chosen due to tiny sizes, morphology, high surface-area, durability, sensitivity, crystallinity, conductivity, photo-catalytic, and distinctive electronic applications (Kaur et al., 2022). Several of metal oxides have been synthesised in order to monitor combustible, reduce gases by monitoring their response. The oxides that exhibit conductivity gas response are Cr_2O_3, Nb_2O_5, ZnO CuO, Nd_2O_3, SrO, TiO_2, WO_3, In_2O_3, V_2O_3, MoO_3, Fe_2O_3, GeO_2, Ta_2O_5, CeO_2, NiO (Kanazawa et al., 2001). Table 1 signifies comparison of numerous metal oxides, fabrication techniques, target gas, operating temperature, and their response.

3.1.1. Metal Sulfides

Sulfide mineral is the origin of most ores and metals because of their readily availability and affordability, they are used for scientific and industrial purposes. These types of materials possess special chemical, electrical, and optical characteristics (Gaiardo et al., 2016). Less energy is required to stimulate metal sulphides gas sensors for monitoring. (Lai, Lu, & Chen, 2012). Mono sulfides/disulfides of metals possess specific properties depending upon the various phases of crystals. Some disulphides like GeS_2 and SnS_2 typically act as semiconductors while MoS_2, SnS_2, and WS_2 are utilized in transistors (Meshram & Abhilash, 2022; Sakhuja, Jha, Chaurasiya, Dixit, & Bhat, 2020). Some disulphides of titanium, hafnium and zirconium show good conductivity (Sakhuja et al., 2020; Yao & Yang, 2022). As the gas physically or chemically come in contact with the sensor, its recognition depends upon its level of absorptivity or adsorption energy. Greater the interaction of gas molecules, stronger will be the response (Sakhuja et al., 2020). In case of MoS_2 sensors, exposure to oxidizing gases such as O_2, NO, NO_2, SO_2, and CO that draw electrons from its surface and decreases the charge density in metal sulphide thus enhances the resistance of sensor (Donarelli & Ottaviano, 2018). On the other hand, exposure of MoS_2 doped sensor to the reducing gases like NH_3 which donates electrons to the MoS_2 nanoworm thin film sensor, results in the increase of charge density (Tyagi et al., 2022). Hence, metal ferrites-based gas sensors are paramount in sensing field. Table 2 signifies the comparison of various metal sulphides, synthesis methodology, target gas, operating temperature, and their response.

Table 2. Comparison of different metal sulphides, fabrication techniques, target gas, operating temperature, and their response

Materials	Synthesis route	Morphology	Gas	Temp./ Conc (ppm)	Response time (s)	Ref.
MoS_2	Hydrothermal process	Thin film	NO_2	RT/ 100	678	(T. Xu et al., 2017)
MoS_2	Hydrothermal process	Nanosheet	NO_2	RT/ 100	1.14	(Ikram et al., 2019)
Lead sulfide	Sputtering process	Nanosheet	Acetone	RT/ 20		(Jahromi, Kazemi, & Sheikhi, 2021)
nanosheets 2D-HfS_2 flakes	Chemical route	Nanoflakes	Methanol	RT/500	12.12	(Das, Sharma, & Sharma, 2019)
Ag/MoS_2 nanorods	Hydrothermal process	Nanorod	Methanol	RT/100	240	(Halvaee, Dehghani, & Mohammadzadeh, 2020)
SnS nanoflakes	Solid state reaction technique	Nanoflakes	Methanol	200 °C /10	-	(Afsar, Rafiq, & Tok, 2017)
ZnS	Thermal evaporation	Nanowires	Acetone	320/100	7	(X. Wang et al., 2012)
SnS_2	Precipitation in aqueous	Nanorod	Acetone	300/10	-	(Giberti et al., 2016)

3.1.2. Metal Ferrites

Metal ferrites, for their distinct physical and chemical characteristics, have attracted a lot of attention. Ferrite compounds include lanthanides and fast transition metals, mostly iron, manganese and Chromium (Srivastava & Yadav, 2012). Due to their exceptionally stable chemical and thermal characteristics, broad resistivity range, availability of a wide range of affordable materials, and adaptable magnetic and mechanical characteristics, metal ferrites are appealing for use in gas sensor (Fallah Vostakola, Salamatinia, & Amini Horri, 2022; Humayun et al., 2022; Srivastava & Yadav, 2012). Typically, ferrites are classified into three groups: N-type, P-type, and mixed (K. Wu, Li, & Zhang, 2019). When O_2 interact with N-type sensor ($ZnFe_2O_4$ and $MgFe_2O_4$), ions are formed by accepting electrons from sensor, consequently enhances the resistance of sensor. The opposite phenomenon occurs, when the sensors interact with hydrogen or acetone gas, here electron density decreases leads to high resistance of sensor. Sensing ability of metal ferrites are significantly enhanced by hydrothermal technique, such as: nano rods of $ZnFe_2O_4$ metal ferrite give quick response when exposed to CH_3COCH_3 gas. Table 3 shows a comparison

Table 3. Comparison of different metal ferrites, fabrication techniques, target gas, operating temperature, and their response

Materials	Synthesis route	Morphology	Gas	Temp./ Conc	Response time (s)	Ref.
$ZnFe_2O_4$/α-Fe_2O_3	sacrificial template	Porous microrods	triethyl-	305 °C /100	–	(Yanwei Li et al., 2018)
$Sr_{0.2}Ni_{0.8}Fe_2O_4$ thin film	Sol–gel spin coating	Thin film	LPG	200 °C /2000	79	(Méndez, Posada, & Ocampo, 2022)
$ZnO/ZnFe_2O_4$	simple and direct pyrolysis	Hollow cube	acetone	250 °C /5	–	(X. Ma et al., 2017)
$CdFe_2O_4$ nanocomposite	Sol–gel auto combustion	Granular nonporous	LPG	25 °C /1000	50	(Kotresh, Ravikiran, Tiwari, & Vijaya Kumari, 2017)
$RGO/ZnFe_2O_4$	solvothermal method	–	acetone	275 °C /10	0.73	(F. Liu et al., 2013)

of numerous ferrites, fabrication techniques, target gas, operating temperature, and their response.

3.1.3. Carbon Based Gas Sensors

The fact that carbon materials have a variety of structural and morphological characteristics at the nanoscale is one of the reasons they are extensively studied in nanomaterials. It is essential to use them as gas-sensing materials since they can exist as nano-tubes, nano-particles, nanowires, and nano-porous forms. Carbon and carbon-based materials serve for several reasons, including their high energy barrier, which contributes to their stability (Llobet, 2013). Single and multi-walled carbon nanotubes (CNTs) are the most notable forms. The particular surface area of these two structures is substantial (Afzali, Mostafavi, Nekooie, & Jahromi, 2019). By altering their synthesis processes, CNTs can produce aligned or unaligned shapes that can directly impact the performance of devices and related applications(Rust et al., 2022). Many of their sensing characteristics have been justified by their excellent mechanical qualities, such as strength and stiffens. As gaseous sensors, graphene and carbon dots have been acknowledged for their ability to identify dangerous elements in the atmosphere (J. Liu, Bao, & Wang, 2022).

Asama *et al.* 2016, fabricated MWCNTs and SWCNTs thin films by the drop casting technique and observed the current response (resistance variation) upon exposure to test gas NO_2 at varying operating temperatures. According to figure 1, the observed current rises when exposed to NO_2 -a strong oxidizing gas- due

Figure 1. Time fluctuation of CNTs sensor time at 25°C, 50°C, 100°C, 150°C and 200°C testing temperature after exposure to NO_2 gas and 5V voltage for (a) MWCNTs and (b) SWCNTs
Reproduced with permission from ref (Naje et al., 2016) Copyright 2016, Springer

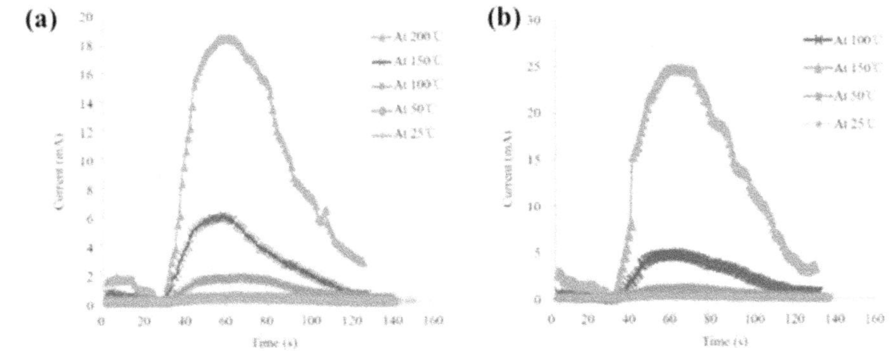

to transfer of electron from CNTs to NO_2. The following explanation can be used to explain how current changes by oxidizing gas: The Fermi levels shift towards the valance band, producing more holes and lowering electrical resistance. The electrical behavior of the carbon nanotubes suggests the charge transfer among the sensing component and test gas; as a result, the primary sensing mechanism is the physisorption of gases in the CNT (Naje, Ibraheem, & Ibrahim, 2016).

3.1.4. Transition Metal Dichalcogenides

Some of the earliest nanostructured transition metal dichalcogenides (TMDs) to be used were MoS_2, $MoSe_2$, and WS_2 (Lv et al., 2015). The behavior of TMDs is pseudocapacitive, and they can function as supercapacitors (Forouzandeh, Kumaravel, & Pillai, 2020). The reason for this is that the transition metal is wedged among two chalcogen sheets and is kept together by weak van der Waals forces (Lv et al., 2015). These weak forces decreases the resistance and consequently increases the sensing and lubricating efficiency of devices (X. Li, Li, Li, Wang, & Zhang, 2017). Table 4 signifies comparison of TMDs, fabrication techniques, target gas, operating temperature, and their response.

Table 4. Comparison of TMDs, fabrication techniques, target gas, operating temperature, and their response

Materials	Morphology	Synthesis route	Temp. /Conc	Gas	Response time (s)	Response/ sensitivity	Ref.
MoS_2-Au	Nanoflakes	Liquid-phase Exfoliation	RT/10	Acetone	~105	15%	(W. Y. Chen, Yen, Xue, Wang, & Stanciu, 2019)
MoS_2	2D	Chemical vapor	RT/25	NO_2	~180	97%	(Zhao et al., 2018)
		deposition					
MoS_2 / rGO	thin film	Soft-lithography patterning	90 °C /5	NO_2	–	25%	(Jung, Kang, Nam, An, & Ku, 2018)
$MoSe_2$	Nanosheets	Gold-assisted	RT/2	NO_2	20	20%	(S. Guo et al., 2019)
		exfoliation					
$WS_{2x}Se_{2-2x}$	Honeycomb-like	Chemical vapor	27 oC /500	NO_2	–	2621%	(Ko et al., 2018)
		deposition					
SnS_2/SnO_2	Nanotubes	In-situ hydrothermal	RT/100	NH_3	21	2.48%	(R. Li et al., 2017)
		sulfuration					

3.2. Graphene-Based Gas Sensors

Graphene, a film made of sp^2 carbon atoms that is one atom thick, was initially reported in 1947 (J. Zhang, Qin, Zeng, & Xie, 2017) and was first prepared in the first year of this century (Kim & Lee, 2014). Researchers from all around the world are immediately drawn to it because of its exceptional morphological traits and unique electrical characteristics. A particularly sensing material for gas detection, graphene can alter in resistance sensibly in response to small amounts of target gases adsorbed on its surface. Graphene nanoribbons (GNRs) have been directly deposited on a multielectrode chip. Platinum electrodes bridged by graphene nanoribbons to form a sensing device. When NRs grow, the chip contained 38 GNR sensors as shown in figure 2a *(Shekhirev et al., 2020).*

It was observed that response of graphene nanoribbons sensor enhanced with temperature, and performance of a sensor is calculated at 180 °C as illustrated in figure 2b–f. Alcohol analytes enhances the resistance of device. The solution-synthesized GNRs' possess unusual gas sensing properties which can be attributed to their extensive electronic band gap in contrast to the CVD-grown graphene

Figure 2. (a) Depiction of the multisensor array. (b–f) Sensor responses of a representative graphene nano ribbon device in the array to selected gas analytes
Reproduced with permission from ref (Shekhirev et al., 2020) Copyright 2020, American Chemical Society

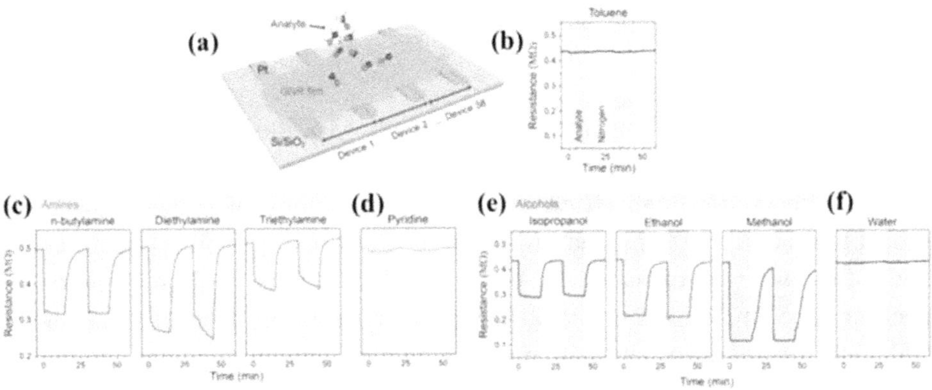

nanoribbons fabricated (Kimouche et al., 2015; L. Yang, Park, Son, Cohen, & Louie, 2007) having peculiar arrangement. Experimentally, toluene has very weak sensor response and weak donor/acceptor characteristics (Figure 2b). Whereas, amine analytes exhibit much greater response as compared to toluene due to nitrogen's ability to donate electrons as it contains lone pair of electrons (Figure 2c). Secondary amine is a stronger electron donor hence, giving maximum response in the amine series (Figure 2c). Pyridine is supposed to be adsorbed in the parallel plane on graphene surface (Voloshina, Mollenhauer, Chiappisi, & Paulus, 2011), weakening the electronic impact of the nitrogen lone pair on the conductivity of graphene nanoribbons and the corresponding sensor response. (Figure 2d). Like amines, alcohols showed robust sensor responses which further got stronger as the molecule size shrunk (Figure 2e). The graphene nanoribbons devices demonstrated a weaker sensor response to H_2O in comparison to the extensively researched rGO sensors (Borini et al., 2013) (Figure 2f).

3.2.1 Gas Sensors Based on Graphene/Metal Oxides Hybrids

These gas sensors exhibit excellent sensitivity, strong stability, cheap, etc., So, appropriate candidates for gas detection such as: ammonia, nitrogen oxides, hydrogen oxide and others (Y. Xu, Sheng, Li, & Shi, 2010). The chemiresistor principle underlies the operation of the metal oxide-based gas sensors. Due to the following factors, creating composites of three dimentional graphene and metal oxides offers a promising way to increase gas-sensing capability (Abideen et al., 2015; Huang et al.,

2017; Xin Liu, Cui, Sun, & Zhang, 2014). Firstly, the sensitivity is increased by the abundance of defective sites present in metal oxides. Secondly, the charge transfer is considerably facilitated by the p-n junction that is created at the intersection of p-type graphene and n-type metal oxides. Thirdly, porous graphene, particularly its 3D version, is an excellent conductor.

3.2.2. Gas Sensors Based on Chemically Modified Graphene

Economic, easy use, high efficacy and functionalization of graphene has been shown to be a viable technique to improve the gas-sensing ability. Modification is done by two type of materials. Firstly, small functional molecules; secondly, conductive polymers. 3D chemically functionalized reduced graphene oxide hydrogel (FRGOH) is prepared by hydrothermal technique to improve gas identifying ability. The hydroquinone introduces hydroxyl groups to 3D RGOH. This modification by small functional molecules enhances the gas sensitivity two-fold as compare to unmodified reduced graphene oxide hydrogel (J. Wu et al., 2016). For a number of reasons, conductive polymers offer important advantages in the field of gas sensing (Chiu & Tang, 2013; Wilson & Baietto, 2009) as they have a high yield and are exhibit cost effective fabrication. Additionally, they can also be synthesized using a solution manufacturing technique. Lastly, they exhibit strong sensitivity even at low temperatures.

3.2.3. Gas Sensors Based on Noble Metal Np-Modified 3D Graphene

Noble metal nanoparticles (NPs) such as Au, Ag, Pd, and Pt, are well-known for being efficient catalysts that enhance functionality of sensors. However, the majority of NPs based sensors need a high operating temperature that restricts their usefulness. Luckily, the decorating of noble metal NPs on graphene makes it possible for them to operate at room temperature (Chu et al., 2011; Yue Li et al., 2015). The composites of 3D graphene and noble metal NPs exhibit incredible synergistic effects.

Zhu et al. presented a biomimetic turbinate-like hydrogen sensor based on 3D laser-induced graphene (LIG) doped by palladium NPs (Zhu et al., 2019). Reaction among molecular hydrogen and the atomic level of three dimensional LIG- palladium are presented in figure 3a-c. The scheme of hydrogen sensing by palladium is attributed to the spillover effect (Figure 3d–f). Typically, molecular hydrogen is disintegrated with palladium NPs and produce PdH_x, that minimize the work function of palladium NPs and causes charge transfer from palladium NPs to n-type laser-induced graphene, minimizing the resistance of laser-induced graphene. In other experiment, the laser-induced graphene without incorporating of palladium NPs

Figure 3. (a–c) Depiction of hydrogen-sensing mechanism of the LIG/Pd sensor: (a) multilayer LIG, (b) LIG/Pd, and (c) H_2 adsorbed on LIG/Pd. (d) Cross-sectional micrograph of LIG/ Pd in dry air, (e) catalytic reaction of H_2 on LIG/Pd, and (f) reaction of H_2 at a higher concentration. Reproduced with permission from ref (Zhu et al., 2019) Copyright 2019, American Chemical Society

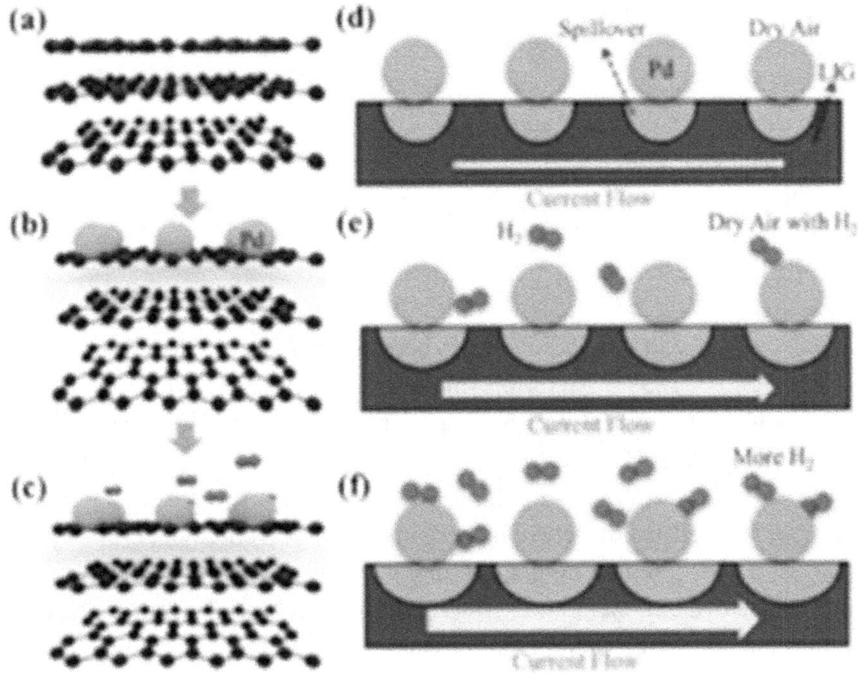

barely responded to hydrogen molecules, while palladium showed efficient sensing ability, representing the good catalytic performance of Pd NPs.

3.2.4. Gas Sensors Based on Graphene/2D Material Hybrids

2D materials have recently received noteworthy attention due to extraordinary properties. Graphene, TMDs, MXenes, boron nitride, phosphorene, and other 2D materials are building blocks for 3D graphene composites that can be used as efficient gas sensors, (Yan et al., 2018) supercapacitors, (Sun et al., 2016) lithium/sodium-ion batteries, (Y. Wang et al., 2018) electromagnetic wave absorption, (Ding, Huang, Li, Zhang, & Wang, 2016)sodium-ion storage, (S. H. Choi, Ko, Lee, & Kang, 2015), etc. Since 2D materials have a high surface area, they have considerable promise for use in gas-sensing applications. Transfer of charge among the adsorbed gas molecules and the two-dimensional materials is main component of gas-sensing mechanism. (W. Choi et al., 2017).

TMDs, like MoS_2, WS_2, and $MoSe_2$, can be prepared efficiently by CVD growth or exfoliation process and have typical layered architectures like graphene (Xianghong Liu, Ma, Pinna, & Zhang, 2017). MoS_2 has attracted significant attention among the various TMDs because of its excellent gas-sensing capabilities, including its high sensitivity and rapid reaction and recovery times (K. Lee, Gatensby, McEvoy, Hallam, & Duesberg, 2013). Moreover, the addition of MoS_2 to three-dimensional graphene improves both the charge transfer that speeds up response times, and the contact surface available for gas adsorption. As a result, MoS_2/graphene hybrids perform better in terms of sensing than pure graphene.

3.3. Cellulose-Based Gas Sensors

With the advent of spot test, cellulose has been extensively employed for analytical purposes (Schiff, 1859). Materials made from cellulose exhibit outstanding qualities such as wide availability, biodegradability, biocompatibility, affordability, or disposability (Pena-Pereira, Lavilla, de la Calle, Romero, & Bendicho, 2023). Moreover, for support purpose being stronger with more capacity, cellulose is frequently employed exclusively or in combinations with biopolymers and synthetic polymers as a substrate in a number of applications, including biosensing and gas sensing devices (Golmohammadi, Morales-Narváez, Naghdi, & Merkoçi, 2017; K.-Y. Lee et al., 2012; Siqueira, Bras, & Dufresne, 2010).

To produce the three-dimensional lattice and multipurpose assembly in hydrogels or in conducting substances, cellulose exhibit variety of derivatives, ranging in size from micro to nano scales. (Teodoro et al., 2021; X. Yang et al., 2021). To create a conducting channel, cavities might be covered by different functional and conducting fillers, like metal-based NPs, metal oxide (Oun, Shankar, & Rhim, 2020), or carbon (Z. Li et al., 2021; L. Zhang, Yu, Zheng, Zhong, & Xue, 2021). Hydrogels or aerogels generated from cellulose are additional matrix options for extremely lightweight applications (Du et al., 2019; Teodoro et al., 2021). Cellulose based conducting substances are specific practical part of strain sensors because their electrical performance when combined with various organic or inorganic fillers is highly responsive to applied strain variations (J. Fan, Zhang, Li, Yang, & Du, 2020).

Insufficient sensitivity, response/recovery rate, reversibility, and selectivity are improved using cellulose-based gas sensors. Due to its more response to different gases, like NO_2, NH_3, CO, CO_2, and other volatile organic composites VOC, cellulose offers a significant deal of promise for use in gas sensors (Tai, Duan, Wang, Wang, & Jiang, 2020). Previous studies have demonstrated how doping technique can increase gas sensing performance. In recent investigations, metal-oxide NPs were used and examined for their gas sensing properties. A new H_2S gas sensor doped with CuO_2 NPs in carboxymethyl cellulose-based materials was published by Hittini et al. (Hittini, Abu-Hani, Reddy, & Mahmoud, 2020).

A profoundly deformable gas sensor with better response was developed by Lee et al. (W. S. Lee & Choi, 2019) by combining multi-walled CNTs with nano-layered transition metal dihalides (WS_2 /MoS_2) on cellulose paper. The sensor can constantly and sensitively sense gas even in extreme distortion (including folds and wrinkles). On cellulose paper, carbon nanostructured tubes and TMDC might be used for the creation of low-priced portable electronics with stable deformability. Figure 4a illustrates how the cellulose paper deforms when bent and twisted. The sensing response of CNTs and CNT-WS_2 can be demonstrated in figure 4b, despite the existence of signal drift and response fluctuations among successive experiences being detected. Comparing reaction of the CNT-WS_2 with the reaction of CNTs under

Figure 4. NO_2 sensing response of CNT-WS_2-cellulose paper. (a) Illustrations of the bent and twisted cellulose paper attached on a polydimethylsiloxane matrix, displaying its ability to be deformed. (b) Temporary sensing response of CNT- and CNT-WS_2-loaded paper on repeated exposure to 10 ppm NO_2 in an oxygen atmosphere. (c, d) NO_2- response of the CNT-WS_2 hybrid based on concentration (0.1 to 0.6 ppm)(c) 1 to 10 ppm (d). Increase in response with the concentration of NO_2 gas. (e) Sensor's sensitivity, that is in agreement to the Langmuir adsorption model (red graph).

Reproduced with permission from ref (W. S. Lee & Choi, 2019) Copyright 2019, American Chemical Society

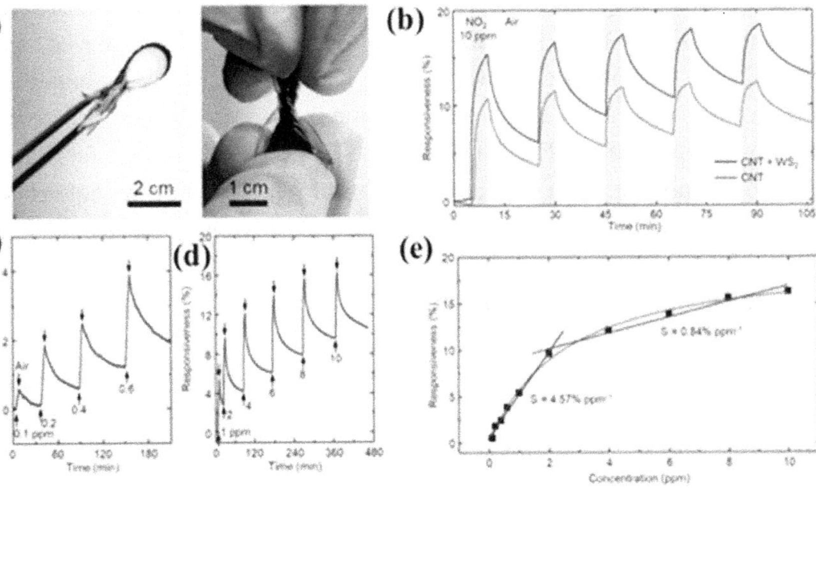

similar, repeated contact of NO_2 (10 ppm), response is enhanced by more than 150%. Increase in response of reactivity in WS_2 and greater transfer of charges to CNTs may be the cause of this enhancement. The CNT/WS_2 based cellulose paper's sensing response to different concentrations of NO_2 (0.1 to 10 ppm) is shown in figure 4c, d. Because of more receptors, response increases with rising NO_2 concentration. Since there are more molecules adsorbed on the CNTs and WS_2, the reaction rises as NO_2 concentration increases. However, there is evidence of signal drift, which is explained by the high adsorption energies of NO_2 (0.7 eV) and WS_2 (0.4 eV) on CNTs (Jijun, Alper, Jie, & Jian Ping, 2002). Figure 4e depicts that sensitivity (specified as the response /ppm of NO_2) ranges 4.57% per ppm at concentration <2 ppm, which is approximately 3-22 times greater relative to formerly reported NO_2 sensors (Mirica, Azzarelli, Weis, Schnorr, & Swager, 2013; G. Yang, Lee, Kim, Ren, & Pearton, 2013).

In order to accomplish quick and precise ammonia detection, Jia et al. (Jia, Yu, Zhang, Dong, & Li, 2016) reported an innovative method that involves the fabrication of a sensor coated for a quartz crystal microbalance (QCM). Positively charged polyethyleneimine along with negative layer of graphene oxide are consecutively constructed over the negative surface of cellulose acetate (CA) nanocomposites via electrostatic approach. Only using electrical signals to differentiate between gas molecules is still difficult. A sensor array (Song et al., 2021) made up of different sensors with diverse chemical surface and specific gas identification patterns are projected to increase selectivity (M. Wang et al., 2021).

Instead of pH indicators like phenolphthalein or Methyl Red that changed colour with pH variations, simple optical pH sensors have been created (Z. Liu, Luo, & Chen, 2004; Van der Schueren et al., 2012). In addition to being employed as dissolved ammonia sensors, sensors created by immersing pH indicators in cellulose esters also showed resistance in the buffer solution of sodium phosphate at pH 5-8 (Waich, Mayr, & Klimant, 2008). As humidity and temperature sensors, cellulose-based conductive polymers were created (Mahadeva, Yun, & Kim, 2011). For using cellulose fibres as a sensor having cyanide ions in aqueous medium, a series of chemosensors were grafted on them (Isaad & El Achari, 2011). These functionalized fabrics changed colour when submerged in aqueous solution of cyanide, and the colour shift was linearly correlated with the concentration of cyanide ions down to 0.01-0.07 M.

3.4. MXene Based Gas Sensor

Taking into account the worldwide degree of pollution and the current serious health risks to the individuals due to noxious and unsafe gases, synthesis of more efficient and quick gas sensors have attained a noteworthy attention (VahidMohammadi, Rosen, & Gogotsi, 2021). However, various gas sensors are accessible, like semiconductor, catalytic combustion, thermal conductivity, electrochemical and optical gas sensors. Yet, some critical constraints, like intense operating temperature, weak selectivity of gas, poor reliability and extended response time substantially confines their practical applications. MXene, the emerging family composed of carbide/nitride 2D nanolayered materials with a graphene like structure, have arisen as brilliant host precursors for the synthesis of brilliant gas sensors because of having different potential features like metallic heat/electrical conduction, vast surface area, remarkable electrochemical performance, exceptional chemical stability, good biocompatibility, low toxicity, structural variation, interlayer spacing tunability, antibacterial action, and hydrophilicity. In addition MXene also shows high selectivity and sensitivity for certain gases (He et al., 2021; Pei et al., 2021; Yuan, Yang, Peng, Li, & Yin, 2018). Moreover, they have extensive applications in energy storage, sensing, biomedical, food and chemical industries industries (Alhabeb et al., 2018; VahidMohammadi et al., 2021).

Hermawan et al. (Hermawan et al., 2020) fabricated CuO/$Ti_3C_2T_x$ MXene heterostructure by using the electrostatic collaboration of CuO with $Ti_3C_2T_x$ in ultrasonic bath and investigated sensor's gas reaction to find toluene gas (Figure 5a). Figure 5b reveals sensor response of CuO, $Ti_3C_2T_x$, and CuO/ $Ti_3C_2T_x$ sample under the exposure of 50 ppm of toluene at various operating temperatures. It was observed that pure $Ti_3C_2T_x$ exhibit nearly no response to toluene gas at all tested temperatures. Conversely, toluene sensing response of CuO NPs showed mountain-hill characteristic, where its response enhanced from 1.12 to 4.8 by rising temperature from 100 to 300 °C, respectively, and reduced slowly to 2.3 by further increase in temperature to 350 °C. Mountain-hill specific response of CuO NPs could be clarified by the gas adsorption-desorption phenomenon.

Figure 5. (a) Schematic representation of manufacturing of CuO NPs/$Ti_3C_2T_x$ and gas sensor instrument; (b) Gas sensing of CuO, $Ti_3C_2T_x$ MXene, and CuO/$Ti_3C_2T_x$ based MXene at various operating temperatures

Reproduced with permission from (Hermawan et al., 2020) Copyright 2020, American Chemical Society

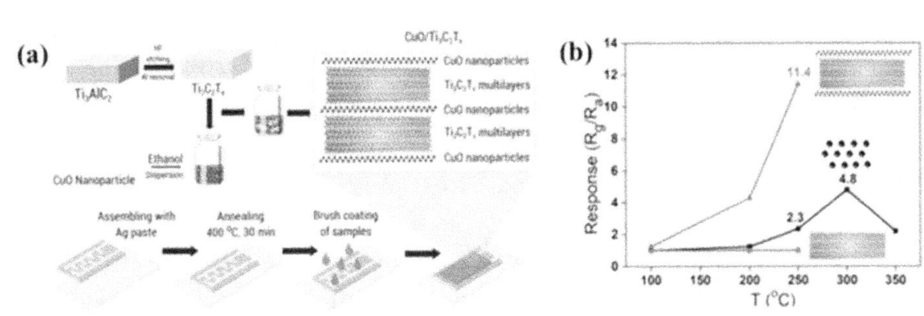

3.5. Acoustic Wave Sensors

Acoustic wave sensors are the part of mass-sensitive sensing materials, depend upon two important synthetic techniques i.e.; quartz crystal microbalance (QCM) and surface acoustic wave (SAW). In QCM, a quartz mineral assists as reverberating unit (Yuwono & Lammers, 2004), where, the amount of the adsorbed target gas alters the vibrational frequency and this frequency change appears as a response of sensor. QCM-based acoustic sensors are not governed by the presence of charge carriers and heat, this ability makes QCM based gas-sensors superior over other gas sensors (M. Yang, He, Hu, Yan, & Cheng, 2011). Conversely, the SAW sensor works on the principle of piezoelectric effect, in which, the vibrant layer is overcoated by chemically active thin layers of material which can adsorb the target gas (Xueli Liu et al., 2018). The gas adsorbs in the reactive layer easily, resulting in a measurable change in the velocity and the attenuation of the spreading wave (L. Fan, Ge, Zhang, Zhang, & Zhu, 2012).

Raj and Singh in their research work applied coating of various reactive layers for example ZnO, SnO_2, TeO_2 and TiO_2 on SAW-doped acoustic sensors and observed that ZnO dropped SAW sensors showed the best results than other reactive layers. The acoustic sensors are weight-dependent on the adsorbed gas, compact and show response to nearly all kinds of gases (Raj et al., 2013). However, no direct physical or chemical reaction occurs with that of adsorbed target gas (Poli, Mer-Calfati, Scorsone, & Saada, 2017). Due to extreme sensitivity of frequency changes, QCM and SAW can work in combination with each other via an organic coating to achieve better performance (Q. Li, Gu, & Wang, 2017).

3.6. Catalytic Sensors

Catalytic gas sensor operates on the basic principle of oxidation of adsorbed gas over the surface of sensor, where heat is liberated as a result of oxidation and thus resistance is increased (Karelin, Baranov, Akbari, Mironov, & Karpova, 2019). Generally, the quantity of heat released has a direct proportion to the adsorbed amount of target gas. Catalytic sensors are extensively applied in the inspection of instantaneous leakage in working gas pipelines (Feng et al., 2019) owing to their reasonable prices, small size, and weight, also being sensitive only to vapors and flammable gases including, hydrogen, methane, and carbon monoxide (Feng et al., 2019) (Karelin et al., 2019) (Feng et al., 2019). When sensor is exposed to the gas, the sensing element usually platinum coil is set to combustion reaction. Subsequently, target gases start burning on the catalyst, most common is TiO_2 (Feng et al., 2019), and produces a specific combustion enthalpy allowing even traces of analytes to be identified in very short response times. Subsequently, a measurable change in the resistance is observed. Still, it is problematic to assess the quantity of combustible mixture of gases in the air.

4. SUMMARY AND OUTLOOK

With the advancements in the nanotechnology, the gas sensing technology has improved significantly. To further develop the gas sensor features, more undertakings should be assessed along with nanostructured materials for gas sensors. The main objective should be to achieve low power, simplified production, and affordable gas sensor that can be employed in numerous environment monitoring. The sensitivity of the gas sensors is currently not any more a trial when contrasted with the selectivity. The gas sensor's response time is a crucial parameter to pay attention to in practical applications. In low-temperature regime, the recovery time at low temperatures is still not sufficient. The need for significant energy to desorb absorbed molecules from the surface of the sample may slow down sensor response times. Though, other studies also recommend UV irradiation as a substitute technique to expedite the recovery process. However, such techniques unfavorably raise costs and prevent the gas sensor from becoming smaller. Stability is also a significant element that should be explored for development of gas sensors.

Though, previous studies reports the long-lasting stability of gas sensors for over months in efficient laboratory environment, which scarcely trails the same fashion in actual operation. Gas sensor research and development are expanding quickly, necessitating the search for new materials as well as the need to enhance currently available materials. The properties of materials can alter at the nanoscale, leading to a wide range of variations in the materials' performances and characteristics, which largely depend on fabrication and structural parameters. In this chapter, efficient gas sensors based on multifunctional nanomaterials such as; metal oxides/sulfides/ferrites, transition metal dichalcogenides CNTs, graphene, cellulose and MXene are outlined comprehensively that explore the path to investigate and optimize gas sensor platforms as well as their better efficiencies. Substantial efforts are also done to meet the challenges in the production of economic multifunctional material-based gas sensor for monitoring in the field of personal healthcare, security warning and environment detection. In addition, more research endeavors are expected to address and work on the stability of gas sensors.

REFERENCES

Abideen, Z. U., Katoch, A., Kim, J.-H., Kwon, Y. J., Kim, H. W., & Kim, S. S. (2015). Excellent gas detection of ZnO nanofibers by loading with reduced graphene oxide nanosheets. *Sensors and Actuators. B, Chemical*, *221*, 1499–1507. doi:10.1016/j.snb.2015.07.120

Afsar, M., Rafiq, M., & Tok, A. (2017). Two-dimensional SnS nanoflakes: Synthesis and application to acetone and alcohol sensors. *RSC Advances*, *7*(35), 21556–21566. doi:10.1039/C7RA03004E

Afzali, M., Mostafavi, A., Nekooie, R., & Jahromi, Z. (2019). A novel voltammetric sensor based on palladium nanoparticles/carbon nanofibers/ionic liquid modified carbon paste electrode for sensitive determination of anti-cancer drug pemetrexed. *Journal of Molecular Liquids*, *282*, 456–465. doi:10.1016/j.molliq.2019.03.041

Alhabeb, M., Maleski, K., Mathis, T. S., Sarycheva, A., Hatter, C. B., Uzun, S., Levitt, A., & Gogotsi, Y. (2018). Selective etching of silicon from Ti3SiC2 (MAX) to obtain 2D titanium carbide (MXene). *Angewandte Chemie*, *130*(19), 5542–5546. doi:10.1002/ange.201802232 PMID:29518271

Ashari, I. A., Widodo, A. P., & Suryono, S. (2019). The Monitoring System for Ammonia Gas (NH3) Hazard Detection in the Livestock Environment uses Inverse Distance Weight Method. Paper presented at the *2019 Fourth International Conference on Informatics and Computing (ICIC)*. IEEE. 10.1109/ICIC47613.2019.8985673

Ayyala, S. K., & Covington, J. A. (2021). Nickel-oxide based thick-film gas sensors for volatile organic compound detection. *Chemosensors (Basel, Switzerland)*, *9*(9), 247. doi:10.3390/chemosensors9090247

Bag, A., & Lee, N. E. (2021). Recent advancements in development of wearable gas sensors. *Advanced Materials Technologies*, *6*(3), 2000883. doi:10.1002/admt.202000883

Blanco-Novoa, O., Fernández-Caramés, T. M., Fraga-Lamas, P., & Castedo, L. (2018). A cost-effective IoT system for monitoring indoor radon gas concentration. *Sensors (Basel)*, *18*(7), 2198. doi:10.339018072198 PMID:29986540

Borini, S., White, R., Wei, D., Astley, M., Haque, S., Spigone, E., Harris, N., Kivioja, J., & Ryhanen, T. (2013). Ultrafast graphene oxide humidity sensors. *ACS Nano*, *7*(12), 11166–11173. doi:10.1021/nn404889b PMID:24206232

Butterfield, D. (2021). *Market review of gas sensors for industrial applications.*

Chen, J., Gu, J., Zhang, R., Mao, Y., & Tian, S. (2019). Freshness evaluation of three kinds of meats based on the electronic nose. *Sensors (Basel), 19*(3), 605. doi:10.339019030605 PMID:30709028

Chen, K., Yuan, D., & Zhao, Y. (2021). Review of optical hydrogen sensors based on metal hydrides: Recent developments and challenges. *Optics & Laser Technology, 137*, 106808. doi:10.1016/j.optlastec.2020.106808

Chen, W. Y., Yen, C.-C., Xue, S., Wang, H., & Stanciu, L. A. (2019). Surface functionalization of layered molybdenum disulfide for the selective detection of volatile organic compounds at room temperature. *ACS Applied Materials & Interfaces, 11*(37), 34135–34143. doi:10.1021/acsami.9b13827 PMID:31453680

Chen, X., Leishman, M., Bagnall, D., & Nasiri, N. (2021). Nanostructured gas sensors: From air quality and environmental monitoring to healthcare and medical applications. *Nanomaterials (Basel, Switzerland), 11*(8), 1927. doi:10.3390/nano11081927 PMID:34443755

Chen, Z., Chen, Z., Song, Z., Ye, W., & Fan, Z. (2019). Smart gas sensor arrays powered by artificial intelligence. *Journal of Semiconductors, 40*(11), 111601. doi:10.1088/1674-4926/40/11/111601

Chiu, S.-W., & Tang, K.-T. (2013). Towards a chemiresistive sensor-integrated electronic nose: A review. *Sensors (Basel), 13*(10), 14214–14247. doi:10.3390131014214 PMID:24152879

Choi, S. H., Ko, Y. N., Lee, J. K., & Kang, Y. C. (2015). 3D MoS2–graphene microspheres consisting of multiple nanospheres with superior sodium ion storage properties. *Advanced Functional Materials, 25*(12), 1780–1788. doi:10.1002/adfm.201402428

Choi, W., Choudhary, N., Han, G. H., Park, J., Akinwande, D., & Lee, Y. H. (2017). Recent development of two-dimensional transition metal dichalcogenides and their applications. *Materials Today, 20*(3), 116–130. doi:10.1016/j.mattod.2016.10.002

Chu, B. H., Nicolosi, J., Lo, C., Strupinski, W., Pearton, S., & Ren, F. (2011). Effect of coated platinum thickness on hydrogen detection sensitivity of graphene-based sensors. *Electrochemical and Solid-State Letters, 14*(7), K43. doi:10.1149/1.3589250

Comini, E., Faglia, G., Sberveglieri, G., Pan, Z., & Wang, Z. L. (2002). Stable and highly sensitive gas sensors based on semiconducting oxide nanobelts. *Applied Physics Letters, 81*(10), 1869–1871. doi:10.1063/1.1504867

Das, S., Sharma, S., & Sharma, S. K. (2019). Facile Synthesis of 2D-HfS 2 Flakes for $\mu $-IDE-Based Methanol Sensor: Fast Detection at Room Temperature. *IEEE Sensors Journal, 19*(20), 9090–9096. doi:10.1109/JSEN.2019.2925027

Ding, X., Huang, Y., Li, S., Zhang, N., & Wang, J. (2016). 3D architecture reduced graphene oxide-MoS2 composite: Preparation and excellent electromagnetic wave absorption performance. *Composites. Part A, Applied Science and Manufacturing, 90*, 424–432. doi:10.1016/j.compositesa.2016.08.006

Donarelli, M., & Ottaviano, L. (2018). 2D materials for gas sensing applications: A review on graphene oxide, MoS2, WS2 and phosphorene. *Sensors (Basel), 18*(11), 3638. doi:10.339018113638 PMID:30373161

Du, H., Liu, W., Zhang, M., Si, C., Zhang, X., & Li, B. (2019). Cellulose nanocrystals and cellulose nanofibrils based hydrogels for biomedical applications. *Carbohydrate Polymers, 209*, 130–144. doi:10.1016/j.carbpol.2019.01.020 PMID:30732792

Fallah Vostakola, M., Salamatinia, B., & Amini Horri, B. (2022). A Review on Recent Progress in the Integrated Green Hydrogen Production Processes. *Energies, 15*(3), 1209. doi:10.3390/en15031209

Fan, J., Zhang, S., Li, F., Yang, Y., & Du, M. (2020). Recent advances in cellulose-based membranes for their sensing applications. *Cellulose (London, England), 27*(16), 9157–9179. doi:10.100710570-020-03445-7 PMID:32934443

Fan, L., Ge, H., Zhang, S.-y., Zhang, H., & Zhu, J. (2012). Optimization of sensitivity induced by surface conductivity and sorbed mass in surface acoustic wave gas sensors. *Sensors and Actuators. B, Chemical, 161*(1), 114–123. doi:10.1016/j.snb.2011.09.077

Farsalinos, K. E., Gillman, I. G., Hecht, S. S., Polosa, R., & Thornburg, J. (2016). *Analytical assessment of e-cigarettes: From contents to chemical and particle exposure profiles*. Elsevier.

Feng, S., Farha, F., Li, Q., Wan, Y., Xu, Y., Zhang, T., & Ning, H. (2019). Review on smart gas sensing technology. *Sensors (Basel), 19*(17), 3760. doi:10.339019173760 PMID:31480359

Forouzandeh, P., Kumaravel, V., & Pillai, S. C. (2020). Electrode materials for supercapacitors: A review of recent advances. *Catalysts, 10*(9), 969. doi:10.3390/catal10090969

Gaiardo, A., Fabbri, B., Guidi, V., Bellutti, P., Giberti, A., Gherardi, S., Vanzetti, L., Malagù, C., & Zonta, G. (2016). Metal sulfides as sensing materials for chemoresistive gas sensors. *Sensors (Basel), 16*(3), 296. doi:10.339016030296 PMID:26927120

Giberti, A., Gaiardo, A., Fabbri, B., Gherardi, S., Guidi, V., Malagù, C., Bellutti, P., Zonta, G., Casotti, D., & Cruciani, G. (2016). Tin (IV) sulfide nanorods as a new gas sensing material. *Sensors and Actuators. B, Chemical*, *223*, 827–833. doi:10.1016/j.snb.2015.10.007

Golmohammadi, H., Morales-Narváez, E., Naghdi, T., & Merkoçi, A. (2017). Nanocellulose in Sensing and Biosensing. *Chemistry of Materials*, *29*(13), 5426–5446. doi:10.1021/acs.chemmater.7b01170

Gomes, J. B., Rodrigues, J. J., Rabêlo, R. A., Kumar, N., & Kozlov, S. (2019). IoT-enabled gas sensors: Technologies, applications, and opportunities. *Journal of Sensor and Actuator Networks*, *8*(4), 57. doi:10.3390/jsan8040057

Guo, K., Yang, P., Guo, D. H., & Liu, Y. (2019). Gas leakage monitoring with mobile wireless sensor networks. *Procedia Computer Science*, *154*, 430–438. doi:10.1016/j.procs.2019.06.061

Guo, S., Yang, D., Zhang, S., Dong, Q., Li, B., Tran, N., Li, Z., Xiong, Y., & Zaghloul, M. E. (2019). Development of a cloud-based epidermal MoSe2 device for hazardous gas sensing. *Advanced Functional Materials*, *29*(18), 1900138. doi:10.1002/adfm.201900138

Halvaee, P., Dehghani, S., & Mohammadzadeh, M. (2020). Room temperature methanol sensors based on rod-shaped nanostructures of MoS 2 functionalized with Ag nanoparticles. *IEEE Sensors Journal*, *21*(4), 4233–4240. doi:10.1109/JSEN.2020.3035753

He, T., Liu, W., Lv, T., Ma, M., Liu, Z., Vasiliev, A., & Li, X. (2021). MXene/SnO2 heterojunction based chemical gas sensors. *Sensors and Actuators. B, Chemical*, *329*, 129275. doi:10.1016/j.snb.2020.129275

Hermawan, A., Zhang, B., Taufik, A., Asakura, Y., Hasegawa, T., Zhu, J., Shi, P., & Yin, S. (2020). CuO nanoparticles/Ti3C2T x MXene hybrid nanocomposites for detection of toluene gas. *ACS Applied Nano Materials*, *3*(5), 4755–4766. doi:10.1021/acsanm.0c00749

Hernández-Gordillo, A., Ruiz-Correa, S., Robledo-Valero, V., Hernández-Rosales, C., & Arriaga, S. (2021). Recent advancements in low-cost portable sensors for urban and indoor air quality monitoring. *Air Quality, Atmosphere & Health*, *14*(12), 1–21. doi:10.100711869-021-01067-x

Hittini, W., Abu-Hani, A. F., Reddy, N., & Mahmoud, S. T. (2020). Cellulose-Copper Oxide hybrid nanocomposites membranes for H2S gas detection at low temperatures. *Scientific Reports*, *10*(1), 2940. doi:10.103841598-020-60069-4 PMID:32076095

Huang, D., Yang, Z., Li, X., Zhang, L., Hu, J., Su, Y., Hu, N., Yin, G., He, D., & Zhang, Y. (2017). Three-dimensional conductive networks based on stacked SiO 2@ graphene frameworks for enhanced gas sensing. *Nanoscale*, *9*(1), 109–118. doi:10.1039/C6NR06465E PMID:27763653

Humayun, M., Ullah, H., Usman, M., Habibi-Yangjeh, A., Tahir, A. A., Wang, C., & Luo, W. (2022). Perovskite-type lanthanum ferrite based photocatalysts: Preparation, properties, and applications. *Journal of Energy Chemistry*, *66*, 314–338. doi:10.1016/j.jechem.2021.08.023

Hunter, G. W., Akbar, S., Bhansali, S., Daniele, M., Erb, P. D., Johnson, K., & Hesketh, P. J. (2020). Editors' choice—Critical review—A critical review of solid state gas sensors. *Journal of the Electrochemical Society*, *167*(3), 037570. doi:10.1149/1945-7111/ab729c

Ikram, M., Liu, Y., Lv, H., Liu, L., Rehman, A. U., Kan, K., & Wang, R. (2019). 3D-multilayer MoS2 nanosheets vertically grown on highly mesoporous cubic In2O3 for high-performance gas sensing at room temperature. *Applied Surface Science*, *466*, 1–11. doi:10.1016/j.apsusc.2018.10.018

Isaad, J., & El Achari, A. (2011). Colorimetric sensing of cyanide anions in aqueous media based on functional surface modification of natural cellulose materials. *Tetrahedron*, *67*(26), 4939–4947. doi:10.1016/j.tet.2011.04.061

Jahromi, H. D., Kazemi, M., & Sheikhi, M. H. (2021). Room temperature and highly sensitive acetone sensor based on lead sulfide nanosheets. *Materials Science and Engineering B*, *267*, 115082. doi:10.1016/j.mseb.2021.115082

Jia, Y., Yu, H., Zhang, Y., Dong, F., & Li, Z. (2016). Cellulose acetate nanofibers coated layer-by-layer with polyethylenimine and graphene oxide on a quartz crystal microbalance for use as a highly sensitive ammonia sensor. *Colloids and Surfaces. B, Biointerfaces*, *148*, 263–269. doi:10.1016/j.colsurfb.2016.09.007 PMID:27619179

Jijun, Z., Alper, B., Jie, H., & Jian Ping, L. (2002). Gas molecule adsorption in carbon nanotubes and nanotube bundles. *Nanotechnology*, *13*(2), 195–200. doi:10.1088/0957-4484/13/2/312

Jung, M. W., Kang, S. M., Nam, K.-H., An, K.-S., & Ku, B.-C. (2018). Highly transparent and flexible NO2 gas sensor film based on MoS2/rGO composites using soft lithographic patterning. *Applied Surface Science*, *456*, 7–12. doi:10.1016/j.apsusc.2018.06.086

Kanazawa, E., Sakai, G., Shimanoe, K., Kanmura, Y., Teraoka, Y., Miura, N., & Yamazoe, N. (2001). Metal oxide semiconductor N2O sensor for medical use. *Sensors and Actuators. B, Chemical*, *77*(1-2), 72–77. doi:10.1016/S0925-4005(01)00675-X

Karelin, A., Baranov, A. M., Akbari, S., Mironov, S., & Karpova, E. (2019). Measurement algorithm for determining unknown flammable gas concentration based on temperature sensitivity of catalytic sensor. *IEEE Sensors Journal*, *19*(11), 4173–4180. doi:10.1109/JSEN.2019.2897626

Kaur, N., Singh, M., & Comini, E. (2022). Materials engineering strategies to control metal oxides nanowires sensing properties. *Advanced Materials Interfaces*, *9*(12), 2101629. doi:10.1002/admi.202101629

Kim, H.-J., & Lee, J.-H. (2014). Highly sensitive and selective gas sensors using p-type oxide semiconductors: Overview. *Sensors and Actuators. B, Chemical*, *192*, 607–627. doi:10.1016/j.snb.2013.11.005

Kim, H.-J., Yoon, J.-W., Choi, K.-I., Jang, H. W., Umar, A., & Lee, J.-H. (2013). Ultraselective and sensitive detection of xylene and toluene for monitoring indoor air pollution using Cr-doped NiO hierarchical nanostructures. *Nanoscale*, *5*(15), 7066–7073. doi:10.1039/c3nr01281f PMID:23807747

Kimouche, A., Ervasti, M. M., Drost, R., Halonen, S., Harju, A., Joensuu, P. M., Sainio, J., & Liljeroth, P. (2015). Ultra-narrow metallic armchair graphene nanoribbons. *Nature Communications*, *6*(1), 10177. doi:10.1038/ncomms10177 PMID:26658960

Ko, K. Y., Lee, S., Park, K., Kim, Y., Woo, W. J., Kim, D., & Lee, Z. (2018). High-Performance Gas Sensor Using a Large-Area WS2 x Se2–2 x Alloy for Low-Power Operation Wearable Applications. *ACS Applied Materials & Interfaces*, *10*(40), 34163–34171. doi:10.1021/acsami.8b10455 PMID:30222310

Kotresh, S., Ravikiran, Y., Tiwari, S., & Vijaya Kumari, S. (2017). Polyaniline–cadmium ferrite nanostructured composite for room-temperature liquefied petroleum gas sensing. *Journal of Electronic Materials*, *46*(8), 5240–5247. doi:10.100711664-017-5535-4

Kroutil, J., Laposa, A., Voves, J., Davydova, M., Nahlik, J., Kulha, P., & Husak, M. (2018). Performance evaluation of low-cost flexible gas sensor array with nanocomposite polyaniline films. *IEEE Sensors Journal*, *18*(9), 3759–3766. doi:10.1109/JSEN.2018.2811461

Lai, C.-H., Lu, M.-Y., & Chen, L.-J. (2012). Metal sulfide nanostructures: Synthesis, properties and applications in energy conversion and storage. *Journal of Materials Chemistry*, *22*(1), 19–30. doi:10.1039/C1JM13879K

Lee, K., Gatensby, R., McEvoy, N., Hallam, T., & Duesberg, G. S. (2013). High-performance sensors based on molybdenum disulfide thin films. *Advanced Materials*, *25*(46), 6699–6702. doi:10.1002/adma.201303230 PMID:24027114

Lee, K.-Y., Tammelin, T., Schulfter, K., Kiiskinen, H., Samela, J., & Bismarck, A. (2012). High Performance Cellulose Nanocomposites: Comparing the Reinforcing Ability of Bacterial Cellulose and Nanofibrillated Cellulose. *ACS Applied Materials & Interfaces*, *4*(8), 4078–4086. doi:10.1021/am300852a PMID:22839594

Lee, W. S., & Choi, J. (2019). Hybrid Integration of Carbon Nanotubes and Transition Metal Dichalcogenides on Cellulose Paper for Highly Sensitive and Extremely Deformable Chemical Sensors. *ACS Applied Materials & Interfaces*, *11*(21), 19363–19371. doi:10.1021/acsami.9b03296 PMID:31062579

Li, Q., Gu, Y., & Wang, N.-F. (2017). Application of random forest classifier by means of a QCM-based e-nose in the identification of Chinese liquor flavors. *IEEE Sensors Journal*, *17*(6), 1788–1794. doi:10.1109/JSEN.2017.2657653

Li, R., Jiang, K., Chen, S., Lou, Z., Huang, T., Chen, D., & Shen, G. (2017). SnO 2/SnS 2 nanotubes for flexible room-temperature NH 3 gas sensors. *RSC Advances*, *7*(83), 52503–52509. doi:10.1039/C7RA10537A

Li, X., Li, X., Li, Z., Wang, J., & Zhang, J. (2017). WS2 nanoflakes based selective ammonia sensors at room temperature. *Sensors and Actuators. B, Chemical*, *240*, 273–277. doi:10.1016/j.snb.2016.08.163

Li, Y., Li, X., Tang, Z., Tang, Z., Yu, J., & Wang, J. (2015). Hydrogen sensing of the mixed-potential-type MnWO4/YSZ/Pt sensor. *Sensors and Actuators. B, Chemical*, *206*, 176–180. doi:10.1016/j.snb.2014.09.055

Li, Y., Luo, N., Sun, G., Zhang, B., Ma, G., Jin, H., Wang, Y., Cao, J., & Zhang, Z. (2018). Facile synthesis of ZnFe2O4/α-Fe2O3 porous microrods with enhanced TEA-sensing performance. *Journal of Alloys and Compounds*, *737*, 255–262. doi:10.1016/j.jallcom.2017.12.068

Li, Z., Wang, J., Xu, Y., Shen, M., Duan, C., Dai, L., & Ni, Y. (2021). Green and sustainable cellulose-derived humidity sensors: A review. *Carbohydrate Polymers*, *270*, 118385. doi:10.1016/j.carbpol.2021.118385 PMID:34364627

Liu, F., Chu, X., Dong, Y., Zhang, W., Sun, W., & Shen, L. (2013). Acetone gas sensors based on graphene-ZnFe2O4 composite prepared by solvothermal method. *Sensors and Actuators. B, Chemical*, *188*, 469–474. doi:10.1016/j.snb.2013.06.065

Liu, J., Bao, S., & Wang, X. (2022). Applications of graphene-based materials in sensors: A review. *Micromachines*, *13*(2), 184. doi:10.3390/mi13020184 PMID:35208308

Liu, X., Cui, J., Sun, J., & Zhang, X. (2014). 3D graphene aerogel-supported SnO 2 nanoparticles for efficient detection of NO 2. *RSC Advances*, *4*(43), 22601–22605. doi:10.1039/c4ra02453b

Liu, X., Ma, T., Pinna, N., & Zhang, J. (2017). Two-dimensional nanostructured materials for gas sensing. *Advanced Functional Materials*, *27*(37), 1702168. doi:10.1002/adfm.201702168

Liu, X., Wang, W., Zhang, Y., Pan, Y., Liang, Y., & Li, J. (2018). Enhanced sensitivity of a hydrogen sulfide sensor based on surface acoustic waves at room temperature. *Sensors (Basel)*, *18*(11), 3796. doi:10.339018113796 PMID:30404204

Liu, Z., Luo, F., & Chen, T. (2004). Phenolphthalein immobilized membrane for an optical pH sensor. *Analytica Chimica Acta*, *510*(2), 189–194. doi:10.1016/j.aca.2004.01.019

Llobet, E. (2013). Gas sensors using carbon nanomaterials: A review. *Sensors and Actuators. B, Chemical*, *179*, 32–45. doi:10.1016/j.snb.2012.11.014

Lv, R., Terrones, H., Elias, A. L., Perea-Lopez, N., Gutierrez, H. R., Cruz-Silva, E., Rajukumar, L. P., Dresselhaus, M. S., & Terrones, M. (2015). Two-dimensional transition metal dichalcogenides: Clusters, ribbons, sheets and more. *Nano Today*, *10*(5), 559–592. doi:10.1016/j.nantod.2015.07.004

Ma, Q., Kutilike, B., Kari, N., Abliz, S., & Yimit, A. (2021). Study on surface sensitization of g-C3N4 by functioned different aggregation behavior porphyrin and its optical properties. *Materials Science in Semiconductor Processing*, *121*, 105316. doi:10.1016/j.mssp.2020.105316

Ma, X., Zhou, X., Gong, Y., Han, N., Liu, H., & Chen, Y. (2017). MOF-derived hierarchical ZnO/ZnFe 2 O 4 hollow cubes for enhanced acetone gas-sensing performance. *RSC Advances*, *7*(55), 34609–34617. doi:10.1039/C7RA04437B

Mahadeva, S. K., Yun, S., & Kim, J. (2011). Flexible humidity and temperature sensor based on cellulose–polypyrrole nanocomposite. *Sensors and Actuators. A, Physical*, *165*(2), 194–199. doi:10.1016/j.sna.2010.10.018

Matindoust, S., Baghaei-Nejad, M., Abadi, M. H. S., Zou, Z., & Zheng, L.-R. (2016). Food quality and safety monitoring using gas sensor array in intelligent packaging. *Sensor Review*, *36*(2), 169–183. doi:10.1108/SR-07-2015-0115

Méndez, E. G., Posada, C. M., & Ocampo, J. M. J. (2022). Statistical analysis of Sr substituted NiFe2O4 thin films for liquefied petroleum gas sensor applications. *Materials Science and Engineering B*, *278*, 115614. doi:10.1016/j.mseb.2022.115614

Meshram, P., & Abhilash. (2022). Strategies for Recycling of Primary and Secondary Resources for Germanium Extraction. *Mining, Metallurgy & Exploration*, *39*(2), 1–19. doi:10.100742461-022-00549-5

Midhuna, A. SD, B. S., & Sreenidhi, P. (2021). *Optimizing Design of Surface Acoustic Wave Gas Sensor.* Paper presented at the 2021 5th International Conference on Trends in Electronics and Informatics (ICOEI).

Mirica, K. A., Azzarelli, J. M., Weis, J. G., Schnorr, J. M., & Swager, T. M. (2013). Rapid prototyping of carbon-based chemiresistive gas sensors on paper. *Proceedings of the National Academy of Sciences of the United States of America*, *110*(35), E3265–E3270. doi:10.1073/pnas.1307251110 PMID:23942132

Naje, A. N., Ibraheem, R. R., & Ibrahim, F. T. (2016). Parametric analysis of NO 2 gas sensor based on carbon nanotubes. *Photonic Sensors*, *6*(2), 153–157. doi:10.100713320-016-0304-1

Oun, A. A., Shankar, S., & Rhim, J.-W. (2020). Multifunctional nanocellulose/metal and metal oxide nanoparticle hybrid nanomaterials. *Critical Reviews in Food Science and Nutrition*, *60*(3), 435–460. doi:10.1080/10408398.2018.1536966 PMID:31131614

Pei, Y., Zhang, X., Hui, Z., Zhou, J., Huang, X., Sun, G., & Huang, W. (2021). Ti3C2TX MXene for sensing applications: Recent progress, design principles, and future perspectives. *ACS Nano*, *15*(3), 3996–4017. doi:10.1021/acsnano.1c00248 PMID:33705113

Pena-Pereira, F., Lavilla, I., de la Calle, I., Romero, V., & Bendicho, C. (2023). Detection of gases and organic vapors by cellulose-based sensors. *Analytical and Bioanalytical Chemistry*, *415*(18), 4039–4060. Advance online publication. doi:10.100700216-023-04649-z PMID:37000210

Persaud, K., & Dodd, G. (1982). Analysis of discrimination mechanisms in the mammalian olfactory system using a model nose. *Nature*, *299*(5881), 352–355. doi:10.1038/299352a0 PMID:7110356

Poli, J.-P., Mer-Calfati, C., Scorsone, E., & Saada, S. (2017). SAW sensor's frequency shift characterization for odor recognition and concentration estimation. *IEEE Sensors Journal*, *17*(21), 7011–7018. doi:10.1109/JSEN.2017.2751666

Raj, V. B., Singh, H., Nimal, A., Tomar, M., Sharma, M., & Gupta, V. (2013). Effect of metal oxide sensing layers on the distinct detection of ammonia using surface acoustic wave (SAW) sensors. *Sensors and Actuators. B, Chemical*, *187*, 563–573. doi:10.1016/j.snb.2013.04.063

Ramu, A., Umar, A., Gopi, S., Algadi, H., Albargi, H., Ibrahim, A. A., Alsaiari, M. A., Wang, Y., & Choi, D. (2021). Tetracyanonickelate (II)/KOH/reduced graphene oxide fabricated carbon felt for mediated electron transfer type electrochemical sensor for efficient detection of N2O gas at room temperature. *Environmental Research, 201*, 111591. doi:10.1016/j.envres.2021.111591 PMID:34186081

Rust, C., Li, H., Gordeev, G., Spari, M., Guttmann, M., Jin, Q., Reich, S., & Flavel, B. S. (2022). Global Alignment of Carbon Nanotubes via High Precision Microfluidic Dead-End Filtration. *Advanced Functional Materials, 32*(10), 2107411. doi:10.1002/adfm.202107411

Sakhuja, N., Jha, R. K., Chaurasiya, R., Dixit, A., & Bhat, N. (2020). 1T-phase titanium disulfide nanosheets for sensing H2S and O2. *ACS Applied Nano Materials, 3*(4), 3382–3394. doi:10.1021/acsanm.0c00127

Schiff, H. (1859). Zur Nachweisung der Harnsäure. *Justus Liebigs Annalen der Chemie, 109*(1), 65–71. doi:10.1002/jlac.18591090107

Shekhirev, M., Lipatov, A., Torres, A., Vorobeva, N. S., Harkleroad, A., Lashkov, A., Sysoev, V., & Sinitskii, A. (2020). Highly selective gas sensors based on graphene nanoribbons grown by chemical vapor deposition. *ACS Applied Materials & Interfaces, 12*(6), 7392–7402. doi:10.1021/acsami.9b13946 PMID:32011111

Siqueira, G., Bras, J., & Dufresne, A. (2010). Cellulosic Bionanocomposites: A Review of Preparation, Properties and Applications. *Polymers, 2*(4), 728–765. doi:10.3390/polym2040728

Song, X., Hu, R., Xu, S., Liu, Z., Wang, J., Shi, Y., Xu, J., Chen, K., & Yu, L. (2021). Highly Sensitive Ammonia Gas Detection at Room Temperature by Integratable Silicon Nanowire Field-Effect Sensors. *ACS Applied Materials & Interfaces, 13*(12), 14377–14384. doi:10.1021/acsami.1c00585 PMID:33750109

Srivastava, R., & Yadav, B. C. (2012). Ferrite materials: Introduction, synthesis techniques, and applications as sensors. *International Journal of Green Nanotechnology, 4*(2), 141–154. doi:10.1080/19430892.2012.676918

Sun, T., Li, Z., Liu, X., Ma, L., Wang, J., & Yang, S. (2016). Facile construction of 3D graphene/MoS2 composites as advanced electrode materials for supercapacitors. *Journal of Power Sources, 331*, 180–188. doi:10.1016/j.jpowsour.2016.09.036

Tai, H., Duan, Z., Wang, Y., Wang, S., & Jiang, Y. (2020). Paper-Based Sensors for Gas, Humidity, and Strain Detections: A Review. *ACS Applied Materials & Interfaces, 12*(28), 31037–31053. doi:10.1021/acsami.0c06435 PMID:32584534

Teodoro, K. B. R., Sanfelice, R. C., Migliorini, F. L., Pavinatto, A., Facure, M. H. M., & Correa, D. S. (2021). A Review on the Role and Performance of Cellulose Nanomaterials in Sensors. *ACS Sensors*, *6*(7), 2473–2496. doi:10.1021/acssensors.1c00473 PMID:34182751

Tyagi, S., Kumar, A., Kumar, A., Gautam, Y. K., Kumar, V., Kumar, Y., & Singh, B. P. (2022). Enhancement in the sensitivity and selectivity of Cu functionalized MoS2 nanoworm thin films for nitrogen dioxide gas sensor. *Materials Research Bulletin*, *150*, 111784. doi:10.1016/j.materresbull.2022.111784

Van der Schueren, L., De Clerck, K., Brancatelli, G., Rosace, G., Van Damme, E., & De Vos, W. (2012). Novel cellulose and polyamide halochromic textile sensors based on the encapsulation of Methyl Red into a sol–gel matrix. *Sensors and Actuators. B, Chemical*, *162*(1), 27–34. doi:10.1016/j.snb.2011.11.077

Voloshina, E., Mollenhauer, D., Chiappisi, L., & Paulus, B. (2011). Theoretical study on the adsorption of pyridine derivatives on graphene. *Chemical Physics Letters*, *510*(4-6), 220–223. doi:10.1016/j.cplett.2011.05.025

Waich, K., Mayr, T., & Klimant, I. (2008). Fluorescence sensors for trace monitoring of dissolved ammonia. *Talanta*, *77*(1), 66–72. doi:10.1016/j.talanta.2008.05.058 PMID:18804600

Wang, M., Wang, T., Luo, Y., He, K., Pan, L., Li, Z., Cui, Z., Liu, Z., Tu, J., & Chen, X. (2021). Fusing Stretchable Sensing Technology with Machine Learning for Human–Machine Interfaces. *Advanced Functional Materials*, *31*(39), 2008807. doi:10.1002/adfm.202008807

Wang, X., Xie, Z., Huang, H., Liu, Z., Chen, D., & Shen, G. (2012). Gas sensors, thermistor and photodetector based on ZnS nanowires. *Journal of Materials Chemistry*, *22*(14), 6845–6850. doi:10.1039/c2jm16523f

Wang, Y., Jin, Y., Li, S., Han, J., Ju, Z., & Jia, M. (2018). Flower-like MoS 2 supported on three-dimensional graphene aerogels as high-performance anode materials for sodium-ion batteries. *Ionics*, *24*(11), 3431–3437. doi:10.100711581-018-2528-0

Wang, Z., Zhu, L., Sun, S., Wang, J., & Yan, W. (2021). One-dimensional nanomaterials in resistive gas sensor: From material design to application. *Chemosensors (Basel, Switzerland)*, *9*(8), 198. doi:10.3390/chemosensors9080198

Wilson, A. D., & Baietto, M. (2009). Applications and advances in electronic-nose technologies. *Sensors (Basel)*, *9*(7), 5099–5148. doi:10.339090705099 PMID:22346690

Wu, J., Tao, K., Zhang, J., Guo, Y., Miao, J., & Norford, L. K. (2016). Chemically functionalized 3D graphene hydrogel for high performance gas sensing. *Journal of Materials Chemistry. A, Materials for Energy and Sustainability*, *4*(21), 8130–8140. doi:10.1039/C6TA01426G

Wu, K., Li, J., & Zhang, C. (2019). Zinc ferrite based gas sensors: A review. *Ceramics International*, *45*(9), 11143–11157. doi:10.1016/j.ceramint.2019.03.086

Xu, T., Pei, Y., Liu, Y., Wu, D., Shi, Z., Xu, J., Tian, Y., & Li, X. (2017). High-response NO2 resistive gas sensor based on bilayer MoS2 grown by a new two-step chemical vapor deposition method. *Journal of Alloys and Compounds*, *725*, 253–259. doi:10.1016/j.jallcom.2017.06.105

Xu, Y., Sheng, K., Li, C., & Shi, G. (2010). Self-assembled graphene hydrogel via a one-step hydrothermal process. *ACS Nano*, *4*(7), 4324–4330. doi:10.1021/nn101187z PMID:20590149

Yamazoe, N., & Miura, N. (1994). Environmental gas sensing. *Sensors and Actuators. B, Chemical*, *20*(2-3), 95–102. doi:10.1016/0925-4005(93)01183-5

Yan, W., Worsley, M. A., Pham, T., Zettl, A., Carraro, C., & Maboudian, R. (2018). Effects of ambient humidity and temperature on the NO2 sensing characteristics of WS2/graphene aerogel. *Applied Surface Science*, *450*, 372–379. doi:10.1016/j.apsusc.2018.04.185

Yang, D., Gopal, R. A., Lkhagvaa, T., & Choi, D. (2021). Metal-oxide gas sensors for exhaled-breath analysis: A review. *Measurement Science & Technology*, *32*(10), 102004. doi:10.1088/1361-6501/ac03e3

Yang, F., Wang, T., Deng, X., Dang, J., Huang, Z., Hu, S., Li, Y., & Ouyang, M. (2021). Review on hydrogen safety issues: Incident statistics, hydrogen diffusion, and detonation process. *International Journal of Hydrogen Energy*, *46*(61), 31467–31488. doi:10.1016/j.ijhydene.2021.07.005

Yang, G., Lee, C., Kim, J., Ren, F., & Pearton, S. J. (2013). Flexible graphene-based chemical sensors on paper substrates. *Physical Chemistry Chemical Physics*, *15*(6), 1798–1801. doi:10.1039/C2CP43717A PMID:23262787

Yang, L., Park, C.-H., Son, Y.-W., Cohen, M. L., & Louie, S. G. (2007). Quasiparticle energies and band gaps in graphene nanoribbons. *Physical Review Letters*, *99*(18), 186801. doi:10.1103/PhysRevLett.99.186801 PMID:17995426

Yang, M., He, J., Hu, X., Yan, C., & Cheng, Z. (2011). CuO nanostructures as quartz crystal microbalance sensing layers for detection of trace hydrogen cyanide gas. *Environmental Science & Technology, 45*(14), 6088–6094. doi:10.1021/es201121w PMID:21699255

Yang, X., Biswas, S. K., Han, J., Tanpichai, S., Li, M.-C., Chen, C., Zhu, S., Das, A. K., & Yano, H. (2021). Surface and Interface Engineering for Nanocellulosic Advanced Materials. *Advanced Materials, 33*(28), 2002264. doi:10.1002/adma.202002264 PMID:32902018

Yao, J., & Yang, G. (2022). 2D layered material alloys: Synthesis and application in electronic and optoelectronic devices. *Advancement of Science, 9*(1), 2103036. doi:10.1002/advs.202103036 PMID:34719873

Yuan, W., Yang, K., Peng, H., Li, F., & Yin, F. (2018). A flexible VOCs sensor based on a 3D Mxene framework with a high sensing performance. *Journal of Materials Chemistry. A, Materials for Energy and Sustainability, 6*(37), 18116–18124. doi:10.1039/C8TA06928J

Yuwono, A. S., & Lammers, P. S. (2004). Odor pollution in the environment and the detection instrumentation. *Agricultural Engineering International: CIGR Journal.*

Zhang, J., Qin, Z., Zeng, D., & Xie, C. (2017). Metal-oxide-semiconductor based gas sensors: Screening, preparation, and integration. *Physical Chemistry Chemical Physics, 19*(9), 6313–6329. doi:10.1039/C6CP07799D PMID:28198897

Zhang, L., Yu, Y., Zheng, S., Zhong, L., & Xue, J. (2021). Preparation and properties of conductive bacterial cellulose-based graphene oxide-silver nanoparticles antibacterial dressing. *Carbohydrate Polymers, 257*, 117671. doi:10.1016/j.carbpol.2021.117671 PMID:33541624

Zhao, Y., Song, J.-G., Ryu, G. H., Ko, K. Y., Woo, W. J., Kim, Y., ... Lee, Z. (2018). Low-temperature synthesis of 2D MoS 2 on a plastic substrate for a flexible gas sensor. *Nanoscale, 10*(19), 9338–9345. doi:10.1039/C8NR00108A PMID:29737989

Zhou, C., Shi, N., Jiang, X., Chen, M., Jiang, J., Zheng, Y., Wu, W., Cui, D., Haick, H., & Tang, N. (2022). Techniques for wearable gas sensors fabrication. *Sensors and Actuators. B, Chemical, 353*, 131133. doi:10.1016/j.snb.2021.131133

Zhu, J., Cho, M., Li, Y., Cho, I., Suh, J.-H., Orbe, D. D., Jeong, Y., Ren, T.-L., & Park, I. (2019). Biomimetic turbinate-like artificial nose for hydrogen detection based on 3D porous laser-induced graphene. *ACS Applied Materials & Interfaces, 11*(27), 24386–24394. doi:10.1021/acsami.9b04495 PMID:31192578

Chapter 7
Photocatalyst for the Degradation of Hazardous and Toxic Compounds

Asma Rafiq
University of the Punjab, Lahore, Pakistan

Saira Riaz
University of the Punjab, Lahore, Pakistan

Shahzad Naseem
https://orcid.org/0000-0002-7814-7577
University of the Punjab, Lahore, Pakistan

ABSTRACT

Every aspect of civilization, including its growth and sustainability on both political and socioeconomic level, is significantly impacted by access to clean water. As a result, maintaining water supplies and treating wastewater have become major global challenges. Water pollution is caused by a variety of gases, microbes, and other poisons that are added to water during rainfall, running water, etc. Studies into the concurrent removal of several contaminants or coexisting contaminants from water by nanoparticles pave the door for a novel route that is not only affordable but also devoid of secondary contamination. Hence, decontaminating water that has been polluted by hazardous and toxic compounds is a topic that generates considerable interest among researchers. This chapter discusses various types of nanomaterials as photocatalysts that can effectively manage contaminated water. By analyzing earlier studies to investigate the toxic compounds present in water, this chapter also highlights the various water treatment approaches for purification of wastewater.

DOI: 10.4018/978-1-6684-8743-3.ch007

1. INTRODUCTION

In the last few decades, increased industrial and agricultural activity had a negative impact on the environment and quality of life due to water pollution (D. A. Reddy, Choi, Lee, Ma, & Kim, 2015). The efficient elimination of various toxic contaminants from industrial wastewater remains a significant challenge. Increasing population expansion and the emergence of several companies have resulted in the environmental release of effluents including hazardous organic pollutants like different dyes (Mahdavi, Zinatloo-Ajabshir, Yousif, & Salavati-Niasari, 2022; Olagunju et al., 2021; Soltaninejad, Ahghari, Taheri-Ledari, & Maleki, 2021). Metal ions are among the main harmful inorganic contaminants, whereas organic pollutants found in water and wastewater include dyes, phenols, and polychlorinated biphenyls (Pan & Kurumada, 2008; D. Zhang, Wei, Zhang, & Qiu, 2015). Increasing freshwater contamination, stricter drinking water regulations, and competing needs for clean water from varied consumers all pose challenges to the increasing global demand for clean water (Elimelech, 2006; Shannon et al., 2009)(1, 2). A severe threat to both human health and ecological systems is posed by the environment's rising amounts of heavy metals. Because mobile and soluble hazardous metal species are not biodegradable, they have a tendency to build up in living things and cause a variety of illnesses and problems.(Jawor & Hoek, 2010).

According to Bora and Dutta's 2014 paper, polyatomic molecules and heavy metals with an inorganic origin typically make up 30% of sewage water pollutants (Kausar et al., 2022). There are several industries that contribute to pollution of water bodies and the environment, such as agriculture, the production of petroleum, steel, and pulp, and the refinement of textiles (Inglezakis, Loizidou, & Grigoropoulou, 2002). The natural abundance of polyatomic molecules and heavy elements leads to a number of ailments in the human body, such as skeletal problems and heart problems (Rasheed, Li, Nabeel, Huang, & Zhou, 2019; Rasheed, Nabeel, Adeel, et al., 2019). Environmental risks are caused by industries that use organic materials that are harmful, such as farming, medicine, dyeing, printing, and the textile industry. According to (Rasheed, Nabeel, Li, & Bilal, 2019; Rasheed, Nabeel, Li, & Zhang, 2019; Rasheed, Nabeel, & Shafi, 2019), 70% of pollutants in sewage-water are organic compounds that are present. Industries often discard biodegradable materials, however because delayed deterioration might lead to bioaccumulation (Rasheed, Li, Fu, et al., 2018). Because of their long-lasting effects and fragility, it is vital to remove organic/inorganic contaminants from drinkable water, particularly heavy metals, and pharmaceuticals. Nonetheless, their modest amounts did not pose a hazard to public health (Jelić, Petrović, & Barceló, 2012; Rasheed, Li, Zhang, et al., 2018). Thus, it is essential to remove pollutants from wastewater before releasing it into the environment.

Some of the nanomaterials employed today include nanoscale metal oxides and chalcogenide semiconductor photocatalysts, polymer nanoparticles (NPs), zeolites, carbon-based nanomaterials, biopolymers, and metal-based NPs (Anjum, Miandad, Waqas, Gehany, & Barakat, 2019; Baruah, Chaudhary, Malik, & Tomer, 2019; Mukhopadhyay et al., 2020). Among these, metal oxide NPs exhibit high reactivity and photolytic characteristics for treatment of contaminated water (Ali et al., 2021; Chaturvedi et al., 2020; Masar et al., 2023) . Because of their high adsorption capability for a variety of pollutants in water and wastewater, these NPs are advantageous (Manikandan et al., 2021). These properties result from possessing special qualities such strong reactivity, selectivity, a large surface area, and a high degree of functionalization (Aragaw, Bogale, & Aragaw, 2021; Araújo, Castro, & Fiúza, 2015). They can be employed to purify wastewater owing their capability to oxidize, precipitate, reduce, and adsorb pollutants (Araújo et al., 2015).

Nanomaterials can be feasibility modified with addition of a variety of functional groups on their surfaces, including carboxyl, epoxide, and carbonyl groups for treatment of effluents (Verma, Asthana, Singh, Prasad, & Susan, 2017). Literature reports the treatment of heavy metals on graphene oxide-based composites for water purification (Nujić & Habuda-Stanić, 2019). One key advantage of NPs is that they are able to mineralize toxicants reducing minimized secondary pollution (Barbero & Vione, 2016). Hua, et al, summarized various mesoporous metal oxides that were effective in the treatment of heavy metals (Hua et al., 2012). Lately, carbon nanotubes have been reported to be highly efficient in adsorption of heavy metals as reported in a comprehensive study provided by Ihsanullah et al (Ihsanullah et al., 2016).

With the potential for coverage of controls on the adsorption behaviors of nanomaterials, research on the elimination of heavy metal ions from wastewater is currently reviving. Furthermore, each of previously reported analyses were focused on a certain kind of nanomaterial. We have looked at recent advancements in increasing the adsorption capacity and selectivity of nanomaterials, such as nanoparticles (NPs), which make them ideal choice for the treatment of wastewater to solve these constraints. Also, we have emphasized the possible significance of nanomaterials in the removal of heavy metals from water and provided suggestions for further research in this area.

2. WATER TREATMENT APPROACHES

2.1. Nanophotocatalysts

Water splitting by photocatalysis is accomplished by a series of chemical and electrochemical processes in a reaction medium. A typical photocatalytic reaction

Figure 1. Schematic illustration of operational factors affecting the photodegradation of organic dyes over semiconductor photocatalysts
Adapted from (Chiu, Chang, Chen, Sone, & Hsu, 2019) Copyrights 2019, MDPI

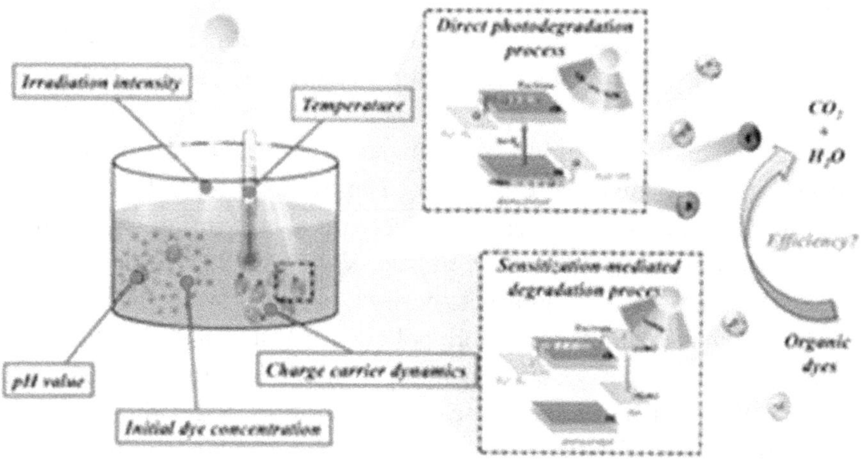

can be classified into various sequential steps, i.e., absorption of a photon, generation of excited charge carriers, charge carriers transfer and/or diffusion, and mass transfer of reactants and products. In the case of charge carriers diffusion and transport, both these processes can occur simultaneously, but for better understanding, they are described separately (Rafiq et al., 2021). Fundamentally, the goal is to keep excited electrons and holes separated at the surface and so that they can participate in reducing and oxidizing reactions, respectively (Akpan & Hameed, 2009). Photocatalytic degradation is influenced by numerous parameters. Figure 1 represents the schematic depiction of the operational factors affecting the photodegradation of dyes.

Nanophotocatalysts are preferred for water treatment applications purposes, as they offer high photoactivity due to large surface-to-volume ratio and morphology-dependent features. These features become prominent at nanoscale in comparison to bulk material due to unique quantum effects and surface-rich properties. Such materials, therefore, behave in a vastly different manner due to enhancement in optoelectronic properties such as especially optical bandgap, mechanical properties, and charge dynamics (Ong, Ng, & Mohammad, 2018). Studies have shown that nanophotocatalysts are more effective towards the oxidation of toxic pollutants due to efficient production of surface charge species (Gómez-Pastora et al., 2017; Yaqoob, Parveen, Umar, & Mohamad Ibrahim, 2020).

2.2. Nano- and Micromotors

Nanotechnology has progressed a lot over the course of the last several decades to an established area of research offering solutions to various environmental and energy issues including wastewater treatment. Particularly, nano/micro motors which convert fuel or non-fuel resources such as sound, magnetic and electric field into useful applications have gained increased interest (Jurado-Sánchez & Wang, 2018). These nano/micro motors offer increased control over speed, movement, and self-mix ability. Remediation of contaminants from polluted sources has become a significant environmental goal to achieve sustainability especially due to the depletion of freshwater sources and increased scrutiny of the discharge of potentially toxic pollutants without proper treatment (Moo & Pumera, 2015).

2.3. Nanomembranes

Nanomembranes are comprised of extraordinary nanofibers which are used to remove undesirable particles in aquatic medium. They offer a high degree of elimination rate for the removal of various pollutants even at minuscule concentrations (Jhaveri & Murthy, 2016). Moreover, nanoparticles are usually immobilized on a support material typically glass films or nanofibers to increase recyclability and recovery of the particles (Yasir, Masar, et al., 2022). There are many reported studies on various types of nanofibers usually fabricated via an electrospinning process and functionalized using different materials to impart additional properties such as increased adsorption capacity and to prevent membrane fouling. Nanomembranes have been used in nanofiltration, ultrafiltration, reverse osmosis etc. Typically, nanomembranes consist of a polymer support material and a composite layer(Yasir, Asabuwa Ngwabebhoh, Šopík, Ali, & Sedlařík, 2022). Usually, the polymeric carbon-based materials such as graphene oxide/CNT are embedded into a support matrix such as polyurethane and cellulose acetate. Utilizing composite material enriches the properties of based material by providing fouling resistance and aqueous phase diffusion. For example, by incorporating CNTs into the nanofibers, additional properties such as antibacterial resistance, prevention of membrane fouling, formation of biofilm, and minimized mechanism failures are imparted (Yaqoob et al., 2020). Additionally, advanced functional materials such as TiO_2, alumina, and zeolites increase the adsorption capacity, change in hydrophilicity, and self-cleaning properties (Bassyouni, Abdel-Aziz, Zoromba, Abdel-Hamid, & Drioli, 2019).

Owing to the necessity of solutions for environmental issues, nanotechnology at a rapid pace led to the development of various nanostructured membranes offering novel and enhanced properties like high selectivity, elimination rate, and larger foul resistance (Bassyouni et al., 2019). In order to synthesize such materials, various techniques have been employed in this regard (Kunduru et al., 2017). However, the trend has been largely shifted towards the development of membranes with catalytic-based degradation in which detoxification of pollutants is not achieved only by adsorption but also by oxidative reactions. In this regard, for practical application, ferrihydrite/iron oxide nanoparticles were immobilized onto a membrane to achieve highly effective degradation of organic pollutants removal in real water (Ibrahim, Hayyan, AlSaadi, Hayyan, & Ibrahim, 2016). So far, various reports have been presented on the immobilization of catalyst NPs onto a suitable membrane (for example, chitosan, cellulose acetate, polyurethane, polysulfone etc.) for dichlorination, remediation of persistent environmental toxicants offering minimized secondary pollution and high rate of degradation due to synergistic catalytic and adsorptive properties (Jawed, Saxena, & Pandey, 2020; Mekaru, Lu, & Tamanoi, 2015). Current research in this field is largely focused on the development of hybrid membranes using bottom-up approaches to achieve multi-functional properties for the treatment of wastewater (Abdullah, Yusof, Lau, Jaafar, & Ismail, 2019).

2.4. Nanosorbents

Nanosorbents offer a high degree of adsorption capacity that make them appropriate for water purification applications. Although the commercial feasibility of nanosorbents is currently lacking but significant progress is being made to achieve sufficient efficiency-cost in comparison to other alternatives (Salim & Ho, 2015). Most used nanosorbents are based on carbon-based materials. Besides that, other different types of nano sorbents also exist such as silica and alumina (L. Yu, Ruan, Xu, Zou, & Hu, 2017). These nanosorbents are usually employed in polymeric membranes or with metal oxide photocatalysts to enhance their efficiency by enabling higher surface sites and adsorption capacity to capture reactant molecules. Nanosorbents such as dendrimers provide functionalization for the removal of toxicants and heavy metals from polluted water (Fuwad, Ryu, Malmstadt, Kim, & Jeon, 2019). In a study, Cu ions were reduced by utilizing a dendrimer-filtration system with high reported efficiency (Shen, Saboe, Sines, Erbakan, & Kumar, 2014). Other advantages of nanosorbents include high biocompatibility, biodegradability, and low risk of environmental pollutants. In the case of synthetic organic dyes, maximum elimination has been reported (Peng et al., 2017). An emerging type of nanosorbents is based on zeolites offering high flexibility in imparting additional functionalization such as the incorporation of Ag and Cu nanoparticles (Giwa et al., 2017). The advantage of zeolites is that they offer a high degree of surface area, are toxic free, and have ease of additional functionalization. Besides zeolites, magnetic nanosorbents are also gaining an increased role in water treatment applications due to their high efficiency and potential recyclability (Sahebi, Sheikhi, & Ramavandi, 2019).

3. NANOMATERIALS AS PHOTOCATALYSTS FOR DEGRADATION OF HAZARDOUS AND TOXIC COMPOUNDS

Increased environmental pollution due to the release of heavy metals is of serious concern due to its potentially harmful impact on humans and animals alike. In this regard, various types of nanomaterials such as iron oxide, graphene oxide, ZnO, and Ag have been reported for the elimination of heavy metals from wastewater (Adel, Ahmed, Elabiad, & Mohamed, 2022; Attatsi & Nsiah, 2020; Dave & Chopda, 2014; Khoso, Haleem, Baig, & Jamal, 2021; Le, Pung, Sreekantan, Matsuda, & Huynh, 2019). In this section, some of the nanomaterials investigated for heavy metals elimination are discussed.

3.1. Elimination of Metal Ions From Wastewater Solutions Using Polymer-Based Composites

In this section, removal of heavy metal ions (hazardous and toxic compounds) via polymer-based composites will be discussed along with the structure and mechanism of the adsorbents and the adsorption process, based on the various counterparts in the composites that are classified as distinct types of polymer-based composites. Among all polymers, polyaniline (PANI) is widely studied polymer among other polymers (figure 2) owing to monomer's cheap cost, chemical and environmental stability,

Figure 2. Progression of the number of published papers between 2010 and 2020 using various polymers for removal of toxic compounds, based on the Scopus database
Adapted from (Hajjaoui, Soufi, Boumya, Abdennouri, & Barka, 2021) Copyrights 2021, MDPI

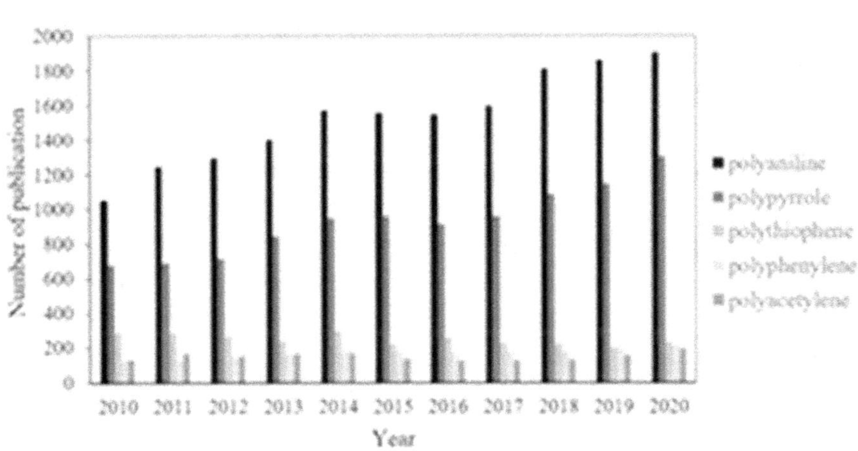

comfortable fabrication route, high electrical conductivity, and biocompatibility. Moreover, it exhibits several amine and imine functional groups that interact with some metal ions because of the strong affinity of nitrogen atoms [20,21].

3.1.1 Elimination of Heavy Metals Over Copolymer Composites

In comparison to using single polymer adsorbent, dual polymer-based adsorbents are commonly preferred due to their high degree of adsorptive capacity, increased structural stability, and prevention of fouling and surface groups (M.-T. Wu, Tsai, Chiu, & Cheng, 2016). Polypyrrole (PPy) has been reported as a highly suitable polymer that can be combined together with other polymeric adsorbents such as PANI (Javadian, 2014; Javadian, Sorkhrodi, & Koutenaei, 2014)and polyacrylonitrile (J. Wang, Pan, He, & Cao, 2013). In one reported study, combined synergistic effects of PANi/PPy polymer-based nanofibers led to a removal efficiency of 99.68% of Co(II) under 11 minutes via a pseudo-second-order model. Similarly, in another study, PPy-PANI composite nanofibers were fabricated by coupling $PPy^{\cdot+}$ and $PANI^{\cdot+}$ radicals via simultaneous polymerization of aniline and pyrrole monomers. The obtained nanofibers were utilized for the elimination of Cr(VI) ions from the wastewater. The adsorption isotherm data correlated well with the Langmuir model and a maximum adsorption capacity of 227 $mg.g^{-1}$ was achieved. Additionally, the nanofibers were found to exhibit high selectivity for Cr(VI) even after several cyclic usage (Bhaumik, Maity, Srinivasu, & Onyango, 2012). Similarly, Checkol et al, fabricated highly efficient nanofiber composite based on poly(3,4 ethylenedioxythiophene)/polystyrene sulfonate (PEDOT/PSS) with additional functionalized with lignin. The prepared nanofibers were able to adsorb Pd^{2+} ions in a neutral solution while applying a negative potential. The adsorbed ions could be released by applying positive potential, indicating a successful system for the selective removal of Pb^{2+} ions (Checkol et al., 2018). Besides above mentioned materials, amphiphilic nanofibers with PMMA cores and PEI shells have also been investigated for treating Cu(II) ions in wastewater with a maximum reported adsorption capacity of 14 $mg.g^{-1}$ obtained under milder operating conditions. Recently, magnetic composites have been employed as well due to their relatively high degree of recyclability and operational ease. For example, Yan et al, synthesized magnetic nanofibrous composites comprised of Fe_3O_4 nanoparticles, poly(acrylic acid), and chitosan for adsorptive removal of Cu(II) from the wastewater (Yan et al., 2012). Obtained findings revealed that the developed composite had higher adsorptive capacity for Cu(II) due to the interaction of Cu(II) with poly(acrylic acid). Moreover, the prepared nanocomposite was able to sustain adsorption capacity at low and high pH. At a pH higher than 4.0, Cu-loaded composite also showed a high adsorption rate for phosphate elimination in the pH range of 4.0-8.0 in the secondary absorption cycle (Yan et al., 2012).

3.1.2. Elimination of Heavy Metals Over Magnetic Polymer Composites

Since many polymer composites with high adsorption capacities, quick adsorption kinetics, and stability have been fabricated for elimination of heavy metal ions. Adsorbent separation from the aqueous solution at a cheap cost, however, is also a crucial element for practical applications. Conventional separation methods like filtration and sedimentation have not evolved because they take too long or the filters blocking. Moreover, the adsorbents might be disposed of along with the sludge from the processing, leading to secondary contamination. Magnetic NPs incorporation in the polymer matrix has gained popularity with the development of magnetic separation (D. H. K. Reddy & Lee, 2013). Until recently, polymers used in magnetic polymer composites for elimination of toxic compounds included natural polymers like chitosan (F.-C. Wu, Tseng, & Juang, 2010), cyclodextrin polymer (Badruddoza, Shawon, Tay, Hidajat, & Uddin, 2013), manmade polymers like polystyrene (M. Chen et al., 2013), PPy (Bhaumik, Leswifi, Maity, Srinivasu, & Onyango, 2011) and poly(1-vinylimidazole) (Takafuji, Ide, Ihara, & Xu, 2004).

Generally, magnetic particles lose some of their magnetism and dispersibility due to their high activity and ease of oxidation in air. Such magnetic nanoparticles can be made less poisonous, have less aggregation, and have longer storage times by covering them with polymers. However, the numerous surface groups on polymers make it possible to further functionalize such composites, producing ternary composites (H. Wang et al., 2015). Fe_3O_4 (Cheng et al., 2005; Covaliu et al., 2011), Fe_2O_3 (Singh, Srivastava, Dutta, & Dutta, 2011), $CoFe_2O_4$ (Qin, Li, Chen, & Jiang, 2009), $NiFe_2O_4$ (Singh, Srivastava, Kalita, & Malhotra, 2012) and $ZnFe_2O_4$ (Lin et al., 2012) are the main magnetic particles inserted into polymers. The homogeneous magnetic cores scattered in the polymer matrix and the magnetic core polymer shell are the two main structural types of the produced magnetic polymer composites (D. H. K. Reddy & Lee, 2013). As a typical example, amine-functionalized magnetite NPs that were easily extracted from water using an external magnet were used to create chitosan magnetic composites. Since chitosan may couple with metal ions and is inactive on the magnetic NPs surface, the magnetic nanoparticles were used to remove metal ions from water. According to the results, the interaction between heavy metal ions and chitosan was reversible, and with the help of ultrasonic radiation, the metal ions were desorbed from chitosan in weak acidic conditions (Liu, Hu, Fang, Zhang, & Zhang, 2009). To improve performance, the magnetic chitosan composites have also undergone other modifications. Chitosan magnetic microspheres were created by Zhou et al. and afterwards treated using thiourea to remove metal ions. In every experiment that was conducted, thiourea showed a greater capacity to bind Hg(II). The sorption activity declined as the temperature rose. With 0.01-0.1 M EDTA, it was possible to regenerate the metal ion-loaded thiourea, having efficiency of >88% (Zhou, Wang, Liu, & Huang, 2009)

3.1.3. Elimination of Heavy Metals Over Polymer/Carbon Material Composites

Carbonous materials like graphene, CNTs and activated carbon (Hwang, Chen, & Wey, 2013), are appropriate for functionalization of in polymer-based composites owing to their high strength, adsorption capacity, aspect and compatibility with the polymeric nanofibers resulting in strong interaction and adhesion (Chowdhury & Balasubramanian, 2014). Most of the polymer/carbon materials are non-covalent due to the inert characteristics of carbon.

Literature reports firstly synthesis of carboxymethyl cellulose (CMC)/CS grafted MWCNTs using N_2-plasma induced grafting approach (Shao, Hu, & Wang, 2010). Obtained CMC grafted MWCNT (MWCNT-g-CMC) was employed in elimination of U(VI) from polluted water. Compared with activated carbon (AC) and CNTs, graphene oxide (GO) is more employed to compose with polymers for enhancing the efficiency of hazardous metal ions' removal (Chowdhury & Balasubramanian, 2014). For instance, GO has been combined with chitosan to prepare hydrogel composites (Chen, Chen, Bai, & Li, 2013; Fan, Luo, Sun, Li, & Qiu, 2013; He, Zhang, & Wang, 2011). He et al. prepared porous GO/chitosan material, in which the inclusion of GO was found to significantly impact the strength and adsorption capacity of the developed material. In case of 5 wt.% GO, the adsorptive capacity of as-prepared was increased ~ 31%, reaching as high as 99 mg/g for Pb2+ ions. (He et al., 2011).

In another report, polyacrylamide (PAM) and reduced graphene oxide (rGO) composites were synthesized via in-situ polymerization reaction. The experimental results suggest that the as-prepared materials had superior adsorption affinity for the Pb(II) heavy metal ions and benzenoid compound (methylene blue). The obtained spontaneous and exothermic adsorption kinetics of PAM/rGO were best fitted to pseudo-second-order model as depicted in figure 3 (Y. Yang et al., 2013).

Figure 3. Fabrication of PAM chains on RGO sheets, and Pb(II) and MB adsorption
Adapted from (Y. Yang et al., 2013) Copyright 2013, American Chemical Society

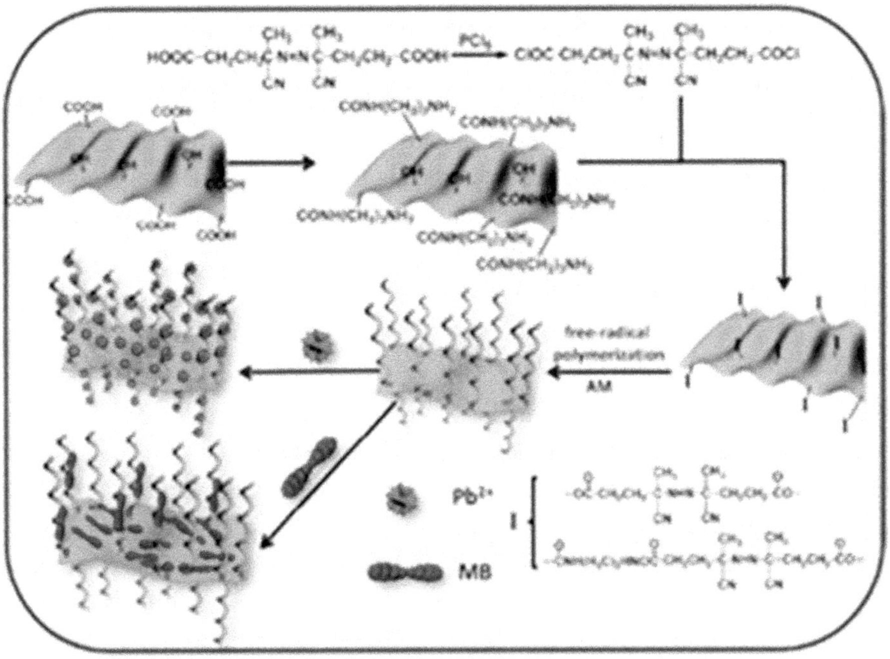

3.1.4. Elimination of Heavy Metals Over Polymer/Clay(Mineral) Composites

3.1.4.1. Polymer/Phosphate Composites

One of the most effective methods for metal ion adsorption is the ion exchange process, which is primarily supported by columbic or electrostatic interaction. Due to their quick kinetics, large capacity, and exceptional thermal stability, phosphates have been one of the most extensively utilised inorganic ion exchanges (Q. Zhang et al., 2009). Phosphates are present as tiny NPs, but due to their poor mechanical strength and high pressure drop, they cannot be used directly in fixed-bed or any other flow-through systems. Such phosphate NPs have been combined with various porous polymers to create hybrids that will ease its use to overcome these challenges (Suzuki et al., 2004; Q. Zhang et al., 2009).

In a highlighted study by Zhang et el., synthesized three zirconium based nanocomposites (nano-ZrP) by encapsulation of macroporous polymer resins comprised of various surface functional groups for elimination of Pb(II). The obtained findings revealed that charged functional groups were more effective in dispersion of nano-ZrP in comparison to the charge neutral -CH_2Cl group. Therefore, ZrP-S and ZrP-CL revealed higher elimination rate for elimination of Pb(II). Additionally, the inclusion of polymeric resins imparted high compressive strength compared to nano-ZrP (Q. Zhang et al., 2011).

Bushra et al., worked on PANI based composite for cation exchange using Ti(IV) arsenophosphate via a simple chemical synthesis route. Different metal ions were separated using the columns of the developed nanocomposite material via sorption. The developed material achieved excellent results for both synthetic mixtures as well as tap water and was able to eliminate Cd(II), Cr(IV), Fe(III), Zn(II), and Cr(IV) ions due to high K_d values (Bushra et al., 2014).

3.1.4.2. Polymer/Clay Composites

Natural clay minerals have garnered a lot of interest in the removal of heavy metal ions. (Kumar, Kalidhasan, Rajesh, & Rajesh, 2012). The development of nanocomposites and the well dispersion were the outcomes of the electrostatic interaction between the polymer and the negatively charged layer clay mineral surfaces. For instance, amidoxime functionalized organobentonite/polyacrylonitrile composite was prepared using in-situ intercalation polymerization and employed as adsorbents for Zn(II), Cu(II), and Cd(II) ions, with maximum elimination of 99.8% for Cu(II), 97.4% for Cd(II) and 98.9% for Zn(II). At 30 °C, the maximal monolayer adsorption capacities for Cu(II), Zn(II), and Cd(II) were determined to be 77.43 mg/g, 65.40 mg/g, and 52.61 mg/g, respectively (G. Zhao et al., 2018).

3.1.4.3. Elimination of Heavy Metals Over Polymer/Metal Composites

Due to their exceptional catalytic and reduction capabilities, metal nanoparticles are regarded as outstanding materials for chemistry and material science. Several studies reported the use of metal NPs for the reduction-based elimination of high-valence heavy metals in the field of wastewater treatment (Jie Li, Chen, Zhang, & Wang, 2015; X. Li, Ai, & Jiang, 2016; Z.-J. Li et al., 2015). Although, these metal nanoparticles are safe for the environment, their evident drawback is that they oxidise in the air and aggregate easily, both of which severely reduce their inherent reducing specificity. Also, it is difficult to recycle nanoparticles and it is unclear if they pose a risk to human health. Several reports have concentrated on the immobilisation of metal NPs onto various supports, such as carbon compounds, clays, and polymers, to overcome these confines.

The large specific surface area polymer-supported metal composites have shown promise as metal nanoparticle stabilisers. Zero valent iron has received the noteworthy attention out of all those metal NPs due to its abundance in the elemental and natural magnetism (Jiansheng Li et al., 2011). Zhao et al. created nanoscale Fe0 using PVA as a support and assessed the removal of Sb(III) and Sb(V) from water. According to the findings, antimony adsorption moved quickly and adhered to a pseudo-second-rate rule. Granular PVA-Fe0 had highest adsorption capacities of 6.99 and 1.65 mg g-1 for Sb(III) and Sb(V), respectively (X. Zhao et al., 2014). Metallic iron NPs supported on PANI nanofibers were easily fabricated at room temperature, according to Bhaumik et al. The composite fibres, which had a diameter of 80 to 150 nm and displayed ferromagnetic activity, removed As quickly and effectively (V) (Bhaumik, Choi, McCrindle, & Maity, 2014).

Even though, copper nanoparticles have received relatively little research, Wu et al. created zero-valent copper-chitosan composites by creating chelating beads from chitosan and tripolyphosphate. The resulting nanoparticles were stable and maintained an acceptable dispersion on the chitosan-tripolyphosphate beads. The findings demonstrated that the synthesised nanocomposites had a better capability for adsorbing Cr(VI) relative to chitosan-tripolyphosphate beads. The three main methods for chromium remediation were thought to be surface sorption, precipitation, and reduction (S.-J. Wu, Liou, & Mi, 2009).

3.2. Carbon-Based Nanomaterials

Due to their relative superior qualities to other materials, such as wide potential window, chemical surface stability, ease of functionalization, and cost-effectiveness, carbon-based nanoparticles were first used in the electronics industry (Dresselhaus & Terrones, 2013; Power, Gorey, Chandra, & Chapman, 2018). The resulting carbon nanostructures (such as graphene, graphene oxide, fullerene, and carbon nanotubes) are recognized as one of the most effective adsorbents for dangerous metals because of their very simple surface functionalization (Bisht, 2019). Carbon-based nanomaterials have become viable alternatives for wastewater purification because of their exceptional qualities, such as a large surface area, ease of physicochemical variation, and potential to remove contaminants (Smith & Rodrigues, 2015). This has generated a great deal of relevance in engineering and science.

Two significant examples of carbon-based nanomaterials—carbon nanotube-based nanomaterials and graphene-based nanomaterials—are presented in this chapter with a focus on their presentation. Carbon nanotubes (CNTs) are easily functionalized with a variety of organic compounds, making them unique in terms of choosing adsorbates and potentially increasing their adsorption capacity. Due to their nanoscale, capacity to interact via pi-pi stacking, huge surface area, hydrogen bonding, and electrostatic interactions, graphene-based nano-adsorbents are exceptional advanced materials for the removal of contaminants from wastewater (Bradder, Ling, Wang, & Liu, 2011).

3.2.1. Carbon Nanotubes

CNTs have grown significantly in significance over the past few decades because of their unique characteristics. Superior thermal, optical, electrical, vibrational, and mechanical qualities are a few of these (Gupta et al., 2016; Popov, 2004). The diameter of the nanotubes can be one to three nanometers, and their length can range from hundreds to thousands of nanometers (G. Yu et al., 2018).

Carbon nanotubes can be classified into two categories: single-walled CNTs and multi-walled CNTs. CNTs provide a lot of comparative benefits when it comes to the treatment of wastewater containing heavy metals (Lu et al., 2016). They basically consist of their high adsorption capacity, large specific surface area, and quick adsorption kinetics (Tang et al., 2012; Yadav & Srivastava, 2017). They consequently exhibit effective adsorption properties for substances like manganese (VII), thallium(I), copper(II), lead(II), and chromium(VI). The CNTs' active sites host a large portion of this adsorption (Agnihotri, Mota, Rostam-Abadi, & Rood, 2006).

3.2.2. Graphene Nanomaterials

A 2D substance called graphene can be employed to eliminate toxic and hazardous compounds from contaminated water (Rajapaksha P, Power, Chandra, & Chapman, 2018). It has several unique qualities, including elasticity, mechanical strength, stiffness, electrical and thermal conductivity, etc. (Novoselov et al., 2012), all of which contribute to its wide range of applications in a variety of settings and fields.

Figure 4. The mechanism of eliminating water pollutants via graphene membrane

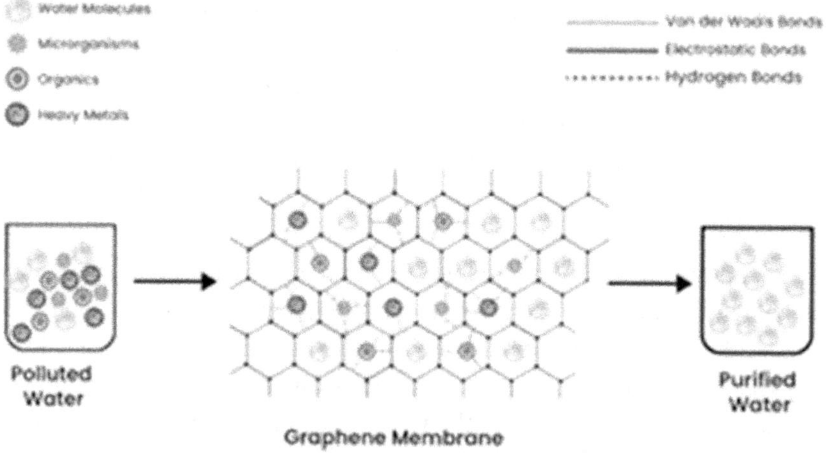

Moreover, GO and RGO, its oxidised counterparts, are widely used in various industries. It is possible for GO to eliminate heavy metals because of its oxygenated moieties (Gao et al., 2011). Due to the layered structure of graphite and the prolonged interlayer separation with polar oxygen groups with a C/O ratio between 1:1 and 4:1, GO is suggested as an acceptable adsorption medium (Nirmala et al., 2022). RGO has more defects than graphene in its native form (Avouris & Dimitrakopoulos, 2012). This is because RGO has less oxygenated moieties. Additionally, GO and graphene membranes exhibit high adsorption rates and capacities. Since organic pollutants are attracted to graphene membranes by non-covalent bonds, the membranes interact strongly with organics (Eduardo, Miguel de la, Ana, & Carlos, 2016). This results in a hydrophobic membrane in the filter (Rajapaksha P et al., 2018). The schematic depiction of eliminating contaminants from graphene is shown in figure 4.

3.3. Nanocomposite Based Nanomaterials

Even if each of the nanomaterial types mentioned thus far frequently has its own inherent benefits, they each also have disadvantages. In contrast to nZVI, which are more likely to experience oxidation, CNTs, for example, are challenging to suspend consistently in a variety of solvents(Hotze, Phenrat, & Lowry, 2010). Moreover, the use of NPs in fixed-bed and flow-through systems typically leads to issues with buildup, insufficient separation, and an unneeded decrease in pressure (Y. Zhang et al., 2016). Such loopholes can be resolved by fabricating hybrid nanocomposites that utilise the benefits provided by a number of distinct nanomaterials (Tounsadi, Khalidi, Abdennouri, & Barka, 2016). Pooseekheaw et al, prepared porous V_2O_5/

Figure 5. Photocatalytic decolorization (a), comparison of the rate constants of methylene blue degradation (b)
Adapted from (Pooseekheaw et al., 2020) Copyright 2020, MDPI.

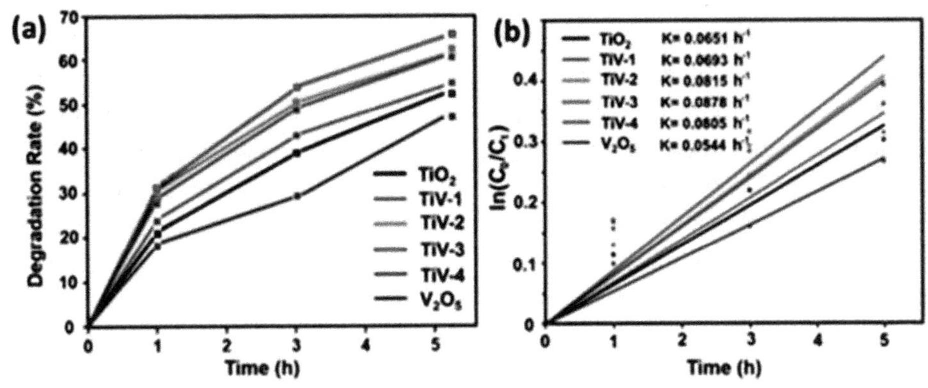

TiO_2 nanoheterostructure films with various atomic ratios of Ti/V using sparking technique. The degradation rate of methylene blue was monitored by examining the photocatalytic activity under visible-light source (figure 5). Calculations reveal that all composite films had more prominent photocatalytic performances relative to pristine TiO_2 which can be attributed to their shortened band gap and heterostructure at the interface between pristine TiO_2 and pristine V_2O_5 (Pooseekheaw et al., 2020). The following section explores nanocomposites of inorganic polymers, as well as the magnetic nanocomposites that can be employed for degradation of pollutants.

3.3.1. Inorganic-Supported Nanocomposites

Due to their high specific surface area, small size, etc., CNTs, activated carbon (AC), and other natural materials are frequently the main inorganic support structures of nanocomposites that are employed for elimination of toxic compounds. One of the most successful, affordable, and simple methods is to use AC as an adsorbent for pollutants in aquatic medium (Parlayici, Eskizeybek, Avcı, & Pehlivan, 2015). A few studies using AC-supported nanocomposites to eliminate heavy metals from wastewater have been published (Jafari Kang, Baghdadi, & Pardakhti, 2016; Jayaweera, Siriwardane, de Silva, & de Silva, 2018; S.-F. Wang, Shen, Zhang, & Tong, 2005). One of the polymers that is prominently used in CNT nanocomposites is chitosan. For instance, Salam et al. [102] created a multi-walled CNTs/chitosan nanocomposite for the efficient removal of Cu(II), Zn(II), Cd(II), and Ni(II) from aquatic environments(Salam, Makki, & Abdelaal, 2011). A poly-amidoamine dendrimer nanocomposite with CNT coating was created by Hayati et al. used to eliminate Zn(II), As(III), and Co(II) in a fixed-bed system (Hayati et al., 2018). The maximum absorption capacities observed for these three ions were 432, 494, and 470 $mg \bullet g^{-1}$, respectively.

3.3.2. Magnetic Nanocomposites

Magnetic nanocomposites are a unique nanomaterial that have recently attracted an increasing amount of attention due to the simplicity with which they may be separated. The majority of magnetic nanocomposites can be produced using the following methods and have as their base components magnetic iron and iron oxides (Lü, Qi, Zhang, Lü, & Zhao, 2018): (1) Surface variation of magnetic iron/iron oxide NPs via functional groups; (2) encasing the iron/iron oxide NPs in different substances, like polyrhodanine, humic acid, polyethylenimine, etc. (Ge et al., 2018; Lü et al., 2018); and (3) coating the iron/iron oxide NPs on various porous materials. Ge et al. (Ge et al., 2018) prepared a Fe@MgO nanocomposite by fusing the advantages of MgO's high adsorption capacity with the potent magnetic properties of nZVI.

Using this nanocomposite led to the greatest removal of Pb(II) and methyl orange in batch adsorption tests, at 1476.4 and 6947.9 mg•g^{-1}, respectively. These results demonstrate the exceptional water purification abilities of this nanocomposite for methyl orange and Pb(II). The methylene orange and Pb (II) adsorption details precisely matched the Langmuir and pseudo-second-order kinetic models, which specify that the adsorption onto the absorbent was a monolayer chemisorption (J. Yang et al., 2019) [4].

4. CONCLUSION AND FUTURE PERCEPTIONS

The fabrication of affordable, effective, and environmentally friendly nanomaterials as photocatalysts opens up new possibilities due to advancements in nanotechnology. Various nanomaterials are being thorough exploited for the elimination of toxic compounds form wastewater. Nanomaterials are good for the filtration of wastewater because of their innovative structural and morphological properties. High specificity, remarkable selectivity, and high adsorption potential make nanomaterials ideal sorbents with the capacity to remove toxic compounds even at low concentrations. This review has covered in depth the significant applications of nanoparticles as photocatalysts, such as polymer-based composites, carbon-based nanomaterials, and nanocomposites-based nanomaterials for eliminations of toxic compounds from wastewater. However, there are still few loopholes that must be considered for wastewater purification approaches. Due to their nano-size, most nanomaterials tend to agglomerate, which reduces their ability to be removed from aqueous solutions. This problem is resolved by using nanocomposites. Nonetheless, there is still a need to focus on commercializing the use of nanomaterials for wastewater purification on an industrial scale. Since nanomaterials are not available in market, efforts should be made to fabricate nanomaterials that can be readily purchased. Green chemistry should be used to manufacture nanomaterials, and their cost should be reasonable. Moreover, various approaches should be developed for removal of multiple contaminants to meet the environmental pollution.

REFERENCES

Abdullah, N., Yusof, N., Lau, W. J., Jaafar, J., & Ismail, A. F. (2019). Recent trends of heavy metal removal from water/wastewater by membrane technologies. *Journal of Industrial and Engineering Chemistry, 76*, 17–38. doi:10.1016/j.jiec.2019.03.029

Adel, M., Ahmed, M. A., Elabiad, M. A., & Mohamed, A. A. (2022). Removal of heavy metals and dyes from wastewater using graphene oxide-based nanomaterials: A critical review. *Environmental Nanotechnology, Monitoring & Management*, *18*, 100719. doi:10.1016/j.enmm.2022.100719

Agnihotri, S., Mota, J. P. B., Rostam-Abadi, M., & Rood, M. J. (2006). Theoretical and Experimental Investigation of Morphology and Temperature Effects on Adsorption of Organic Vapors in Single-Walled Carbon Nanotubes. *The Journal of Physical Chemistry B*, *110*(15), 7640–7647. doi:10.1021/jp060040a PMID:16610854

Akpan, U. G., & Hameed, B. H. (2009). Parameters affecting the photocatalytic degradation of dyes using TiO2-based photocatalysts: A review. *Journal of Hazardous Materials*, *170*(2), 520–529. doi:10.1016/j.jhazmat.2009.05.039 PMID:19505759

Ali, H., Guler, A. C., Masar, M., Urbanek, P., Urbanek, M., Skoda, D., Suly, P., Machovsky, M., Galusek, D., & Kuritka, I. (2021). Solid-State Synthesis of Direct Z-Scheme Cu2O/WO3 Nanocomposites with Enhanced Visible-Light Photocatalytic Performance. *Catalysts*, *11*(2), 293. doi:10.3390/catal11020293

Anjum, M., Miandad, R., Waqas, M., Gehany, F., & Barakat, M. A. (2019). Remediation of wastewater using various nano-materials. *Arabian Journal of Chemistry*, *12*(8), 4897–4919. doi:10.1016/j.arabjc.2016.10.004

Aragaw, T. A., Bogale, F. M., & Aragaw, B. A. (2021). Iron-based nanoparticles in wastewater treatment: A review on synthesis methods, applications, and removal mechanisms. *Journal of Saudi Chemical Society*, *25*(8), 101280. doi:10.1016/j.jscs.2021.101280

Araújo, R., Castro, A. C. M., & Fiúza, A. (2015). The Use of Nanoparticles in Soil and Water Remediation Processes. *Materials Today: Proceedings*, *2*(1), 315–320. doi:10.1016/j.matpr.2015.04.055

Attatsi, I. K., & Nsiah, F. (2020). Application of silver nanoparticles toward Co(II) and Pb(II) ions contaminant removal in groundwater. *Applied Water Science*, *10*(6), 152. doi:10.100713201-020-01240-0

Avouris, P., & Dimitrakopoulos, C. (2012). Graphene: Synthesis and applications. *Materials Today*, *15*(3), 86–97. doi:10.1016/S1369-7021(12)70044-5

Badruddoza, A. Z. M., Shawon, Z. B. Z., Tay, W. J. D., Hidajat, K., & Uddin, M. S. (2013). Fe3O4/cyclodextrin polymer nanocomposites for selective heavy metals removal from industrial wastewater. *Carbohydrate Polymers*, *91*(1), 322–332. doi:10.1016/j.carbpol.2012.08.030 PMID:23044139

Barbero, N., & Vione, D. (2016). Why Dyes Should Not Be Used to Test the Photocatalytic Activity of Semiconductor Oxides. *Environmental Science & Technology*, *50*(5), 2130–2131. doi:10.1021/acs.est.6b00213 PMID:26886856

Baruah, A., Chaudhary, V., Malik, R., & Tomer, V. K. (2019). 17 - Nanotechnology Based Solutions for Wastewater Treatment. In A. Ahsan & A. F. Ismail (Eds.), *Nanotechnology in Water and Wastewater Treatment* (pp. 337–368). Elsevier. doi:10.1016/B978-0-12-813902-8.00017-4

Bassyouni, M., Abdel-Aziz, M. H., Zoromba, M. S., Abdel-Hamid, S. M. S., & Drioli, E. (2019). A review of polymeric nanocomposite membranes for water purification. *Journal of Industrial and Engineering Chemistry*, *73*, 19–46. doi:10.1016/j.jiec.2019.01.045

Bhaumik, M., Choi, H. J., McCrindle, R. I., & Maity, A. (2014). Composite nanofibers prepared from metallic iron nanoparticles and polyaniline: High performance for water treatment applications. *Journal of Colloid and Interface Science*, *425*, 75–82. doi:10.1016/j.jcis.2014.03.031 PMID:24776666

Bhaumik, M., Leswifi, T. Y., Maity, A., Srinivasu, V. V., & Onyango, M. S. (2011). Removal of fluoride from aqueous solution by polypyrrole/Fe3O4 magnetic nanocomposite. *Journal of Hazardous Materials*, *186*(1), 150–159. doi:10.1016/j.jhazmat.2010.10.098 PMID:21112695

Bhaumik, M., Maity, A., Srinivasu, V. V., & Onyango, M. S. (2012). Removal of hexavalent chromium from aqueous solution using polypyrrole-polyaniline nanofibers. *Chemical Engineering Journal*, *181-182*, 323–333. doi:10.1016/j.cej.2011.11.088

Bisht, M. (2019). Nanomaterials for Removal of Toxic Metals Ions from the Water. In A. H. Bhat, I. Khan, M. Jawaid, F. O. Suliman, H. Al-Lawati, & S. M. Al-Kindy (Eds.), *Nanomaterials for Healthcare, Energy and Environment* (pp. 159–174). Springer Singapore. doi:10.1007/978-981-13-9833-9_8

Bradder, P., Ling, S. K., Wang, S., & Liu, S. (2011). Dye Adsorption on Layered Graphite Oxide. *Journal of Chemical & Engineering Data*, *56*(1), 138–141. doi:10.1021/je101049g

Bushra, R., Shahadat, M., Ahmad, A., Nabi, S. A., Umar, K., Oves, M., Raeissi, A. S., & Muneer, M. (2014). Synthesis, characterization, antimicrobial activity and applications of polyanilineTi(IV)arsenophosphate adsorbent for the analysis of organic and inorganic pollutants. *Journal of Hazardous Materials*, *264*, 481–489. doi:10.1016/j.jhazmat.2013.09.044 PMID:24238807

Chaturvedi, V. K., Kushwaha, A., Maurya, S., Tabassum, N., Chaurasia, H., & Singh, M. P. (2020). Wastewater Treatment Through Nanotechnology: Role and Prospects. In A. K. Upadhyay, R. Singh, & D. P. Singh (Eds.), *Restoration of Wetland Ecosystem: A Trajectory Towards a Sustainable Environment* (pp. 227–247). Springer Singapore. doi:10.1007/978-981-13-7665-8_14

Checkol, F., Elfwing, A., Greczynski, G., Mehretie, S., Inganäs, O., & Admassie, S. (2018). Highly Stable and Efficient Lignin-PEDOT/PSS Composites for Removal of Toxic Metals. *Advanced Sustainable Systems*, *2*(1), 1700114. doi:10.1002/adsu.201700114

Chen, M., Jiang, W., Wang, F., Shen, P., Ma, P., Gu, J., Mao, J., & Li, F. (2013). Synthesis of highly hydrophobic floating magnetic polymer nanocomposites for the removal of oils from water surface. *Applied Surface Science*, *286*, 249–256. doi:10.1016/j.apsusc.2013.09.059

Chen, Y., Chen, L., Bai, H., & Li, L. (2013). Graphene oxide–chitosan composite hydrogels as broad-spectrum adsorbents for water purification. *Journal of Materials Chemistry. A, Materials for Energy and Sustainability*, *1*(6), 1992–2001. doi:10.1039/C2TA00406B

Cheng, F.-Y., Su, C.-H., Yang, Y.-S., Yeh, C.-S., Tsai, C.-Y., Wu, C.-L., ... Shieh, D.-B. (2005). Characterization of aqueous dispersions of Fe3O4 nanoparticles and their biomedical applications. *Biomaterials*, *26*(7), 729–738. doi:10.1016/j.biomaterials.2004.03.016 PMID:15350777

Chiu, Y.-H., Chang, T.-F. M., Chen, C.-Y., Sone, M., & Hsu, Y.-J. (2019). Mechanistic Insights into Photodegradation of Organic Dyes Using Heterostructure Photocatalysts. *Catalysts*, *9*(5), 430. Advance online publication. doi:10.3390/catal9050430

Chowdhury, S., & Balasubramanian, R. (2014). Recent advances in the use of graphene-family nanoadsorbents for removal of toxic pollutants from wastewater. *Advances in Colloid and Interface Science*, *204*, 35–56. doi:10.1016/j.cis.2013.12.005 PMID:24412086

Covaliu, C. I., Berger, D., Matei, C., Diamandescu, L., Vasile, E., Cristea, C., Ionita, V., & Iovu, H. (2011). Magnetic nanoparticles coated with polysaccharide polymers for potential biomedical applications. *Journal of Nanoparticle Research*, *13*(11), 6169–6180. doi:10.100711051-011-0452-6

Dave, P. N., & Chopda, L. V. (2014). Application of Iron Oxide Nanomaterials for the Removal of Heavy Metals. *Journal of Nanotechnology*, *398569*, 1–14. doi:10.1155/2014/398569

Dresselhaus, M. S., & Terrones, M. (2013). Carbon-Based Nanomaterials From a Historical Perspective. *Proceedings of the IEEE, 101*(7), 1522–1535. doi:10.1109/JPROC.2013.2261271

Eduardo, E. P.-R. Miguel de la, L.-A., Ana, L. M.-H., & Carlos, V.-S. (2016). Graphene Materials to Remove Organic Pollutants and Heavy Metals from Water: Photocatalysis and Adsorption. In C. Wenbin (Ed.), Semiconductor Photocatalysis. Rijeka: IntechOpen.

Elimelech, M. (2006). The global challenge for adequate and safe water. *Journal of Water Supply: Research & Technology - Aqua, 55*(1), 3–10. doi:10.2166/aqua.2005.064

Fan, L., Luo, C., Sun, M., Li, X., & Qiu, H. (2013). Highly selective adsorption of lead ions by water-dispersible magnetic chitosan/graphene oxide composites. *Colloids and Surfaces. B, Biointerfaces, 103*, 523–529. doi:10.1016/j.colsurfb.2012.11.006 PMID:23261576

Fuwad, A., Ryu, H., Malmstadt, N., Kim, S. M., & Jeon, T.-J. (2019). Biomimetic membranes as potential tools for water purification: Preceding and future avenues. *Desalination, 458*, 97–115. doi:10.1016/j.desal.2019.02.003

Gao, W., Majumder, M., Alemany, L. B., Narayanan, T. N., Ibarra, M. A., Pradhan, B. K., & Ajayan, P. M. (2011). Engineered Graphite Oxide Materials for Application in Water Purification. *ACS Applied Materials & Interfaces, 3*(6), 1821–1826. doi:10.1021/am200300u PMID:21568266

Ge, L., Wang, W., Peng, Z., Tan, F., Wang, X., Chen, J., & Qiao, X. (2018). Facile fabrication of Fe@MgO magnetic nanocomposites for efficient removal of heavy metal ion and dye from water. *Powder Technology, 326*, 393–401. doi:10.1016/j.powtec.2017.12.003

Giwa, A., Hasan, S. W., Yousuf, A., Chakraborty, S., Johnson, D. J., & Hilal, N. (2017). Biomimetic membranes: A critical review of recent progress. *Desalination, 420*, 403–424. doi:10.1016/j.desal.2017.06.025

Gómez-Pastora, J., Dominguez, S., Bringas, E., Rivero, M. J., Ortiz, I., & Dionysiou, D. D. (2017). Review and perspectives on the use of magnetic nanophotocatalysts (MNPCs) in water treatment. *Chemical Engineering Journal, 310*, 407–427. doi:10.1016/j.cej.2016.04.140

Gupta, V. K., Moradi, O., Tyagi, I., Agarwal, S., Sadegh, H., Shahryari-Ghoshekandi, R., Makhlouf, A. S. H., Goodarzi, M., & Garshasbi, A. (2016). Study on the removal of heavy metal ions from industry waste by carbon nanotubes: Effect of the surface modification: a review. *Critical Reviews in Environmental Science and Technology*, *46*(2), 93–118. doi:10.1080/10643389.2015.1061874

Hajjaoui, H., Soufi, A., Boumya, W., Abdennouri, M., & Barka, N. (2021). Polyaniline/Nanomaterial Composites for the Removal of Heavy Metals by Adsorption: A Review. *Journal of Composites Science*, *5*(9), 233. doi:10.3390/jcs5090233

Hayati, B., Maleki, A., Najafi, F., Gharibi, F., McKay, G., Gupta, V. K., Harikaranahalli Puttaiah, S., & Marzban, N. (2018). Heavy metal adsorption using PAMAM/CNT nanocomposite from aqueous solution in batch and continuous fixed bed systems. *Chemical Engineering Journal*, *346*, 258–270. doi:10.1016/j.cej.2018.03.172

He, Y. Q., Zhang, N. N., & Wang, X. D. (2011). Adsorption of graphene oxide/chitosan porous materials for metal ions. *Chinese Chemical Letters*, *22*(7), 859–862. doi:10.1016/j.cclet.2010.12.049

Hotze, E. M., Phenrat, T., & Lowry, G. V. (2010). Nanoparticle Aggregation: Challenges to Understanding Transport and Reactivity in the Environment. *Journal of Environmental Quality*, *39*(6), 1909–1924. doi:10.2134/jeq2009.0462 PMID:21284288

Hua, M., Zhang, S., Pan, B., Zhang, W., Lv, L., & Zhang, Q. (2012). Heavy metal removal from water/wastewater by nanosized metal oxides: A review. *Journal of Hazardous Materials*, *211-212*, 317–331. doi:10.1016/j.jhazmat.2011.10.016 PMID:22018872

Hwang, L.-L., Chen, J.-C., & Wey, M.-Y. (2013). The properties and filtration efficiency of activated carbon polymer composite membranes for the removal of humic acid. *Desalination*, *313*, 166–175. doi:10.1016/j.desal.2012.12.019

Ibrahim, R. K., Hayyan, M., AlSaadi, M. A., Hayyan, A., & Ibrahim, S. (2016). Environmental application of nanotechnology: Air, soil, and water. *Environmental Science and Pollution Research International*, *23*(14), 13754–13788. doi:10.100711356-016-6457-z PMID:27074929

Ihsanullah, A., Abbas, A., Al-Amer, A. M., Laoui, T., Al-Marri, M. J., Nasser, M. S., Khraisheh, M., & Atieh, M. A. (2016). Heavy metal removal from aqueous solution by advanced carbon nanotubes: Critical review of adsorption applications. *Separation and Purification Technology*, *157*, 141–161. doi:10.1016/j.seppur.2015.11.039

Inglezakis, V. J., Loizidou, M. D., & Grigoropoulou, H. P. (2002). Equilibrium and kinetic ion exchange studies of Pb2+, Cr3+, Fe3+ and Cu2+ on natural clinoptilolite. *Water Research*, *36*(11), 2784–2792. doi:10.1016/S0043-1354(01)00504-8 PMID:12146866

Jafari Kang, A., Baghdadi, M., & Pardakhti, A. (2016). Removal of cadmium and lead from aqueous solutions by magnetic acid-treated activated carbon nanocomposite. *Desalination and Water Treatment*, *57*(40), 18782–18798. doi:10.1080/19443994.2015.1095123

Javadian, H. (2014). Application of kinetic, isotherm and thermodynamic models for the adsorption of Co(II) ions on polyaniline/polypyrrole copolymer nanofibers from aqueous solution. *Journal of Industrial and Engineering Chemistry*, *20*(6), 4233–4241. doi:10.1016/j.jiec.2014.01.026

Javadian, H., Sorkhrodi, F. Z., & Koutenaei, B. B. (2014). Experimental investigation on enhancing aqueous cadmium removal via nanostructure composite of modified hexagonal type mesoporous silica with polyaniline/polypyrrole nanoparticles. *Journal of Industrial and Engineering Chemistry*, *20*(5), 3678–3688. doi:10.1016/j.jiec.2013.12.066

Jawed, A., Saxena, V., & Pandey, L. M. (2020). Engineered nanomaterials and their surface functionalization for the removal of heavy metals: A review. *Journal of Water Process Engineering*, *33*, 101009. doi:10.1016/j.jwpe.2019.101009

Jawor, A., & Hoek, E. M. V. (2010). Removing Cadmium Ions from Water via Nanoparticle-Enhanced Ultrafiltration. *Environmental Science & Technology*, *44*(7), 2570–2576. doi:10.1021/es902310e PMID:20230021

Jayaweera, H. D. A. C., Siriwardane, I., de Silva, K. M. N., & de Silva, R. M. (2018). Synthesis of multifunctional activated carbon nanocomposite comprising biocompatible flake nano hydroxyapatite and natural turmeric extract for the removal of bacteria and lead ions from aqueous solution. *Chemistry Central Journal*, *12*(1), 18. doi:10.118613065-018-0384-7 PMID:29468333

Jelić, A., Petrović, M., & Barceló, D. (2012). Pharmaceuticals in Drinking Water. In D. Barceló (Ed.), *Emerging Organic Contaminants and Human Health* (pp. 47–70). Springer Berlin Heidelberg. doi:10.1007/698_2011_133

Jhaveri, J. H., & Murthy, Z. V. P. (2016). A comprehensive review on anti-fouling nanocomposite membranes for pressure driven membrane separation processes. *Desalination*, *379*, 137–154. doi:10.1016/j.desal.2015.11.009

Jurado-Sánchez, B., & Wang, J. (2018). Micromotors for environmental applications: A review. *Environmental Science. Nano*, *5*(7), 1530–1544. doi:10.1039/C8EN00299A

Kausar, F., Bagheri, A. R., Rasheed, T., Bilal, M., Rizwan, K., Nguyen, T. A., & Iqbal, H. M. N. (2022). Chapter 7 - Nanomaterials for removal of heavy metals from wastewater. In A. Denizli, N. Ali, M. Bilal, A. Khan, & T. A. Nguyen (Eds.), Nano-Biosorbents for Decontamination of Water, Air, and Soil Pollution (pp. 135-161): Elsevier.

Khoso, W. A., Haleem, N., Baig, M. A., & Jamal, Y. (2021). Synthesis, characterization and heavy metal removal efficiency of nickel ferrite nanoparticles (NFN's). *Scientific Reports*, *11*(1), 3790. doi:10.103841598-021-83363-1 PMID:33589710

Kumar, A. S. K., Kalidhasan, S., Rajesh, V., & Rajesh, N. (2012). Application of Cellulose-Clay Composite Biosorbent toward the Effective Adsorption and Removal of Chromium from Industrial Wastewater. *Industrial & Engineering Chemistry Research*, *51*(1), 58–69. doi:10.1021/ie201349h

Kunduru, K. R., Nazarkovsky, M., Farah, S., Pawar, R. P., Basu, A., & Domb, A. J. (2017). 2 - Nanotechnology for water purification: applications of nanotechnology methods in wastewater treatment. In A. M. Grumezescu (Ed.), *Water Purification* (pp. 33–74). Academic Press. doi:10.1016/B978-0-12-804300-4.00002-2

Le, A. T., Pung, S.-Y., Sreekantan, S., Matsuda, A., & Huynh, D. P. (2019). Mechanisms of removal of heavy metal ions by ZnO particles. *Heliyon*, *5*(4), e01440. doi:10.1016/j.heliyon.2019.e01440 PMID:31008388

Li, J., Chen, C., Zhang, R., & Wang, X. (2015). Nanoscale Zero-Valent Iron Particles Supported on Reduced Graphene Oxides by Using a Plasma Technique and Their Application for Removal of Heavy-Metal Ions. *Chemistry, an Asian Journal*, *10*(6), 1410–1417. doi:10.1002/asia.201500242 PMID:25917859

Li, J., Li, H., Zhu, Y., Hao, Y., Sun, X., & Wang, L. (2011). Dual roles of amphiphilic triblock copolymer P123 in synthesis of α-Fe nanoparticle/ordered mesoporous silica composites. *Applied Surface Science*, *258*(2), 657–661. doi:10.1016/j.apsusc.2011.07.037

Li, X., Ai, L., & Jiang, J. (2016). Nanoscale zerovalent iron decorated on graphene nanosheets for Cr(VI) removal from aqueous solution: Surface corrosion retard induced the enhanced performance. *Chemical Engineering Journal*, *288*, 789–797. doi:10.1016/j.cej.2015.12.022

Li, Z.-J., Wang, L., Yuan, L.-Y., Xiao, C.-L., Mei, L., Zheng, L.-R., Zhang, J., Yang, J.-H., Zhao, Y.-L., Zhu, Z.-T., Chai, Z.-F., & Shi, W.-Q. (2015). Efficient removal of uranium from aqueous solution by zero-valent iron nanoparticle and its graphene composite. *Journal of Hazardous Materials*, *290*, 26–33. doi:10.1016/j.jhazmat.2015.02.028 PMID:25734531

Lin, Y., Yao, W., Cheng, Y., Qian, H., Wang, X., Ding, Y., Wu, W., & Jiang, X. (2012). Multifold enhanced T2 relaxation of ZnFe2O4 nanoparticles by jamming them inside chitosan nanospheres. *Journal of Materials Chemistry*, *22*(12), 5684–5693. doi:10.1039/c2jm15133b

Liu, X., Hu, Q., Fang, Z., Zhang, X., & Zhang, B. (2009). Magnetic Chitosan Nanocomposites: A Useful Recyclable Tool for Heavy Metal Ion Removal. *Langmuir*, *25*(1), 3–8. doi:10.1021/la802754t PMID:19032060

Lu, H., Wang, J., Stoller, M., Wang, T., Bao, Y., & Hao, H. (2016). An Overview of Nanomaterials for Water and Wastewater Treatment. *Advances in Materials Science and Engineering*, *4964828*, 1–10. Advance online publication. doi:10.1155/2016/4964828

Lü, T., Qi, D., Zhang, D., Lü, Y., & Zhao, H. (2018). A facile method for emulsified oil-water separation by using polyethylenimine-coated magnetic nanoparticles. *Journal of Nanoparticle Research*, *20*(4), 88. doi:10.100711051-018-4193-7

Mahdavi, K., Zinatloo-Ajabshir, S., Yousif, Q. A., & Salavati-Niasari, M. (2022). Enhanced photocatalytic degradation of toxic contaminants using Dy2O3-SiO2 ceramic nanostructured materials fabricated by a new, simple and rapid sonochemical approach. *Ultrasonics Sonochemistry*, *82*, 105892. doi:10.1016/j.ultsonch.2021.105892 PMID:34959201

Manikandan, S., Karmegam, N., Subbaiya, R., Karthiga Devi, G., Arulvel, R., Ravindran, B., & Kumar Awasthi, M. (2021). Emerging nano-structured innovative materials as adsorbents in wastewater treatment. *Bioresource Technology*, *320*, 124394. doi:10.1016/j.biortech.2020.124394 PMID:33220545

Masar, M., Ali, H., Guler, A. C., Urbanek, M., Urbanek, P., Hanulikova, B., Pistekova, H., Annusova, A., Machovsky, M., & Kuritka, I. (2023). Multifunctional bandgap-reduced ZnO nanocrystals for photocatalysis, self-cleaning, and antibacterial glass surfaces. *Colloids and Surfaces. A, Physicochemical and Engineering Aspects*, *656*, 130447. doi:10.1016/j.colsurfa.2022.130447

Mekaru, H., Lu, J., & Tamanoi, F. (2015). Development of mesoporous silica-based nanoparticles with controlled release capability for cancer therapy. *Advanced Drug Delivery Reviews*, *95*, 40–49. doi:10.1016/j.addr.2015.09.009 PMID:26434537

Moo, J. G. S., & Pumera, M. (2015). Chemical Energy Powered Nano/Micro/Macromotors and the Environment. *Chemistry (Weinheim an der Bergstrasse, Germany)*, *21*(1), 58–72. doi:10.1002/chem.201405011 PMID:25410790

Mukhopadhyay, R., Bhaduri, D., Sarkar, B., Rusmin, R., Hou, D., Khanam, R., Sarkar, S., Kumar Biswas, J., Vithanage, M., Bhatnagar, A., & Ok, Y. S. (2020). Clay–polymer nanocomposites: Progress and challenges for use in sustainable water treatment. *Journal of Hazardous Materials*, *383*, 121125. doi:10.1016/j.jhazmat.2019.121125 PMID:31541959

Nirmala, N., Shriniti, V., Aasresha, K., Arun, J., Gopinath, K. P., Dawn, S. S., Sheeladevi, A., Priyadharsini, P., Birindhadevi, K., Chi, N. T. L., & Pugazhendhi, A. (2022). Removal of toxic metals from wastewater environment by graphene-based composites: A review on isotherm and kinetic models, recent trends, challenges and future directions. *The Science of the Total Environment*, *840*, 156564. doi:10.1016/j.scitotenv.2022.156564 PMID:35690214

Novoselov, K. S., Fal′ko, V. I., Colombo, L., Gellert, P. R., Schwab, M. G., & Kim, K. (. (2012). A roadmap for graphene. *Nature*, *490*(7419), 192–200. doi:10.1038/nature11458 PMID:23060189

Nujić, M., & Habuda-Stanić, M. (2019). Toxic Metal Ions in Drinking Water and Effective Removal Using Graphene Oxide Nanocomposite. In M. Naushad (Ed.), *A New Generation Material Graphene: Applications in Water Technology* (pp. 373–395). Springer International Publishing. doi:10.1007/978-3-319-75484-0_15

Olagunju, M. O., Zahran, E. M., Reed, J. M., Zeynaloo, E., Shukla, D., Cohn, J. L., Surnar, B., Dhar, S., Bachas, L. G., & Knecht, M. R. (2021). Halide Effects in BiVO4/BiOX Heterostructures Decorated with Pd Nanoparticles for Photocatalytic Degradation of Rhodamine B as a Model Organic Pollutant. *ACS Applied Nano Materials*, *4*(3), 3262–3272. doi:10.1021/acsanm.1c00481

Ong, C. B., Ng, L. Y., & Mohammad, A. W. (2018). A review of ZnO nanoparticles as solar photocatalysts: Synthesis, mechanisms and applications. *Renewable & Sustainable Energy Reviews*, *81*, 536–551. doi:10.1016/j.rser.2017.08.020

Pan, G., & Kurumada, K.-I. (2008). Hybrid gel reinforced with coating layer for removal of phenol from aqueous solution. *Chemical Engineering Journal*, *138*(1), 194–199. doi:10.1016/j.cej.2007.06.025

Parlayici, S., Eskizeybek, V., Avcı, A., & Pehlivan, E. (2015). Removal of chromium (VI) using activated carbon-supported-functionalized carbon nanotubes. *Journal of Nanostructure in Chemistry*, *5*(3), 255–263. doi:10.100740097-015-0156-z

Peng, F., Xu, T., Wu, F., Ma, C., Liu, Y., Li, J., Zhao, B., & Mao, C. (2017). Novel biomimetic enzyme for sensitive detection of superoxide anions. *Talanta, 174*, 82–91. doi:10.1016/j.talanta.2017.05.028 PMID:28738660

Pooseekheaw, P., Thongpan, W., Panthawan, A., Kantarak, E., Sroila, W., & Singjai, P. (2020). Porous V2O5/TiO2 Nanoheterostructure Films with Enhanced Visible-Light Photocatalytic Performance Prepared by the Sparking Method. *Molecules (Basel, Switzerland)*, *25*(15), 3327. Advance online publication. doi:10.3390/molecules25153327 PMID:32707984

Popov, V. N. (2004). Carbon nanotubes: Properties and application. *Materials Science and Engineering R Reports*, *43*(3), 61–102. doi:10.1016/j.mser.2003.10.001

Power, A. C., Gorey, B., Chandra, S., & Chapman, J. (2018). Carbon nanomaterials and their application to electrochemical sensors: a review. *7*(1), 19-41. 10.1515/ntrev-2017-0160

Qin, R., Li, F., Chen, M., & Jiang, W. (2009). Preparation of chitosan–ethylenediaminetetraacetate-enwrapped magnetic CoFe2O4 nanoparticles via zero-length emulsion crosslinking method. *Applied Surface Science*, *256*(1), 27–32. doi:10.1016/j.apsusc.2009.07.032

Rafiq, A., Ikram, M., Ali, S., Niaz, F., Khan, M., Khan, Q., & Maqbool, M. (2021). Photocatalytic degradation of dyes using semiconductor photocatalysts to clean industrial water pollution. *Journal of Industrial and Engineering Chemistry*, *97*, 111–128. doi:10.1016/j.jiec.2021.02.017

Rajapaksha, P. P., Power, A., Chandra, S., & Chapman, J. (2018). Graphene, electrospun membranes and granular activated carbon for eliminating heavy metals, pesticides and bacteria in water and wastewater treatment processes. *Analyst (London)*, *143*(23), 5629–5645. doi:10.1039/C8AN00922H PMID:30357153

Rasheed, T., Li, C., Fu, L., Nabeel, F., Yu, C., Gong, L., & Zhou, Y. (2018). Development and characterization of newly engineered chemosensor with intracellular monitoring potentialities and lowest detection of toxic elements. *Journal of Molecular Liquids*, *272*, 440–449. doi:10.1016/j.molliq.2018.09.112

Rasheed, T., Li, C., Nabeel, F., Huang, W., & Zhou, Y. (2019). Self-assembly of alternating copolymer vesicles for the highly selective, sensitive and visual detection and quantification of aqueous Hg2+. *Chemical Engineering Journal*, *358*, 101–109. doi:10.1016/j.cej.2018.09.216

Rasheed, T., Li, C., Zhang, Y., Nabeel, F., Peng, J., Qi, J., Gong, L., & Yu, C. (2018). Rhodamine-based multianalyte colorimetric probe with potentialities as on-site assay kit and in biological systems. *Sensors and Actuators. B, Chemical, 258*, 115–124. doi:10.1016/j.snb.2017.11.100

Rasheed, T., Nabeel, F., Adeel, M., Rizwan, K., Bilal, M., & Iqbal, H. M. N. (2019). Carbon nanotubes-based cues: A pathway to future sensing and detection of hazardous pollutants. *Journal of Molecular Liquids, 292*, 111425. doi:10.1016/j.molliq.2019.111425

Rasheed, T., Nabeel, F., Li, C., & Bilal, M. (2019). Rhodamine-assisted fluorescent strategy for the sensitive and selective in-field mapping of environmental pollutant Hg(II) with potential bioimaging. *Journal of Luminescence, 208*, 519–526. doi:10.1016/j.jlumin.2019.01.032

Rasheed, T., Nabeel, F., Li, C., & Zhang, Y. (2019). Rhodol assisted alternating copolymer based chromogenic vesicles for the aqueous detection and quantification of hydrazine via switch-on strategy. *Journal of Molecular Liquids, 274*, 461–469. doi:10.1016/j.molliq.2018.11.014

Rasheed, T., Nabeel, F., & Shafi, S. (2019). Chromogenic vesicles for aqueous detection and quantification of Hg2+/Cu2+ in real water samples. *Journal of Molecular Liquids, 282*, 489–498. doi:10.1016/j.molliq.2019.03.048

Reddy, D. A., Choi, J., Lee, S., Ma, R., & Kim, T. K. (2015). Self-assembled macro porous ZnS–graphene aerogels for photocatalytic degradation of contaminants in water. *RSC Advances, 5*(24), 18342–18351. doi:10.1039/C4RA16494F

Reddy, D. H. K., & Lee, S.-M. (2013). Application of magnetic chitosan composites for the removal of toxic metal and dyes from aqueous solutions. *Advances in Colloid and Interface Science, 201-202*, 68–93. doi:10.1016/j.cis.2013.10.002 PMID:24182685

Sahebi, S., Sheikhi, M., & Ramavandi, B. (2019). A new biomimetic aquaporin thin film composite membrane for forward osmosis: Characterization and performance assessment. *Desalination and Water Treatment, 148*, 42–50. doi:10.5004/dwt.2019.23748

Salam, M. A., Makki, M. S. I., & Abdelaal, M. Y. A. (2011). Preparation and characterization of multi-walled carbon nanotubes/chitosan nanocomposite and its application for the removal of heavy metals from aqueous solution. *Journal of Alloys and Compounds, 509*(5), 2582–2587. doi:10.1016/j.jallcom.2010.11.094

Salim, W., & Ho, W. S. W. (2015). Recent developments on nanostructured polymer-based membranes. *Current Opinion in Chemical Engineering*, *8*, 76–82. doi:10.1016/j.coche.2015.03.003

Shannon, M. A., Bohn, P. W., Elimelech, M., Georgiadis, J. G., Mariñas, B. J., & Mayes, A. M. (2009). Science and technology for water purification in the coming decades. In *Nanoscience and Technology* (pp. 337–346). Co-Published with Macmillan Publishers Ltd. doi:10.1142/9789814287005_0035

Shao, D., Hu, J., & Wang, X. (2010). Plasma Induced Grafting Multiwalled Carbon Nanotube with Chitosan and Its Application for Removal of UO, Cu2+, and Pb2+ from Aqueous Solutions. *Plasma Processes and Polymers*, *7*(12), 977–985. doi:10.1002/ppap.201000062

Shen, Y., Saboe, P. O., Sines, I. T., Erbakan, M., & Kumar, M. (2014). Biomimetic membranes: A review. *Journal of Membrane Science*, *454*, 359–381. doi:10.1016/j.memsci.2013.12.019

Singh, J., Srivastava, M., Dutta, J., & Dutta, P. K. (2011). Preparation and properties of hybrid monodispersed magnetic α-Fe2O3 based chitosan nanocomposite film for industrial and biomedical applications. *International Journal of Biological Macromolecules*, *48*(1), 170–176. doi:10.1016/j.ijbiomac.2010.10.016 PMID:21056054

Singh, J., Srivastava, M., Kalita, P., & Malhotra, B. D. (2012). A novel ternary NiFe2O4/CuO/FeO-chitosan nanocomposite as a cholesterol biosensor. *Process Biochemistry (Barking, London, England)*, *47*(12), 2189–2198. doi:10.1016/j.procbio.2012.08.012

Smith, S. C., & Rodrigues, D. F. (2015). Carbon-based nanomaterials for removal of chemical and biological contaminants from water: A review of mechanisms and applications. *Carbon*, *91*, 122–143. doi:10.1016/j.carbon.2015.04.043

Soltaninejad, V., Ahghari, M. R., Taheri-Ledari, R., & Maleki, A. (2021). Bifunctional PVA/ZnO/AgI/Chlorophyll Nanocomposite Film: Enhanced Photocatalytic Activity for Degradation of Pollutants and Antimicrobial Property under Visible-Light Irradiation. *Langmuir*, *37*(15), 4700–4713. doi:10.1021/acs.langmuir.1c00501 PMID:33821643

Suzuki, T. M., Kobayashi, S., Pacheco Tanaka, D. A., Llosa Tanco, M. A., Nagase, T., & Onodera, Y. (2004). Separation and concentration of trace Pb(II) by the porous resin loaded with α-zirconium phosphate crystals. *Reactive & Functional Polymers*, *58*(2), 131–138. doi:10.1016/j.reactfunctpolym.2003.11.012

Takafuji, M., Ide, S., Ihara, H., & Xu, Z. (2004). Preparation of Poly(1-vinylimidazole)-Grafted Magnetic Nanoparticles and Their Application for Removal of Metal Ions. *Chemistry of Materials*, *16*(10), 1977–1983. doi:10.1021/cm030334y

Tang, W.-W., Zeng, G.-M., Gong, J.-L., Liu, Y., Wang, X.-Y., Liu, Y.-Y., Liu, Z.-F., Chen, L., Zhang, X.-R., & Tu, D.-Z. (2012). Simultaneous adsorption of atrazine and Cu (II) from wastewater by magnetic multi-walled carbon nanotube. *Chemical Engineering Journal*, *211-212*, 470–478. doi:10.1016/j.cej.2012.09.102

Tounsadi, H., Khalidi, A., Abdennouri, M., & Barka, N. (2016). Activated carbon from Diplotaxis Harra biomass: Optimization of preparation conditions and heavy metal removal. *Journal of the Taiwan Institute of Chemical Engineers*, *59*, 348–358. doi:10.1016/j.jtice.2015.08.014

Verma, R., Asthana, A., Singh, A. K., Prasad, S., & Susan, M. A. B. H. (2017). Novel glycine-functionalized magnetic nanoparticles entrapped calcium alginate beads for effective removal of lead. *Microchemical Journal*, *130*, 168–178. doi:10.1016/j.microc.2016.08.006

Wang, H., Yuan, X., Wu, Y., Chen, X., Leng, L., Wang, H., Li, H., & Zeng, G. (2015). Facile synthesis of polypyrrole decorated reduced graphene oxide–Fe3O4 magnetic composites and its application for the Cr(VI) removal. *Chemical Engineering Journal*, *262*, 597–606. doi:10.1016/j.cej.2014.10.020

Wang, J., Pan, K., He, Q., & Cao, B. (2013). Polyacrylonitrile/polypyrrole core/shell nanofiber mat for the removal of hexavalent chromium from aqueous solution. *Journal of Hazardous Materials*, *244-245*, 121–129. doi:10.1016/j.jhazmat.2012.11.020 PMID:23246947

Wang, S.-F., Shen, L., Zhang, W.-D., & Tong, Y.-J. (2005). Preparation and Mechanical Properties of Chitosan/Carbon Nanotubes Composites. *Biomacromolecules*, *6*(6), 3067–3072. doi:10.1021/bm050378v PMID:16283728

Wu, F.-C., Tseng, R.-L., & Juang, R.-S. (2010). A review and experimental verification of using chitosan and its derivatives as adsorbents for selected heavy metals. *Journal of Environmental Management*, *91*(4), 798–806. doi:10.1016/j.jenvman.2009.10.018 PMID:19917518

Wu, M.-T., Tsai, Y.-L., Chiu, C.-W., & Cheng, C.-C. (2016). Synthesis, characterization, and highly acid-resistant properties of crosslinking β-chitosan with polyamines for heavy metal ion adsorption. *RSC Advances*, *6*(106), 104754–104762. doi:10.1039/C6RA21993D

Wu, S.-J., Liou, T.-H., & Mi, F.-L. (2009). Synthesis of zero-valent copper-chitosan nanocomposites and their application for treatment of hexavalent chromium. *Bioresource Technology*, *100*(19), 4348–4353. doi:10.1016/j.biortech.2009.04.013 PMID:19414251

Yadav, D. K., & Srivastava, S. (2017). Carbon nanotubes as adsorbent to remove heavy metal ion (Mn+7) in wastewater treatment. *Materials Today: Proceedings*, *4*(2, 2, Part A), 4089–4094. doi:10.1016/j.matpr.2017.02.312

Yan, H., Yang, L., Yang, Z., Yang, H., Li, A., & Cheng, R. (2012). Preparation of chitosan/poly(acrylic acid) magnetic composite microspheres and applications in the removal of copper(II) ions from aqueous solutions. *Journal of Hazardous Materials*, *229-230*, 371–380. doi:10.1016/j.jhazmat.2012.06.014 PMID:22749139

Yang, J., Hou, B., Wang, J., Tian, B., Bi, J., Wang, N., Li, X., & Huang, X. (2019). Nanomaterials for the Removal of Heavy Metals from Wastewater. *Nanomaterials (Basel, Switzerland)*, *9*(3), 424. Advance online publication. doi:10.3390/nano9030424 PMID:30871096

Yang, Y., Xie, Y., Pang, L., Li, M., Song, X., Wen, J., & Zhao, H. (2013). Preparation of Reduced Graphene Oxide/Poly(acrylamide) Nanocomposite and Its Adsorption of Pb(II) and Methylene Blue. *Langmuir*, *29*(34), 10727–10736. doi:10.1021/la401940z PMID:23895359

Yaqoob, A. A., Parveen, T., Umar, K., & Mohamad Ibrahim, M. N. (2020). Role of Nanomaterials in the Treatment of Wastewater: A Review. *Water (Basel)*, *12*(2), 495. Advance online publication. doi:10.3390/w12020495

Yasir, M., Asabuwa Ngwabebhoh, F., Šopík, T., Ali, H., & Sedlařík, V. (2022). Electrospun polyurethane nanofibers coated with polyaniline/polyvinyl alcohol as ultrafiltration membranes for the removal of ethinylestradiol hormone micropollutant from aqueous phase. *Journal of Environmental Chemical Engineering*, *10*(3), 107811. doi:10.1016/j.jece.2022.107811

Yasir, M., Masar, M., Sopik, T., Ali, H., Urbanek, M., Antos, J., Machovsky, M., & Kuritka, I. (2022). ZnO nanowires and nanorods based ZnO/WO3/Pt heterojunction for efficient photocatalytic degradation of estriol (E3) hormone. *Materials Letters*, *319*, 132291. doi:10.1016/j.matlet.2022.132291

Yu, G., Lu, Y., Guo, J., Patel, M., Bafana, A., Wang, X., Qiu, B., Jeffryes, C., Wei, S., Guo, Z., & Wujcik, E. K. (2018). Carbon nanotubes, graphene, and their derivatives for heavy metal removal. *Advanced Composites and Hybrid Materials*, *1*(1), 56–78. doi:10.100742114-017-0004-3

Yu, L., Ruan, S., Xu, X., Zou, R., & Hu, J. (2017). One-dimensional nanomaterial-assembled macroscopic membranes for water treatment. *Nano Today*, *17*, 79–95. doi:10.1016/j.nantod.2017.10.012

Zhang, D., Wei, A., Zhang, J., & Qiu, R. (2015). The photocatalytic interaction of Cr(VI) ions and phenol on polymer-modified TiO2 under visible light irradiation. *Kinetics and Catalysis*, *56*(5), 569–573. doi:10.1134/S0023158415050195

Zhang, Q., Pan, B., Zhang, S., Wang, J., Zhang, W., & Lv, L. (2011). New insights into nanocomposite adsorbents for water treatment: A case study of polystyrene-supported zirconium phosphate nanoparticles for lead removal. *Journal of Nanoparticle Research*, *13*(10), 5355–5364. doi:10.100711051-011-0521-x

Zhang, Q., Pan, B., Zhang, W., Pan, B., Lv, L., Wang, X., Wu, J., & Tao, X. (2009). Selective removal of Pb(II), Cd(II), and Zn(II) ions from waters by an inorganic exchanger Zr(HPO3S)2. *Journal of Hazardous Materials*, *170*(2), 824–828. doi:10.1016/j.jhazmat.2009.05.038 PMID:19493618

Zhang, Y., Wu, B., Xu, H., Liu, H., Wang, M., He, Y., & Pan, B. (2016). Nanomaterials-enabled water and wastewater treatment. *NanoImpact*, *3-4*, 22–39. doi:10.1016/j.impact.2016.09.004

Zhao, G., Huang, X., Tang, Z., Huang, Q., Niu, F., & Wang, X. (2018). Polymer-based nanocomposites for heavy metal ions removal from aqueous solution: A review. *Polymer Chemistry*, *9*(26), 3562–3582. doi:10.1039/C8PY00484F

Zhao, X., Dou, X., Mohan, D., Pittman, C. U. Jr, Ok, Y. S., & Jin, X. (2014). Antimonate and antimonite adsorption by a polyvinyl alcohol-stabilized granular adsorbent containing nanoscale zero-valent iron. *Chemical Engineering Journal*, *247*, 250–257. doi:10.1016/j.cej.2014.02.096

Zhou, L., Wang, Y., Liu, Z., & Huang, Q. (2009). Characteristics of equilibrium, kinetics studies for adsorption of Hg(II), Cu(II), and Ni(II) ions by thiourea-modified magnetic chitosan microspheres. *Journal of Hazardous Materials*, *161*(2), 995–1002. doi:10.1016/j.jhazmat.2008.04.078 PMID:18538924

Chapter 8
Study of Interesting Applications of Nanomaterials

Himanshi Tyagi
Manipal University Jaipur, India

ABSTRACT

'Nanomaterials' is a word is made by two words; first is 'nano,' which means very small; and second one is 'material,' which means compound. Nanomaterials are based on nanoscience and nanotechnology. They deal with the synthesis, characterization, and application of nanostructure materials. Nanomaterials atoms size range 10^{-9}m to 10^{-7}m (1 to 100nm). They have interesting and useful behavior. Nano structures are bridges between molecules and infinite bulk systems. Individual nanostructure includes quantum dots, nano wires, and nanotubes, etc. While a collection of nanostructures involves arrays, assemblies, and super lattices of the nanostructures. The uniqueness of nanomaterials structure consists of chemistry of mechanism, dynamics on the basis of nanoscience which is the chemistry (science) only applied for nanomaterials.

INTRODUCTION TO NANOMATERIALS

"Nanomaterials" word is made by two words first one is *nano* that means very small and second one is *material* that mean compound. Nanomaterials based on nanoscience and nanotechnology. They deal with the synthesis, characterization and application of nanostructure materials. Nanomaterials atoms size range 10^{-9}m to 10^{-7}m (1 to 100nm) (Ju et al., 2017; Som et al., 2014; Travis et al., 2016). They have interesting and useful behavior. Nano structures are bridge between molecules and infinite bulk systems. Individual nanostructure includes quantum dots, nano wires

DOI: 10.4018/978-1-6684-8743-3.ch008

and nanotubes etc. While collection of nanostructures involve arrays, assemblies and super lattices of the nanostructures. The uniqueness of nanomaterials structure consist of chemistry of mechanism, dynamics on the basis of nanoscience which is the chemistry (science) only applied for nanomaterials (Chen et al., 2015).

Some nanomaterials occur naturally but of particular interest are engineered nanomaterials (EN). Nanomaterials are found in such things as sunscreens, cosmetic, sporting goods, strain-resistant clothing, tires, electronics, as well as many other everyday items and used in medicine for purposes of diagnosis, imaging and drug delivery (Katyayan & Agrawal, 2018; Noto et al., 2013). Nano coatings and nano composites are finding uses in diverse consumer products, such as windows, sports equipment, bicycle and automobiles.

Nanomaterials properties have differed on the area wise like that means on the surface properties is are different from the bulk counterparts. This difference observed because in changes in physical attributes (size, shape and specific surface area). We know very well surface have less molecule compared with that the surface atoms from a negligible part of the total number of atoms. The smaller a particle becomes the more the ratio of surface atoms increase (Shi et al., 2008).

Classification of Nanomaterials

Nanomaterials are extremely small in size with zero dimensions. They have categories in four types:-

1. 0 D Nanomaterials
2. 1 D Nanomaterials
3. 2 D Nanomaterials
4. 3 D Nanomaterials

In 0 D nanomaterials include spheres filaments and clusters. 1 D nanoscale includes surface films, strands and fibers. In 2 D nanomaterials or 3 D include films, plates and particles (Gonçalves et al., 2015; Katyayan & Agrawal, 2018).

Properties of Nanomaterials

Nanomaterials have different properties in between atoms and bulk materials. On surface type of materials the property of nanomaterials have large fraction of surface atoms, high surface energy spatial confinement and reduced performance which do not exist in the corresponding bulk materials because of their small dimensions, nanomaterials have extremely large surface area to volume ratio. The entire material will be affected by the surface properties of nanomaterials. It also brings the quantum effects.

Applications of Nanomaterials

Nanomaterials have wide range of applications the field of fuel cells, batteries, food industry and medicine etc. In medical field on drug delivery nano sized active materials is kept inside a biomaterials matrix. It prevents the low or high dosage and side effects. They are also used in cancer therapy.

Nanomaterials are also used to develop miniature video camera attached to a blind person's glasses. In electronics industry memory circuits, diodes, and micro sensor chips are made of nano tube composites are found to be superior quantity. Transistors with P-N junction can be made by nano wires (Battisha et al., 2002; Fu et al., 2009; Katyayan & Agrawal, 2018; Marí et al., 2013).

Forming of Nanomaterials

Some nanomaterials are occurring naturally. But EN which are designed for, and already being used in many commercial products and processes. They are found in many cosmetics products like Sunscreens, sporting goods, stain- resistant clothing, tires, and many other essentials for our daily life (Katyayan & Agrawal, 2019).

Nanomaterials are used in commercials use. The range of commercial products available today is very vast including stain- resistant and textiles industry. They are also used in U-V blocking coatings on glass bottles which are used for protection of beverages from damager by sunlight.

Importance of Nanomaterials

Nanomaterials have unusal mechanical, optical, electrical, and magnetic properties.

i. Nanophase ceramics are useful because they are more malleable at high temperature as compared with coarse- grained ceramics.
ii. Gas tight materials are produced by nanosized metallic powders. Cold welding properties combined with the ductility make them suitable for metal- metal bonding in the electronic industry.
iii. Micro particle have special atomic structures with discrete electronic states, which give special properties in addition to the super paramagnetism nature. They are used in ferro fluids for high density information storage and magnetic refrigeration (Lemański & Dereń, 2014).

Examples of Nanomaterials

Nanomaterials (gold, carbon, metals) with variety of morphologies are depicted (Katyayan & Agrawal, 2019).

Examples are as follows-

1) Au nanoparticle
2) Buck minster fullerene
3) FePt nanosphere
4) Titanium nanoflower
5) Silver nanocubes
6) SnO_2 nanoflower

SYNTHESIS AND PROCESSING OF NANOMATERIALS

Methods

There are many methods to create nanostructures. They are arranged by equilibrium thermodynamics such as methods of self- organization and self assembly. Synthesized materials can be arranged into useful shapes so that finally the material can be applied to a certain application.

Grinding of Materials

Materials is only not prepared by assembly of cluster but by the decomposition of coarse- grained structure severe plastic deformation.

The major advantage is the scaling up to quantities of materials for various application (Deng et al., 2010).

1. Milling media and atmosphere.
2. To form powder mixture with coarsening the nano crystalline microstructure.

Mechanical mill is using high energy shaker, planetary ball or tumber mills. The transformation of energy from the refractory or steel balls depends on the rotational or vibrational speed, size and number of the balls.

Wet Chemical Synthesis

There are two main postulate regarding this-

1. Top down method

First single crystal are etched in an aqueous solution for forming nanomaterials. Ex- The synthesis of porous silicon by electro chemical ehching.

2. Bottom up method

Consisting of sol- gel method, precipitation etc. where materials containing the desired precursors are mixed in a controlled fashion to form a colloidal solution (Terashi et al., 2008).

Gas- Phase Synthesis

In gas- phase synthesis they allow elegant way to control process parameters in order to be able to produce size, shape and chemical composition controlled nanostructure.

i. Homogeneous chemical vapor deposition (CVD)

Particles from gas phase towards a cold surface due to thermophoretic forces. Any can either be from the cold surface to give nano-powder. After the diffusion deposited yield is known as “particulate films”.

ii. Heterogenous CVD

The formation of solid with the help of substrate surface which are used for catalyses the reaction and formed a dense film (Xiao et al., 2017).

Advantage of Gas Phase Synthesis

- Control of size is excellent, shape and chemical composition.
- Pure materials can be produced.
- Multi component systems are relatively easy to form.
- On reaction mechanism control is very easy.

Most of the synthesis routes are based on the formation of small lattices that can be use in the formation of nano particles. This process is also known as condensation. Condensation mean the solid crystals of nano materials is convert into with help of gas phase synthesis in gas phase is utilized to form gas particles, this can be obtained by physical and chemical methods (Li et al., 2015).

Chemical Vapor Condensation

Hot wall reactor in the chemical vapor condensation in the CVC process using source of GPC is replaced by evaporator. The process of CVC advanced the parameter field during the synthesis in order to suppress film forms and enhance homogeneous nucleus particle from the gas flow. CVC both particle and film formation can be obtained. When we change the gas flow rate the pressure difference occur in the main chamber. And also the temperature is also increases. Result in the fertile production of nano sized particles of the metals and ceramics instead of thin films occur in CVD processing (Katyayan & Agrawal,).

Properties of Nanomaterials

In the atoms and the bulk materials nanomaterials have structural difference. And also many bulk and nanomaterials atoms have same properties. This is mainly to the size of nanometer.

- First large size of surface atoms
- High surface energy
- In the bulk material reduce defect (Impuririte)

Because of their small size of particles dimensions have extremely large surface area and volume ratio. When we compared the length of nanomaterials whole molecule will be affected by the surface properties of nanomaterials. This system may enhance or modify the properties of the bulk materials. Nature these are quantum effects.

The band theory for the electron charges Spectra quite differently from their bulk and in turn will modify the electronic and optical properties of the materials for example lasers and light emitting diodes (LED) from both of the quantum dots and quantum wires are very useful in the future. High density information storage using quantum dot devices is also a fast developing area. Nanomaterials and nanostructures favor self purification. Due to their nanometer size, nano materials are already known to have many novle properties (Yang et al., 2012).

Optical Properties

Optical optical properties are very important property of nanomaterials. This based on detector, laser, sensors, imaging, fast food, display, solar cell, photocatalysis, electrochemistry and biomedicine. They also depend size, shape, surface area and other variables including doping and interaction with the surrounding environment or other nanostructures. In semiconductor when simple change in their size they alter the optical properties of the nanoparticles.

Electrical Properties

In electrical properties of nanoparticles discuss about fundamentals of electrical conductivity in nanotubes and nano rods, carbon nanotubes, photoconductivity of nanorods, electrical conductivity of nanocomposite composites. Main point is decreasing diameter of the wire, the number of electron wave modes contributing to the electrical conductivity is becoming increasingly smaller by well defined quantized steps (Ruzimuradov et al.,).

Mechanical Properties

Nanomaterials have different type of mechanical properties including bike metallic and ceramic metal materials, influence of process porosity, influence of queen size, superplasticity, filled polymer composites, particle field polymers, polymer based nanocomposites filled with platelets, carbon nanotube based composites. The mechanical properties of nanomaterials only applicable of size less than 100 nanometer (Katyayan & Agrawal, 2020; Kuwabara, 2012).

REFERENCES

Battisha, I. K., Speghini, A., Polizzi, S., Agnoli, F., & Bettinelli, M. (2002). Molten chloride synthesis, structural characterisation and luminescence spectroscopy of ultrafine Eu3+-doped BaTiO3 and SrTiO3. *Materials Letters*, *57*(1), 183–187. doi:10.1016/S0167-577X(02)00727-9

Chen, Q., De Marco, N., Yang, Y., Song, T.-B., Chen, C.-C., Zhao, H., Hong, Z., Zhou, H., & Yang, Y. (2015). Under the spotlight: The organic–inorganic hybrid halide perovskite for optoelectronic applications. *Nano Today*, *10*(3), 355–396. doi:10.1016/j.nantod.2015.04.009

Deng, Z., Dai, Y., Chen, W., Pei, X., & Liao, J. (2010). Synthesis and Characterization of Bowl-Like Single-Crystalline BaTiO(3) Nanoparticles. *Nanoscale Research Letters*, *5*(7), 1217–1221. doi:10.100711671-010-9629-7 PMID:20596350

Fu, J., Zhang, Q., Li, Y., & Wang, H. (2009). Preparation and photoluminescence characteristics of a new promising red NUV phosphor CaTiO3:Eu3+. *Journal of Alloys and Compounds*, *485*(1–2), 418–421. doi:10.1016/j.jallcom.2009.05.128

Gonçalves, R. F., Moura, A. P., Godinho, M. J., Longo, E., Machado, M. A. C., de Castro, D. A., Siu Li, M. S., & Marques, A. P. A. (2015). Crystal growth and photoluminescence of europium-doped strontium titanate prepared by a microwave hydrothermal method. *Ceramics International*, *41*(3), 3549–3554. doi:10.1016/j.ceramint.2014.11.018

Ju, M.-G., Dai, J., Ma, L., & Zeng, X. C. (2017). Perovskite Chalcogenides with Optimal Bandgap and Desired Optical Absorption for Photovoltaic Devices. *Advanced Energy Materials*, *7*(18), 1700216. doi:10.1002/aenm.201700216

Katyayan, S., & Agrawal, S. (2018a). Effect of rare earth doping on optical and spectroscopic characteristics of BaZrO3:Eu3+,Tb3+ perovskites. *Methods and Applications in Fluorescence*, *6*(3), 035002. doi:10.1088/2050-6120/aaae0b PMID:29422457

Katyayan, S., & Agrawal, S. (2018b, September 10). Molten salt synthesis and photoluminescent studies of ATiO3:Eu2+, Yb2+ (A = Ca, Ba, Sr) perovskites phosphors. *Journal of Photonics for Energy*, *8*(3), 1. doi:10.1117/1.JPE.8.036001

Katyayan, S., & Agrawal, S. (2018c). Facile molten salt synthesis, structural, morphological and optical studies of ASiO3: Eu2+, Er3+ (A = Ca, Ba, Sr) perovskites. *Journal of Materials Science Materials in Electronics*, *29*(19), 16609–16629. doi:10.100710854-018-9754-0

Katyayan, S., & Agrawal, S. (2019a). Optical Behavior of CaSiO3:Eu2+, Er3+, BaSiO3:Eu2+, Er3+and SrSiO3:Eu2+, Er3+ Phosphors. *Journal of the Minerals Metals & Materials Society*, *71*(8), 2899–2905. doi:10.100711837-019-03545-1

Katyayan, S., & Agrawal, S. (2019b). Optical behavior and TL kinetics of Eu3+ and Tb3+ doped zirconate thermoluminescent phosphors. *Optical and Quantum Electronics*, *51*(8), 277. Advance online publication. doi:10.100711082-019-1993-7

Katyayan, S., & Agrawal, S. (2020). Low temperature salt melt synthesis, thermal and optical studies of AZrO3:Eu3+,Tb3+ perovskite phosphors. *Optik (Stuttgart)*, *204*, 164125. doi:10.1016/j.ijleo.2019.164125

Kuwabara, M. (2012). Innovative nanocrystal-based technologies for ceramic devices with novel electronic functions. *Journal of Advanced Ceramics*, *1*(2), 79–88. doi:10.100740145-012-0016-y

Lemański, K., & Dereń, P. J. (2014). Spectroscopic properties of Dy3+ ions in CaTiO3 nano-perovskites. *Journal of Luminescence*, *145*, 661–664. doi:10.1016/j.jlumin.2013.08.048

Li, Y., Guo, W., Hao, H., Wang, L., Su, Q., Jin, S., Qin, L., Gao, W., Liu, G., & Hu, Z. (2015). Enhancing photoelectrical performance of dye–sensitized solar cell by doping SrTiO 3:Sm 3+ @SiO 2 core–shell nanoparticles in the photoanode. *Electrochimica Acta, 173*, 656–664. doi:10.1016/j.electacta.2015.05.116

Marí, B., Singh, K. C., Cembrero-Coca, P., Singh, I., Singh, D., & Chand, S. (2013). Red emitting MTiO3 (M=Ca or Sr) phosphors doped with Eu3+ or Pr3+ with some cations as co-dopants. *Displays, 34*(4), 346–351. doi:10.1016/j.displa.2013.07.003

Noto, L. L., Pitale, S. S., Gusowki, M. A., Terblans, J. J., Ntwaeaborwa, O. M., & Swart, H. C. (2013). Afterglow enhancement with In3+ codoping in CaTiO3:Pr3+ red phosphor. *Powder Technology, 237*, 141–146. doi:10.1016/j.powtec.2013.01.029

Shi, Y., Wang, C.-H., & Jing, X. P. (2008)... *Journal of the Electrochemical Society, 155*(6), J148–J151. doi:10.1149/1.2898897

Som, S., Kunti, A. K., Kumar, V., Kumar, V., Dutta, S., Chowdhury, M., Sharma, S. K., Terblans, J. J., & Swart, H. C. (2014). Defect correlated fluorescent quenching and electron phonon coupling in the spectral transition of Eu3+ in CaTiO3 for red emission in display application. *Journal of Applied Physics, 115*(19), 193101. doi:10.1063/1.4876316

Terashi, Y., Purwanto, A., Wang, W., Iskandar, F., & Okuyama, K. (2008). Role of urea addition in the preparation of tetragonal BaTiO3 nanoparticles using flame-assisted spray pyrolysis. *Journal of the European Ceramic Society, 28*(13), 2573–2580. doi:10.1016/j.jeurceramsoc.2008.03.002

Travis, W., Glover, E. N. K., Bronstein, H., Scanlon, D. O., & Palgrave, R. G. (2016). On the application of the tolerance factor to inorganic and hybrid halide perovskites: A revised system. *Chemical Science (Cambridge), 7*(7), 4548–4556. doi:10.1039/C5SC04845A PMID:30155101

Xiao, Z., Zhang, J., Jin, L., Xia, Y., Lei, L., Wang, H., Zhang, J., & Xu, S. (2017). Enhanced up-conversion luminescence intensity in single-crystal SrTiO3: Er3+ nanocubes by codoping with Yb3+ ions. *Journal of Alloys and Compounds, 724*, 139–145. doi:10.1016/j.jallcom.2017.07.024

Yang, J., Geng, B., Ye, Y., & Yu, X. (2012). Stick-like titania precursor route to MTiO3 (M = Sr, Ba, and Ca) polyhedra. *CrystEngComm, 14*(8), 2959–2965. doi:10.1039/c2ce06532k

Compilation of References

Abbad, S., Guergouri, K., Gazaout, S., Djebabra, S., Zertal, A., Barille, R., & Zaabat, M. (2020). Effect of silver doping on the photocatalytic activity of TiO2 nanapowders synthesized by sol-gel route. *Journal of Environmental Chemical Engineering*, *8*(3), 103718. doi:10.1016/j.jece.2020.103718

Abdelhaleem, A., & Chu, M. (2019). Insights into peroxymonosulfate activation for carbofuran degradation under visible LED via a double-component photocatalyst of Fe (III) impregnated N-doped TiO2. *Chemosphere*, *237*, 124487. doi:10.1016/j.chemosphere.2019.124487 PMID:31398605

Abdullah, N., Yusof, N., Lau, W. J., Jaafar, J., & Ismail, A. F. (2019). Recent trends of heavy metal removal from water/wastewater by membrane technologies. *Journal of Industrial and Engineering Chemistry*, *76*, 17–38. doi:10.1016/j.jiec.2019.03.029

Abideen, Z. U., Katoch, A., Kim, J.-H., Kwon, Y. J., Kim, H. W., & Kim, S. S. (2015). Excellent gas detection of ZnO nanofibers by loading with reduced graphene oxide nanosheets. *Sensors and Actuators. B, Chemical*, *221*, 1499–1507. doi:10.1016/j.snb.2015.07.120

Adel, M., Ahmed, M. A., Elabiad, M. A., & Mohamed, A. A. (2022). Removal of heavy metals and dyes from wastewater using graphene oxide-based nanomaterials: A critical review. *Environmental Nanotechnology, Monitoring & Management*, *18*, 100719. doi:10.1016/j.enmm.2022.100719

Afsar, M., Rafiq, M., & Tok, A. (2017). Two-dimensional SnS nanoflakes: Synthesis and application to acetone and alcohol sensors. *RSC Advances*, *7*(35), 21556–21566. doi:10.1039/C7RA03004E

Afzali, M., Mostafavi, A., Nekooie, R., & Jahromi, Z. (2019). A novel voltammetric sensor based on palladium nanoparticles/carbon nanofibers/ionic liquid modified carbon paste electrode for sensitive determination of anti-cancer drug pemetrexed. *Journal of Molecular Liquids*, *282*, 456–465. doi:10.1016/j.molliq.2019.03.041

Agnihotri, S., Mota, J. P. B., Rostam-Abadi, M., & Rood, M. J. (2006). Theoretical and Experimental Investigation of Morphology and Temperature Effects on Adsorption of Organic Vapors in Single-Walled Carbon Nanotubes. *The Journal of Physical Chemistry B*, *110*(15), 7640–7647. doi:10.1021/jp060040a PMID:16610854

Ahmad, H., Kamarudin, S. K., Minggu, L. J., & Kassim, M. (2015). Hydrogen from photo-catalytic water splitting process: A review. *Renewable & Sustainable Energy Reviews, 43*, 599–610. doi:10.1016/j.rser.2014.10.101

Ahmed, M., & Xinxin, G. (2016). A review of metal oxynitrides for photocatalysis. *Inorganic Chemistry Frontiers, 3*(5), 578–590. doi:10.1039/C5QI00202H

Ajay, S., Panicker, J. S., Manjumol, K. A., & Subramanian, P. P. (2022). Photocatalytic activity of biogenic silver nanoparticles synthesized using Coleus Vettiveroids. *Inorganic Chemistry Communications, 144*, 109926. doi:10.1016/j.inoche.2022.109926

Akpan, U. G., & Hameed, B. H. (2009). Parameters affecting the photocatalytic degradation of dyes using TiO2-based photocatalysts: A review. *Journal of Hazardous Materials, 170*(2), 520–529. doi:10.1016/j.jhazmat.2009.05.039 PMID:19505759

Al–Ekabi, H., Serpone, N., Pelizzetti, E., Minero, C., Fox, M. A., & Draper, R. B. (1989). Kinetic studies in heterogeneous photocatalysis. Titania-mediated degradation of 4-chlorophenol in air-equilibrated aqueous media. *Langmuir, 5*(1), 250–255. doi:10.1021/la00085a048

Alhabeb, M., Maleski, K., Mathis, T. S., Sarycheva, A., Hatter, C. B., Uzun, S., Levitt, A., & Gogotsi, Y. (2018). Selective etching of silicon from Ti3SiC2 (MAX) to obtain 2D titanium carbide (MXene). *Angewandte Chemie, 130*(19), 5542–5546. doi:10.1002/ange.201802232 PMID:29518271

Alhaddad, M., Navarro, R. M., Hussein, M. A., & Mohamed, R. M. (2020). Bi_2O_3/ g-C_3N_4 nanocomposites as proficient photocatalysts for hydrogen generation from aqueous glycerol solutions beneath visible light. *Ceramics International, 46*(16), 24873–24881. doi:10.1016/j.ceramint.2020.06.271

Ali, H., Guler, A. C., Masar, M., Urbanek, P., Urbanek, M., Skoda, D., Suly, P., Machovsky, M., Galusek, D., & Kuritka, I. (2021). Solid-State Synthesis of Direct Z-Scheme Cu2O/WO3 Nanocomposites with Enhanced Visible-Light Photocatalytic Performance. *Catalysts, 11*(2), 293. doi:10.3390/catal11020293

Al-Sheddi, E. S., Farshori, N. N., Al-Oqail, M. M., Al-Massarani, S. M., Saquib, Q., Wahab, R., Musarrat, J., Al-Khedhairy, A. A., & Siddiqui, M. A. (2018). Anticancer potential of green synthesized silver nanoparticles using extract of nepeta deflersiana against human cervical cancer cells (HeLA). *Bioinorganic Chemistry and Applications, Nov*(1), 9390784. doi:10.1155/2018/9390784

Alwhibi, M. S., Soliman, D. A., al khaldy, H., Alonaizan, A., Abdulhaq Marraiki, N., El-Zaidy, M., & AlSubeie, M. S. (2020). Green biosynthesis of silver nanoparticle using Commiphora myrrh extract and evaluation of their antimicrobial activity and colon cancer cells viability. *Journal of King Saud University. Science, 32*(8), 3372–3379. doi:10.1016/j.jksus.2020.09.024

Amani, A., Begdelo, J. M., Yaghoubi, H., & Motallebinia, S. (2019). Multifunctional magnetic nanoparticles for controlled release of anticancer drug, breast cancer cell targeting, MRI/ fluorescence imaging, and anticancer drug delivery. *Journal of Drug Delivery Science and Technology*, *49*, 534–546. doi:10.1016/j.jddst.2018.12.034

Amiri, S., & Anbia, M. (2023). Enhanced degradation of diazinon in aqueous solution using C-TiO2/ g-C3N4 nanocomposite under visible light: Synthesis, characterization, kinetics, and mechanism studies. *Materials Research Bulletin*, *165*, 112289. doi:10.1016/j.materresbull.2023.112289

Anjum, M., Miandad, R., Waqas, M., Gehany, F., & Barakat, M. A. (2019). Remediation of wastewater using various nano-materials. *Arabian Journal of Chemistry*, *12*(8), 4897–4919. doi:10.1016/j.arabjc.2016.10.004

Ansari, S., Khan, M., Ansari, M., & Cho, M. (2013). Nitrogen-doped titanium dioxide (N-doped TiO2) for visible light photocatalysis. *New Journal of Chemistry*, *00*, 1–3.

Aragaw, B. A. (2020). Reduced graphene oxide-intercalated graphene oxide nano-hybrid for enhanced photoelectrochemical water reduction. *Journal of Nanostructure in Chemistry*, *10*(1), 9–18. doi:10.100740097-019-00324-x

Aragaw, T. A., Bogale, F. M., & Aragaw, B. A. (2021). Iron-based nanoparticles in wastewater treatment: A review on synthesis methods, applications, and removal mechanisms. *Journal of Saudi Chemical Society*, *25*(8), 101280. doi:10.1016/j.jscs.2021.101280

Araújo, R., Castro, A. C. M., & Fiúza, A. (2015). The Use of Nanoparticles in Soil and Water Remediation Processes. *Materials Today: Proceedings*, *2*(1), 315–320. doi:10.1016/j.matpr.2015.04.055

Aravind, M., Ahmad, A., Ahmad, I., Amalanathan, M., Naseem, K., Mary, S. M. M., Parvathiraja, C., Hussain, S., Algarni, T. S., Pervaiz, M., & Zuber, M. (2021). Critical green routing synthesis of silver NPs using jasmine flower extract for biological activities and photocatalytical degradation of methylene blue. *Journal of Environmental Chemical Engineering*, *9*(1), 104877. doi:10.1016/j.jece.2020.104877

Arboleda, D. M., Santillán, J. M. J., Arce, V. B., Fernández van Raap, M. B., Muraca, D., Fernández, M. A., Torres Sanchez, R. M., Schinca, D. C., & Scaffardi, L. B. (2018). A simple and "green" technique to synthesize long-term stability colloidal Ag nanoparticles: Fs laser ablation in a biocompatible aqueous medium. *Materials Characterization*, *140*, 320–332. doi:10.1016/j.matchar.2018.04.021

Aryan, R., Ruby, & Mehata, M. S. (2021). Green synthesis of silver nanoparticles using Kalanchoe pinnata leaves (life plant) and their antibacterial and photocatalytic activities. *Chemical Physics Letters*, *778*, 138760. doi:10.1016/j.cplett.2021.138760

Ashari, I. A., Widodo, A. P., & Suryono, S. (2019). The Monitoring System for Ammonia Gas (NH3) Hazard Detection in the Livestock Environment uses Inverse Distance Weight Method. Paper presented at the *2019 Fourth International Conference on Informatics and Computing (ICIC)*. IEEE. 10.1109/ICIC47613.2019.8985673

Ashique, S., Upadhyay, A., Hussain, A., Bag, S., Chaterjee, D., Rihan, M., Mishra, N., Bhatt, S., Puri, V., Sharma, A., Prasher, P., Singh, S. K., Chellappan, D. K., Gupta, G., & Dua, K. (2022). Green biogenic silver nanoparticles, therapeutic uses, recent advances, risk assessment, challenges, and future perspectives. *Journal of Drug Delivery Science and Technology*, *77*, 103876. doi:10.1016/j.jddst.2022.103876

Attatsi, I. K., & Nsiah, F. (2020). Application of silver nanoparticles toward Co(II) and Pb(II) ions contaminant removal in groundwater. *Applied Water Science*, *10*(6), 152. doi:10.100713201-020-01240-0

Avouris, P., & Dimitrakopoulos, C. (2012). Graphene: Synthesis and applications. *Materials Today*, *15*(3), 86–97. doi:10.1016/S1369-7021(12)70044-5

Awad, M. A. A., Hendi, A., Ortashi, K. M., Alzahrani, B., Soliman, D., Alanazi, A., Alenazi, W., Taha, R. M., Ramadan, R., El-Tohamy, M., AlMasoud, N., & Alomar, T. S. (2021). Biogenic synthesis of silver nanoparticles using Trigonella foenum-graecum seed extract: Characterization, photocatalytic and antibacterial activities. *Sensors and Actuators. A, Physical*, *323*, 112670. doi:10.1016/j.sna.2021.112670

Ayyala, S. K., & Covington, J. A. (2021). Nickel-oxide based thick-film gas sensors for volatile organic compound detection. *Chemosensors (Basel, Switzerland)*, *9*(9), 247. doi:10.3390/chemosensors9090247

Badruddoza, A. Z. M., Shawon, Z. B. Z., Tay, W. J. D., Hidajat, K., & Uddin, M. S. (2013). Fe3O4/cyclodextrin polymer nanocomposites for selective heavy metals removal from industrial wastewater. *Carbohydrate Polymers*, *91*(1), 322–332. doi:10.1016/j.carbpol.2012.08.030 PMID:23044139

Baek, J. H., Kim, B. J., Han, G. S., Hwang, S. W., Kim, D. R., Cho, I. S., & Jung, H. S. (2017). $BiVO_4/WO_3/SnO_2$ double-heterojunction photoanode with enhanced charge separation and visible-transparency for bias-free solar water-splitting with a perovskite solar cell. *ACS Applied Materials & Interfaces*, *9*(2), 1479–1486. doi:10.1021/acsami.6b12782 PMID:27989115

Bag, A., & Lee, N. E. (2021). Recent advancements in development of wearable gas sensors. *Advanced Materials Technologies*, *6*(3), 2000883. doi:10.1002/admt.202000883

Balamurugan, V., Muruganandam, L., & Natarajan, R. (2022). Photocatalytic activity of silver nanoparticles synthesized using cadiospermum halicacabum and moringa oleifera leaf extract. *Materials Today: Proceedings*, *62*, 5365–5370. doi:10.1016/j.matpr.2022.03.540

Balandeh, M., Mezzetti, A., Tacca, A., Leonardi, S., Marra, G., Divitini, G., Ducati, C., Meda, L., & Di Fonzo, F. (2015). Quasi-1D hyperbranched WO_3 nanostructures for low-voltage photoelectrochemical water splitting. *Journal of Materials Chemistry. A, Materials for Energy and Sustainability*, *3*(11), 6110–6117. doi:10.1039/C4TA06786J

Baranowska-Wojak, E., Szwajgeir, D., Oleszczuk, P., & Winiarsk-Mieczan, A. (2019). Effects of Titanium Dioxide Nanoparticles Exposure to Human Health-A Review. *Biological Trace Element Research*, *193*(1), 118–129. doi:10.100712011-019-01706-6 PMID:30982201

Barbero, N., & Vione, D. (2016). Why Dyes Should Not Be Used to Test the Photocatalytic Activity of Semiconductor Oxides. *Environmental Science & Technology*, *50*(5), 2130–2131. doi:10.1021/acs.est.6b00213 PMID:26886856

Bard, A. J. (1979). Photoelectrochemistry and heterogeneous photo-catalysis at semiconductors. *Journal of Photochemistry*, *10*(1), 59–75. doi:10.1016/0047-2670(79)80037-4

Barreto, J. C., Smith, G. S., Strobel, N. H. P., McQuillin, P. A., & Miller, T. A. (1995). Terephthalic acid: A dosimeter for the detection of hydroxyl radicals in vitro. *Life Sciences*, *56*(4), 89–96. doi:10.1016/0024-3205(94)00925-2 PMID:7823778

Baruah, A., Chaudhary, V., Malik, R., & Tomer, V. K. (2019). 17 - Nanotechnology Based Solutions for Wastewater Treatment. In A. Ahsan & A. F. Ismail (Eds.), *Nanotechnology in Water and Wastewater Treatment* (pp. 337–368). Elsevier. doi:10.1016/B978-0-12-813902-8.00017-4

Bassyouni, M., Abdel-Aziz, M. H., Zoromba, M. S., Abdel-Hamid, S. M. S., & Drioli, E. (2019). A review of polymeric nanocomposite membranes for water purification. *Journal of Industrial and Engineering Chemistry*, *73*, 19–46. doi:10.1016/j.jiec.2019.01.045

Basumatary, S., Kumar, K. J., Daimari, J., Mondal, A., Kalita, S., Dey, K. S., & Deka, A. K. (2023). Biosynthesis of silver nanoparticles using Antidesma acidum leaf extract: Its application in textile organic dye degradation. *Environmental Nanotechnology, Monitoring & Management*, *19*, 100769. doi:10.1016/j.enmm.2022.100769

Battisha, I. K., Speghini, A., Polizzi, S., Agnoli, F., & Bettinelli, M. (2002). Molten chloride synthesis, structural characterisation and luminescence spectroscopy of ultrafine Eu3+-doped BaTiO3 and SrTiO3. *Materials Letters*, *57*(1), 183–187. doi:10.1016/S0167-577X(02)00727-9

Bello, M., Raman, A., & Asghar, A. (2019). A review on approaches for addressing the limitations of Fenton oxidation for recacitrant wastewater. *Process Safety and Environmental Protection*, *126*, 119–140. doi:10.1016/j.psep.2019.03.028

Bhaumik, M., Choi, H. J., McCrindle, R. I., & Maity, A. (2014). Composite nanofibers prepared from metallic iron nanoparticles and polyaniline: High performance for water treatment applications. *Journal of Colloid and Interface Science*, *425*, 75–82. doi:10.1016/j.jcis.2014.03.031 PMID:24776666

Bhaumik, M., Leswifi, T. Y., Maity, A., Srinivasu, V. V., & Onyango, M. S. (2011). Removal of fluoride from aqueous solution by polypyrrole/Fe3O4 magnetic nanocomposite. *Journal of Hazardous Materials*, *186*(1), 150–159. doi:10.1016/j.jhazmat.2010.10.098 PMID:21112695

Bhaumik, M., Maity, A., Srinivasu, V. V., & Onyango, M. S. (2012). Removal of hexavalent chromium from aqueous solution using polypyrrole-polyaniline nanofibers. *Chemical Engineering Journal*, *181-182*, 323–333. doi:10.1016/j.cej.2011.11.088

Bi, S., & Ahmad, N. (2022). Green synthesis of palladium nanoparticles and their biomedical applications. *Materials Today: Proceedings*, *62*, 3172–3177. doi:10.1016/j.matpr.2022.03.441

Bisht, M. (2019). Nanomaterials for Removal of Toxic Metals Ions from the Water. In A. H. Bhat, I. Khan, M. Jawaid, F. O. Suliman, H. Al-Lawati, & S. M. Al-Kindy (Eds.), *Nanomaterials for Healthcare, Energy and Environment* (pp. 159–174). Springer Singapore. doi:10.1007/978-981-13-9833-9_8

Biswas, S. K., & Baeg, J. O. (2013). A facile one-step synthesis of single crystalline hierarchical WO_3 with enhanced activity for photoelectrochemical solar water oxidation. *International Journal of Hydrogen Energy*, *38*(8), 3177–3188. doi:10.1016/j.ijhydene.2012.12.114

Blanco-Novoa, O., Fernández-Caramés, T. M., Fraga-Lamas, P., & Castedo, L. (2018). A cost-effective IoT system for monitoring indoor radon gas concentration. *Sensors (Basel)*, *18*(7), 2198. doi:10.339018072198 PMID:29986540

Borini, S., White, R., Wei, D., Astley, M., Haque, S., Spigone, E., Harris, N., Kivioja, J., & Ryhanen, T. (2013). Ultrafast graphene oxide humidity sensors. *ACS Nano*, *7*(12), 11166–11173. doi:10.1021/nn404889b PMID:24206232

Bose, S., Maity, S., & Sarkar, A. (2022). Nano-materials as biosensor for heavy metal detection. *Food, Medical, and Environmental Applications of Nanomaterials*, 493–526. doi:10.1016/B978-0-12-822858-6.00018-2

Bradder, P., Ling, S. K., Wang, S., & Liu, S. (2011). Dye Adsorption on Layered Graphite Oxide. *Journal of Chemical & Engineering Data*, *56*(1), 138–141. doi:10.1021/je101049g

Brattain, W. H., & Garrett, C. G. B. (1955). Experiments on the interface between germanium and an electrolyte. *The Bell System Technical Journal*, *34*(1), 129–176. doi:10.1002/j.1538-7305.1955.tb03766.x

Bullett, D. W. (1983). Bulk and surface electron states in WO_3 and tungsten bronzes. *Journal of Physics. C. Solid State Physics*, *16*(11), 2197–2207. doi:10.1088/0022-3719/16/11/022

Bushra, R., Shahadat, M., Ahmad, A., Nabi, S. A., Umar, K., Oves, M., Raeissi, A. S., & Muneer, M. (2014). Synthesis, characterization, antimicrobial activity and applications of polyanilineTi(IV) arsenophosphate adsorbent for the analysis of organic and inorganic pollutants. *Journal of Hazardous Materials*, *264*, 481–489. doi:10.1016/j.jhazmat.2013.09.044 PMID:24238807

Butterfield, D. (2021). *Market review of gas sensors for industrial applications.*

Cahen, D., Hodes, G., & Manassen, J. (1976). Tungsten trioxide as a photoanode for a photoelectrochemical cell (PEC). *Nature*, 260–312.

Çetin Uyanıkgil, E. Ö., & Selcen Salmanoğlu, D. (2020). Metalik nanopartiküllerin hedeflendirilmesi. *Ege Tıp Dergisi*, *59*(1), 71–81. doi:10.19161/etd.698596

Chae, S. Y., Jung, H. J., Jeon, H. S., Min, B. K., Hwang, Y. J., & Joo, O. S. (2014). Morphology control of one-dimensional heterojunctions for highly efficient photoanodes used for solar water splitting. *Journal of Materials Chemistry. A, Materials for Energy and Sustainability*, *2*(29), 11408–11416. doi:10.1039/c4ta00702f

Chae, S. Y., Lee, C. S., Jung, H., Joo, O. S., Min, B. K., Kim, J. H., & Hwang, Y. J. (2017). Insight into charge separation in $WO_3/BiVO_4$ heterojunction for solar water splitting. *ACS Applied Materials & Interfaces, 9*(23), 19780–19790. doi:10.1021/acsami.7b02486 PMID:28530789

Chandra, D., Saito, K., Yui, T., & Yagi, M. (2013). Crystallization of tungsten trioxide having small mesopores: Highly efficient photoanode for visible-light-driven water oxidation. *Angewandte Chemie International Edition, 52*(48), 12606–12609. doi:10.1002/anie.201306004 PMID:24123708

Chatchai, P., Murakami, Y., Kishioka, S. Y., Nosaka, A. Y., & Nosaka, Y. (2009). Efficient photocatalytic activity of water oxidation over $WO_3/BiVO_4$ composite under visible light irradiation. *Electrochimica Acta, 54*(3), 1147–1152. doi:10.1016/j.electacta.2008.08.058

Chaturvedi, V. K., Kushwaha, A., Maurya, S., Tabassum, N., Chaurasia, H., & Singh, M. P. (2020). Wastewater Treatment Through Nanotechnology: Role and Prospects. In A. K. Upadhyay, R. Singh, & D. P. Singh (Eds.), *Restoration of Wetland Ecosystem: A Trajectory Towards a Sustainable Environment* (pp. 227–247). Springer Singapore. doi:10.1007/978-981-13-7665-8_14

Chavali, M. S., & Nikolova, M. P. (2019). Metal oxide nanoparticles and their applications in nanotechnology. *SN Applied Sciences, 1*(6), 1–30. doi:10.1007/s42452-019-0592-3

Chavan, N., Dharmaraj, D., Sarap, S., & Surve, C. (2022). Magnetic nanoparticles – A new era in nanotechnology. *Journal of Drug Delivery Science and Technology, 77*, 103899. doi:10.1016/j.jddst.2022.103899

Checkol, F., Elfwing, A., Greczynski, G., Mehretie, S., Inganäs, O., & Admassie, S. (2018). Highly Stable and Efficient Lignin-PEDOT/PSS Composites for Removal of Toxic Metals. *Advanced Sustainable Systems, 2*(1), 1700114. doi:10.1002/adsu.201700114

Che, G. L., Lakshmi, B. B., Martin, C. R., & Fisher, E. R. (1999). Metal-nanocluster-filled carbon nanotubes: Catalytic properties and possible applications in electrochemical energy storage and production. *Langmuir, 15*(3), 750–758. doi:10.1021/la980663i

Cheng, F.-Y., Su, C.-H., Yang, Y.-S., Yeh, C.-S., Tsai, C.-Y., Wu, C.-L., ... Shieh, D.-B. (2005). Characterization of aqueous dispersions of Fe3O4 nanoparticles and their biomedical applications. *Biomaterials, 26*(7), 729–738. doi:10.1016/j.biomaterials.2004.03.016 PMID:15350777

Chen, J., Gu, J., Zhang, R., Mao, Y., & Tian, S. (2019). Freshness evaluation of three kinds of meats based on the electronic nose. *Sensors (Basel), 19*(3), 605. doi:10.339019030605 PMID:30709028

Chen, K., Yuan, D., & Zhao, Y. (2021). Review of optical hydrogen sensors based on metal hydrides: Recent developments and challenges. *Optics & Laser Technology, 137*, 106808. doi:10.1016/j.optlastec.2020.106808

Chen, M., Jiang, W., Wang, F., Shen, P., Ma, P., Gu, J., Mao, J., & Li, F. (2013). Synthesis of highly hydrophobic floating magnetic polymer nanocomposites for the removal of oils from water surface. *Applied Surface Science, 286*, 249–256. doi:10.1016/j.apsusc.2013.09.059

Chen, Q., De Marco, N., Yang, Y., Song, T.-B., Chen, C.-C., Zhao, H., Hong, Z., Zhou, H., & Yang, Y. (2015). Under the spotlight: The organic–inorganic hybrid halide perovskite for optoelectronic applications. *Nano Today*, *10*(3), 355–396. doi:10.1016/j.nantod.2015.04.009

Chen, Q., Li, J., Zhou, B., Long, M., Chen, H., Liu, Y., Cai, W., & Shangguan, W. (2012). Preparation of well-aligned WO_3 nanoflake arrays vertically grown on tungsten substrate as photoanode for photoelectrochemical water splitting. *Electrochemistry Communications*, *20*, 153–156. doi:10.1016/j.elecom.2012.03.043

Chen, W. Y., Yen, C.-C., Xue, S., Wang, H., & Stanciu, L. A. (2019). Surface functionalization of layered molybdenum disulfide for the selective detection of volatile organic compounds at room temperature. *ACS Applied Materials & Interfaces*, *11*(37), 34135–34143. doi:10.1021/acsami.9b13827 PMID:31453680

Chen, X., Leishman, M., Bagnall, D., & Nasiri, N. (2021). Nanostructured gas sensors: From air quality and environmental monitoring to healthcare and medical applications. *Nanomaterials (Basel, Switzerland)*, *11*(8), 1927. doi:10.3390/nano11081927 PMID:34443755

Chen, X., Liu, Q., Wu, Q., Du, P., Zhu, J., Dai, S., & Yang, S. (2016). Incorporating graphitic carbon nitride (g-C_3N_4) quantum dots into bulk heterojunction polymer solar cells leads to efficiency enhancement. *Advanced Functional Materials*, *26*(11), 1719–1728. doi:10.1002/adfm.201505321

Chen, X., Zhang, Z., Chi, L., Nair, A. K., Shangguan, W., & Jiang, Z. (2016). Recent advances in visible light driven photoelectrochemical water splitting: Catalyst nanostructures and reaction systems. *Nano-Micro Letters*, *8*(1), 1–12. doi:10.100740820-015-0063-3 PMID:30464988

Chen, Y., Chen, L., Bai, H., & Li, L. (2013). Graphene oxide–chitosan composite hydrogels as broad-spectrum adsorbents for water purification. *Journal of Materials Chemistry. A, Materials for Energy and Sustainability*, *1*(6), 1992–2001. doi:10.1039/C2TA00406B

Chen, Z., Chen, Z., Song, Z., Ye, W., & Fan, Z. (2019). Smart gas sensor arrays powered by artificial intelligence. *Journal of Semiconductors*, *40*(11), 111601. doi:10.1088/1674-4926/40/11/111601

Chen, Z., Ning, G., Ma, G., Meng, Q., Zhang, Y., Gao, J., Jin, M., Chen, Z., Yuan, M., Wang, X., Liu, J.-M., & Zhou, G. (2017). Effective silicon nanowire arrays/WO_3 core/shell photoelectrode for neutral pH water splitting. *Nanotechnology*, *28*(27), 275401. doi:10.1088/1361-6528/aa749d PMID:28531092

Chiarello, G. L., Aguirre, M. H., & Selli, E. (2010). Hydrogen production by photocatalytic steam reforming of methanol on noble metal modified TiO_2. *Journal of Catalysis*, *273*(2), 182–190. doi:10.1016/j.jcat.2010.05.012

Chiarello, G. L., Bernareggi, M., Pedroni, M., Magni, M., Pietralunga, S. M., Tagliaferri, A., Vassallo, E., & Selli, E. (2017). Enhanced photopromoted electron transfer over a bilayer WO_3 n-n heterojunction prepared by RF diode sputtering. *Journal of Materials Chemistry. A, Materials for Energy and Sustainability*, *5*(25), 12977–12989. doi:10.1039/C7TA03887A

Chiu, S.-W., & Tang, K.-T. (2013). Towards a chemiresistive sensor-integrated electronic nose: A review. *Sensors (Basel)*, *13*(10), 14214–14247. doi:10.3390131014214 PMID:24152879

Chiu, Y.-H., Chang, T.-F. M., Chen, C.-Y., Sone, M., & Hsu, Y.-J. (2019). Mechanistic Insights into Photodegradation of Organic Dyes Using Heterostructure Photocatalysts. *Catalysts*, *9*(5), 430. Advance online publication. doi:10.3390/catal9050430

Choi, S. H., Ko, Y. N., Lee, J. K., & Kang, Y. C. (2015). 3D MoS2–graphene microspheres consisting of multiple nanospheres with superior sodium ion storage properties. *Advanced Functional Materials*, *25*(12), 1780–1788. doi:10.1002/adfm.201402428

Choi, W., Choudhary, N., Han, G. H., Park, J., Akinwande, D., & Lee, Y. H. (2017). Recent development of two-dimensional transition metal dichalcogenides and their applications. *Materials Today*, *20*(3), 116–130. doi:10.1016/j.mattod.2016.10.002

Choi, Y. W., Kim, S., Seong, M., Yoo, H., & Choi, J. (2015). NH_4-doped anodic WO_3 prepared through anodization and subsequent NH_4OH treatment for water splitting. *Applied Surface Science*, *324*, 414–418. doi:10.1016/j.apsusc.2014.10.059

Chowdhury, S., & Balasubramanian, R. (2014). Recent advances in the use of graphene-family nanoadsorbents for removal of toxic pollutants from wastewater. *Advances in Colloid and Interface Science*, *204*, 35–56. doi:10.1016/j.cis.2013.12.005 PMID:24412086

Chu, B. H., Nicolosi, J., Lo, C., Strupinski, W., Pearton, S., & Ren, F. (2011). Effect of coated platinum thickness on hydrogen detection sensitivity of graphene-based sensors. *Electrochemical and Solid-State Letters*, *14*(7), K43. doi:10.1149/1.3589250

Comini, E., Faglia, G., Sberveglieri, G., Pan, Z., & Wang, Z. L. (2002). Stable and highly sensitive gas sensors based on semiconducting oxide nanobelts. *Applied Physics Letters*, *81*(10), 1869–1871. doi:10.1063/1.1504867

Covaliu, C. I., Berger, D., Matei, C., Diamandescu, L., Vasile, E., Cristea, C., Ionita, V., & Iovu, H. (2011). Magnetic nanoparticles coated with polysaccharide polymers for potential biomedical applications. *Journal of Nanoparticle Research*, *13*(11), 6169–6180. doi:10.100711051-011-0452-6

Cravanzola, S., Cesano, F., Geziano, F., & Scarano, D. (2017). Sulfur-doped TiO2: Structure and surface properties. *Catalysis*, *7*(7), 214.

Cui, Y., Barford, J. P., & Renneberg, R. (2007). Development of an interference-free biosensor for glucose-6-phosphate using a bienzyme-based Clark-type electrode. *Sensors and Actuators. B, Chemical*, *123*(2), 696–700. doi:10.1016/j.snb.2006.10.004

Cui, Y., Nengzi, L., Gou, J., Huang, Y., Li, B., & Cheng, X. (2020). Fabrication of dual Z-scheme MIL-53(Fe)/α-Bi_2O_3/g-C_3N_4 ternary composite with enhanced visible light photocatalytic performance. *Separation and Purification Technology*, *232*, 115959. doi:10.1016/j.seppur.2019.115959

Dağlıoğlu, C. (2018). A Comparative Study of Passive and Active Tumor Targeting Using Nanoparticles as Drug Delivery Systems. *Academic Platform - Journal of Engineering and Science*, *6*(1), 1–7. doi:10.21541/apjes.349889

Dahl, M., Liu, Y., & Yin, X. (2014). Composite Titanium Dioxide Nanomaterials. *Chemical Reviews*, *114*(19), 9853–9889. doi:10.1021/cr400634p PMID:25011918

Das, I., & Ansari, S. A. (2009). Nanomaterials in science and technology. *Journal of Scientific and Industrial Research*, *68*(8), 657–667.

Das, S., Samanta, S., Maji, S. K., Samanta, P. K., Dutta, A. K., Srivastava, D. N., Adhikary, B., & Biswas, P. (2013). Visible-light-driven synthesis of 2-substituted benzothiazoles using CdS Tet. *Lett.*, *54*(3), 1090–1096. doi:10.1016/j.tetlet.2012.12.044

Das, S., Sharma, S., & Sharma, S. K. (2019). Facile Synthesis of 2D-HfS 2 Flakes for $\mu $-IDE-Based Methanol Sensor: Fast Detection at Room Temperature. *IEEE Sensors Journal*, *19*(20), 9090–9096. doi:10.1109/JSEN.2019.2925027

Dave, P. N., & Chopda, L. V. (2014). Application of Iron Oxide Nanomaterials for the Removal of Heavy Metals. *Journal of Nanotechnology*, *398569*, 1–14. doi:10.1155/2014/398569

Debska, J., Kot-Wasik, A., & Namiesnik, J. (2004). Fate and analysis of pharmaceutical residues in aquatic environment. *critical. Critical Reviews in Analytical Chemistry*, *34*(1), 51–67. doi:10.1080/10408340490273753

Dede, S., & Altay, F. (2018). Biosensors from the First Generation to Nano-biosensors. *International Advanced Researches and Engineering Journal*, *02*(02), 200–207.

Deng, Y., Tang, L., Zeng, G., Zhu, Z., Yan, M., Zhou, Y., Wang, J., Liu, Y., & Wang, J. (2017). Insight into highly efficient simultaneous photocatalytic removal of Cr(VI) and 2,4-diclorophenol under visible light irradiation by phosphorus doped porous ultrathin g-C_3N_4 nano-sheets from aqueous media: Performance and reaction mechanism. *Applied Catalysis B: Environmental*, *203*, 343–354. doi:10.1016/j.apcatb.2016.10.046

Deng, Z., Dai, Y., Chen, W., Pei, X., & Liao, J. (2010). Synthesis and Characterization of Bowl-Like Single-Crystalline BaTiO(3) Nanoparticles. *Nanoscale Research Letters*, *5*(7), 1217–1221. doi:10.100711671-010-9629-7 PMID:20596350

Devi, L., & Kavitha, R. (2013). A review of non metal ion doped titania for the catalytic degradation of organic pollutants under UV/solar light: Role of photogenerated charge carrier dynamics in enhancing the activity. *Applied Catalysis B: Environmental*, *140-141*, 559–587. doi:10.1016/j.apcatb.2013.04.035

Di Valentin, C., & Pacchioni, G. (2014). Spectroscopic properties of doped and defective semiconducting oxides from hybrid density functional calculations. *Accounts of Chemical Research*, *47*(11), 3233–3241. doi:10.1021/ar4002944 PMID:24828320

Diaz-Sanchez, M., Renones, P., Mena-Palomo, I., Lopez-Coleazo, E., Fresno, F., Oropeza, F., & Gomez-Ruiz, S. (2021). Ionic liquid assisted synthesis of F-doped titanium dioxide nanomaterials with high surface area for multifunctional catalytic and photocatalytic applications. *Applied Catalysis A, General*, *613*, 118029. doi:10.1016/j.apcata.2021.118029

Diehl, R., Brandt, G., & Saije, E. (1978). The crystal structure of triclinic WO_3. *Acta Crystallographica. Section B, Structural Crystallography and Crystal Chemistry, 34*(4), 1105–1111. doi:10.1107/S0567740878005014

Dilshad, E., Bibi, M., Sheikh, N. A., Tamrin, K. F., Mansoor, Q., Maqbool, Q., & Nawaz, M. (2020). Synthesis of functional silver nanoparticles and microparticles with modifiers and evaluation of their antimicrobial, anticancer, and antioxidant activity. *Journal of Functional Biomaterials, 11*(4), 76. doi:10.3390/jfb11040076 PMID:33113975

Ding, X., Huang, Y., Li, S., Zhang, N., & Wang, J. (2016). 3D architecture reduced graphene oxide-MoS2 composite: Preparation and excellent electromagnetic wave absorption performance. *Composites. Part A, Applied Science and Manufacturing, 90*, 424–432. doi:10.1016/j.compositesa.2016.08.006

Donarelli, M., & Ottaviano, L. (2018). 2D materials for gas sensing applications: A review on graphene oxide, MoS2, WS2 and phosphorene. *Sensors (Basel), 18*(11), 3638. doi:10.339018113638 PMID:30373161

Dresselhaus, M. S., & Terrones, M. (2013). Carbon-Based Nanomaterials From a Historical Perspective. *Proceedings of the IEEE, 101*(7), 1522–1535. doi:10.1109/JPROC.2013.2261271

Dubale, A. A., Su, W. N., Tamirat, A. G., Pan, C. J., Aragaw, B. A., Chen, H. M., Chen, C. H., & Hwang, B. J. (2014). The synergetic effect of graphene on Cu_2O nanowire arrays as a highly efficient hydrogen evolution photocathode in water splitting. *Journal of Materials Chemistry. A, Materials for Energy and Sustainability, 2*(43), 18383–18397. doi:10.1039/C4TA03464C

Du, H., Liu, W., Zhang, M., Si, C., Zhang, X., & Li, B. (2019). Cellulose nanocrystals and cellulose nanofibrils based hydrogels for biomedical applications. *Carbohydrate Polymers, 209*, 130–144. doi:10.1016/j.carbpol.2019.01.020 PMID:30732792

Duonghong, D., Borgarello, E., & Graetzel, M. (1981). Dynamics of light-induced water cleavage in colloidal systems. *Journal of the American Chemical Society, 103*(16), 4685–4690. doi:10.1021/ja00406a004

Dutta, A. K., Maji, S. K., & Adhikary, B. (2014). γ–Fe_2O_3 nanoparticles: An easily recoverable effective photo-catalyst for the degradation of rose bengal and methylene blue dyes in the waste-water treatment plant. *Materials Research Bulletin, 49*, 28–34. doi:10.1016/j.materresbull.2013.08.024

Dutta, A. K., Maji, S. K., Srivastava, D. N., Mondal, A., Biswas, P., Paul, P., & Adhikary, B. (2012). Synthesis of FeS and FeSe Nanoparticles from a Single Source Precursor: A Study of Their Photocatalytic Activity, Peroxidase-Like Behavior, and Electrochemical Sensing of H_2O_2. *ACS Applied Materials & Interfaces, 4*(4), 1919–1927. doi:10.1021/am300408r PMID:22448782

Eduardo, E. P.-R. Miguel de la, L.-A., Ana, L. M.-H., & Carlos, V.-S. (2016). Graphene Materials to Remove Organic Pollutants and Heavy Metals from Water: Photocatalysis and Adsorption. In C. Wenbin (Ed.), Semiconductor Photocatalysis. Rijeka: IntechOpen.

Elimelech, M. (2006). The global challenge for adequate and safe water. *Journal of Water Supply: Research & Technology - Aqua, 55*(1), 3–10. doi:10.2166/aqua.2005.064

El-Naka, M. A., El-Dissouky, A., Ali, G. Y., Ebrahim, S., & Shokry, A. (2023). Garlic capped silver nanoparticles for rapid detection of cholesterol. *Talanta, 253*, 123908. doi:10.1016/j.talanta.2022.123908 PMID:36087411

Fallah Vostakola, M., Salamatinia, B., & Amini Horri, B. (2022). A Review on Recent Progress in the Integrated Green Hydrogen Production Processes. *Energies, 15*(3), 1209. doi:10.3390/en15031209

Fan, G., Ma, Z., Li, X., & Deng, L. (2021). Coupling of Bi_2O_3 nanoparticles with g-C_3N_4 for enhanced photocatalytic degradation of methylene blue. *Ceramics International, 47*(4), 5758–5766. doi:10.1016/j.ceramint.2020.10.162

Fan, J., Zhang, S., Li, F., Yang, Y., & Du, M. (2020). Recent advances in cellulose-based membranes for their sensing applications. *Cellulose (London, England), 27*(16), 9157–9179. doi:10.100710570-020-03445-7 PMID:32934443

Fan, L., Ge, H., Zhang, S.-y., Zhang, H., & Zhu, J. (2012). Optimization of sensitivity induced by surface conductivity and sorbed mass in surface acoustic wave gas sensors. *Sensors and Actuators. B, Chemical, 161*(1), 114–123. doi:10.1016/j.snb.2011.09.077

Fan, L., Luo, C., Sun, M., Li, X., & Qiu, H. (2013). Highly selective adsorption of lead ions by water-dispersible magnetic chitosan/graphene oxide composites. *Colloids and Surfaces. B, Biointerfaces, 103*, 523–529. doi:10.1016/j.colsurfb.2012.11.006 PMID:23261576

Fan, N., Li, P., Wang, J., Gongsun, X., Xue, L., Bai, J., Morovvati, H., & Goorani, S. (2022). Novel formulation, characterization, cytotoxicity, antioxidant, and anti-lung cancer activities of silver nanoparticles green-formulated by plant extract. *Inorganic Chemistry Communications, 143*, 109707. doi:10.1016/j.inoche.2022.109707

Fan, X., Gao, B., Wang, T., Huang, X., Gong, H., Xue, H., Guo, H., Song, L., Xia, W., & He, J. (2016). Layered double hydroxide modified WO_3 nanorod arrays for enhanced photoelectrochemical water splitting. *Applied Catalysis A, General, 528*, 52–58. doi:10.1016/j.apcata.2016.09.014

Farsalinos, K. E., Gillman, I. G., Hecht, S. S., Polosa, R., & Thornburg, J. (2016). *Analytical assessment of e-cigarettes: From contents to chemical and particle exposure profiles*. Elsevier.

Fatima, T., Bansal, S., Husain, S., & Khanuja, M. (2022). Biosensors. *Electrochemical Sensors: From Working Electrodes to Functionalization and Miniaturized Devices*, 1–30. doi:10.1016/B978-0-12-823148-7.00001-5

Feng, C., Wang, S., & Geng, B. (2014). Ti(IV) doped WO_3 nanocuboids: Fabrication and enhanced visible-light-driven photocatalytic performance. *Nanoscale, 3*(9), 3695. doi:10.1039/c1nr10460h PMID:21785781

Feng, S., Farha, F., Li, Q., Wan, Y., Xu, Y., Zhang, T., & Ning, H. (2019). Review on smart gas sensing technology. *Sensors (Basel), 19*(17), 3760. doi:10.339019173760 PMID:31480359

Fina, F., Callear, S. K., Carins, G. M., & Irvine, J. T. S. (2015). Structural investigation of graphitic carbon nitride via XRD and neutron diffraction. *Chemistry of Materials*, *27*(7), 2612–2618. doi:10.1021/acs.chemmater.5b00411

Forouzandeh, P., Kumaravel, V., & Pillai, S. C. (2020). Electrode materials for supercapacitors: A review of recent advances. *Catalysts*, *10*(9), 969. doi:10.3390/catal10090969

Fugishima, A., Zhang, X., & Tyrk, D. (2008). TiO2 Photocatalysis and related surface phenomena. *Surface Science Reports*, *63*(12), 515–582. doi:10.1016/j.surfrep.2008.10.001

Fu, J., Zhang, Q., Li, Y., & Wang, H. (2009). Preparation and photoluminescence characteristics of a new promising red NUV phosphor CaTiO3:Eu3+. *Journal of Alloys and Compounds*, *485*(1–2), 418–421. doi:10.1016/j.jallcom.2009.05.128

Fujishima, A., Honda, K. (1972). Electrochemical photolysis of water at a semiconductor electrode. *Nature, 238,* 238037a0.

Fujishima, A., & Honda, K. (1972). Electrochemical photolysis of water at a semiconductor electrode. *Nature*, *238*(5358), 37–38. doi:10.1038/238037a0 PMID:12635268

Fuwad, A., Ryu, H., Malmstadt, N., Kim, S. M., & Jeon, T.-J. (2019). Biomimetic membranes as potential tools for water purification: Preceding and future avenues. *Desalination*, *458*, 97–115. doi:10.1016/j.desal.2019.02.003

Gaiardo, A., Fabbri, B., Guidi, V., Bellutti, P., Giberti, A., Gherardi, S., Vanzetti, L., Malagù, C., & Zonta, G. (2016). Metal sulfides as sensing materials for chemoresistive gas sensors. *Sensors (Basel)*, *16*(3), 296. doi:10.339016030296 PMID:26927120

Gaillard, N., Chang, Y. C., De Angelis, A., Higgins, S., & Braun, A. (2013). A nanocompositephotoelectrode made of 2.2 eV band gap copper tungstate (CuWO4) and multi-wall carbon nanotubes for solar-assisted water splitting. *International Journal of Hydrogen Energy*, *38*(8), 3166–3176. doi:10.1016/j.ijhydene.2012.12.104

Gao, B., Chen, G. Z., & Li Puma, G. (2009). Carbon nanotubes/titanium dioxide (CNTs/TiO2) nanocomposites prepared by conventional and novel surfactant wrapping sol–gel methods exhibiting enhanced photocatalytic activity. *Applied Catalysis B: Environmental*, *89*(3-4), 503–509. doi:10.1016/j.apcatb.2009.01.009

Gao, W., Majumder, M., Alemany, L. B., Narayanan, T. N., Ibarra, M. A., Pradhan, B. K., & Ajayan, P. M. (2011). Engineered Graphite Oxide Materials for Application in Water Purification. *ACS Applied Materials & Interfaces*, *3*(6), 1821–1826. doi:10.1021/am200300u PMID:21568266

Ge, L., Peng, Z., Wang, W., Tan, F., Wang, X., Su, B., Qiao, X., & Wong, P. K. (2018). g-C_3N_4/MgO nanosheets: Light-independent, metal-poisoning-free catalysts for the activation of hydrogen peroxide to degrade organics. *Journal of Materials Chemistry. A, Materials for Energy and Sustainability*, *6*(34), 16421–16429. doi:10.1039/C8TA05488F

Ge, L., Wang, W., Peng, Z., Tan, F., Wang, X., Chen, J., & Qiao, X. (2018). Facile fabrication of Fe@MgO magnetic nanocomposites for efficient removal of heavy metal ion and dye from water. *Powder Technology, 326*, 393–401. doi:10.1016/j.powtec.2017.12.003

Gerischer, H. (1966). Electrochemical behavior of semiconductors under illumination. *Journal of the Electrochemical Society, 113*(11), 1174–1182. doi:10.1149/1.2423779

Ghosh, S., Hajra, P., Sariket, D., Ray, D., Baduri, S., & Bhattacharya, C. (2022). *Modifications of $BiVO_4$ semiconductors for oxidation of water and detoxification of organic wastes: photoelectrochemical applications of semiconductors*. IGI Global. doi:10.4018/978-1-6684-4553-2.ch001

Ghosh, S., Laha, D., Hajra, P., Sariket, D., Ray, D., Baduri, S., Sahoo, H. S., & Bhattacharya, C. (2023). Development of transition metal incorporated bismuth-based oxide semiconductors as potential candidates for solar assisted wáter splitting applications. *ChemElectroChem, 10*(6), e202201062. doi:10.1002/celc.202201062

Giberti, A., Gaiardo, A., Fabbri, B., Gherardi, S., Guidi, V., Malagù, C., Bellutti, P., Zonta, G., Casotti, D., & Cruciani, G. (2016). Tin (IV) sulfide nanorods as a new gas sensing material. *Sensors and Actuators. B, Chemical, 223*, 827–833. doi:10.1016/j.snb.2015.10.007

Giwa, A., Hasan, S. W., Yousuf, A., Chakraborty, S., Johnson, D. J., & Hilal, N. (2017). Biomimetic membranes: A critical review of recent progress. *Desalination, 420*, 403–424. doi:10.1016/j.desal.2017.06.025

Golmohammadi, H., Morales-Narváez, E., Naghdi, T., & Merkoçi, A. (2017). Nanocellulose in Sensing and Biosensing. *Chemistry of Materials, 29*(13), 5426–5446. doi:10.1021/acs.chemmater.7b01170

Gomes, J. B., Rodrigues, J. J., Rabêlo, R. A., Kumar, N., & Kozlov, S. (2019). IoT-enabled gas sensors: Technologies, applications, and opportunities. *Journal of Sensor and Actuator Networks, 8*(4), 57. doi:10.3390/jsan8040057

Gómez-Pastora, J., Dominguez, S., Bringas, E., Rivero, M. J., Ortiz, I., & Dionysiou, D. D. (2017). Review and perspectives on the use of magnetic nanophotocatalysts (MNPCs) in water treatment. *Chemical Engineering Journal, 310*, 407–427. doi:10.1016/j.cej.2016.04.140

Gonçalves, R. F., Moura, A. P., Godinho, M. J., Longo, E., Machado, M. A. C., de Castro, D. A., Siu Li, M. S., & Marques, A. P. A. (2015). Crystal growth and photoluminescence of europium-doped strontium titanate prepared by a microwave hydrothermal method. *Ceramics International, 41*(3), 3549–3554. doi:10.1016/j.ceramint.2014.11.018

Gonçalves, R. H., Leite, L. D. T., & Leite, E. R. (2012). Colloidal WO_3 nanowires as a versatile route to prepare a photoanode for solar water splitting. *ChemSusChem, 5*(12), 2341–2347. doi:10.1002/cssc.201200484 PMID:23139181

Gondal, M. A., Suliman, M. A., Dastageer, M. A., Chuah, G. K., Basheer, C., Yang, D., & Suwaiyan, A. (2016). Visible light photocatalytic degradation of herbicide (Atrazine) using surface plasmon resonance induced in mesoporous Ag-WO_3/SBA-15 composite. *Journal of Molecular Catalysis A Chemical*, *425*, 208–216. doi:10.1016/j.molcata.2016.10.015

Gong, K. P., Du, F., Xia, Z. H., Durstock, M., & Dai, L. M. (2009). Nitrogen-doped carbon nanotube arrays with high electrocatalytic activity for oxygen reduction. *Science*, *323*(5915), 760–764. doi:10.1126cience.1168049 PMID:19197058

Gooding, J. J. (2005). Nanostructuring electrodes with carbon nanotubes: A review on electrochemistry and applications for sensing. *Electrochimica Acta*, *50*(15), 3049–3060. doi:10.1016/j.electacta.2004.08.052

Grätzel, M. (2005). Solar energy conversion by dye-sensitized photovoltaic cells. *Inorganic Chemistry*, *44*(20), 6841–6851. doi:10.1021/ic0508371 PMID:16180840

Guo, K., Yang, P., Guo, D. H., & Liu, Y. (2019). Gas leakage monitoring with mobile wireless sensor networks. *Procedia Computer Science*, *154*, 430–438. doi:10.1016/j.procs.2019.06.061

Guo, S., Yang, D., Zhang, S., Dong, Q., Li, B., Tran, N., Li, Z., Xiong, Y., & Zaghloul, M. E. (2019). Development of a cloud-based epidermal MoSe2 device for hazardous gas sensing. *Advanced Functional Materials*, *29*(18), 1900138. doi:10.1002/adfm.201900138

Guo, Y., Quan, X., Lu, N., Zhao, H., & Chen, S. (2007). High photocatalytic capability of self-assembled nanoporous WO_3 with preferential orientation of (002) planes. *Environmental Science & Technology*, *41*(12), 4422–4427. doi:10.1021/es062546c PMID:17626446

Gupta, V. K., Moradi, O., Tyagi, I., Agarwal, S., Sadegh, H., Shahryari-Ghoshekandi, R., Makhlouf, A. S. H., Goodarzi, M., & Garshasbi, A. (2016). Study on the removal of heavy metal ions from industry waste by carbon nanotubes: Effect of the surface modification: a review. *Critical Reviews in Environmental Science and Technology*, *46*(2), 93–118. doi:10.1080/10643389.2015.1061874

Habibullah, G., Viktorova, J., & Ruml, T. (2021). Current Strategies for Noble Metal Nanoparticle Synthesis. *Nanoscale Research Letters, 16*(1), 1–12. doi:10.1186/s11671-021-03480-8

Hai, N. D., Dat, N. M., Huong, L. M., Tai, L. T., Thinh, D. B., Nam, N. T. H., Dat, N. T., Phong, M. T., & Hieu, N. H. (2022). Phytosynthesis of silver nanoparticles using Mangifera indica leaves extract at room temperature: Formation mechanism, catalytic reduction, colorimetric sensing, and antimicrobial activity. *Colloids and Surfaces. B, Biointerfaces*, *220*, 112974. doi:10.1016/j.colsurfb.2022.112974

Hains, A. W., Liang, Z., Woodhouse, M. A., & Gregg, B. A. (2010). Molecular semiconductors in organic photovoltaic cells. *Chemical Reviews*, *110*(11), 6689–6735. doi:10.1021/cr9002984 PMID:20184362

Hajjaoui, H., Soufi, A., Boumya, W., Abdennouri, M., & Barka, N. (2021). Polyaniline/Nanomaterial Composites for the Removal of Heavy Metals by Adsorption: A Review. *Journal of Composites Science*, *5*(9), 233. doi:10.3390/jcs5090233

Hajra, P., Kundu, S., Maity, A., & Bhattacharya, C. (2019). Facile photoelectrochemical water oxidation on Co^{2+}-adsorbed $BiVO_4$ thin films synthesized from aqueous solutions. *Chemical Engineering Journal, 374*, 1221–1230. doi:10.1016/j.cej.2019.06.014

Ha, L. P. P., Vinh, T. H. T., Thuy, N. T. B., Thi, C. M., & Van Viet, P. (2021). Visible-light-driven photocatalysis of anisotropic silver nanoparticles decorated on ZnO nanorods: Synthesis and characterizations. *Journal of Environmental Chemical Engineering, 9*(2), 105103. doi:10.1016/j.jece.2021.105103

Halvaee, P., Dehghani, S., & Mohammadzadeh, M. (2020). Room temperature methanol sensors based on rod-shaped nanostructures of MoS 2 functionalized with Ag nanoparticles. *IEEE Sensors Journal, 21*(4), 4233–4240. doi:10.1109/JSEN.2020.3035753

Hariharan, D., Christy, A. J., Pitchaiya, S., Sagadevan, S., Thangamaniyandi, P., Devan, U., & Nehru, L. (2019). Green hydrothermal synthesis of gold and palladium doped titanium dioxide nanoparticles for multifunctional performance. *Journal of Materials Science Materials in Electronics, 30*(13), 12812–12819. doi:10.100710854-019-01647-9

Hasan, J., Li, H., Tian, G., & Qin, C. (2020). Fabrication of Cr2S3-GO-TiO2 composite with high visible-light-driven photocatalytic activity on degradation of organic dyes. *Chemical Physics, 539*, 110950. doi:10.1016/j.chemphys.2020.110950

Hasija, V., Raizada, P., Sudhaik, A., Sharma, K., Kumar, A., Singh, P., Jonnalagadda, S. B., & Thakur, V. K. (2019). Recent advances in noble metal free doped graphitic carbon nitride based nanohybrids for photocatalysis of organic contaminants in water: A review. *Applied Materials Today, 15*, 494–524. doi:10.1016/j.apmt.2019.04.003

Hayati, B., Maleki, A., Najafi, F., Gharibi, F., McKay, G., Gupta, V. K., Harikaranahalli Puttaiah, S., & Marzban, N. (2018). Heavy metal adsorption using PAMAM/CNT nanocomposite from aqueous solution in batch and continuous fixed bed systems. *Chemical Engineering Journal, 346*, 258–270. doi:10.1016/j.cej.2018.03.172

He, R., Zhou, J., Fu, H., Zhang, S., & Jiang, C. (2018). Room-temperature in situ fabrication of $Bi_2O_3/g\text{-}C_3N_4$ direct Z-scheme photocatalyst with enhanced photocatalytic activity. *Applied Surface Science, 430*, 273–282. doi:10.1016/j.apsusc.2017.07.191

Hermawan, A., Zhang, B., Taufik, A., Asakura, Y., Hasegawa, T., Zhu, J., Shi, P., & Yin, S. (2020). CuO nanoparticles/Ti3C2T x MXene hybrid nanocomposites for detection of toluene gas. *ACS Applied Nano Materials, 3*(5), 4755–4766. doi:10.1021/acsanm.0c00749

Hernández-Gordillo, A., Ruiz-Correa, S., Robledo-Valero, V., Hernández-Rosales, C., & Arriaga, S. (2021). Recent advancements in low-cost portable sensors for urban and indoor air quality monitoring. *Air Quality, Atmosphere & Health, 14*(12), 1–21. doi:10.100711869-021-01067-x

He, T., Liu, W., Lv, T., Ma, M., Liu, Z., Vasiliev, A., & Li, X. (2021). MXene/SnO2 heterojunction based chemical gas sensors. *Sensors and Actuators. B, Chemical, 329*, 129275. doi:10.1016/j.snb.2020.129275

He, Y. Q., Zhang, N. N., & Wang, X. D. (2011). Adsorption of graphene oxide/chitosan porous materials for metal ions. *Chinese Chemical Letters*, *22*(7), 859–862. doi:10.1016/j.cclet.2010.12.049

Hittini, W., Abu-Hani, A. F., Reddy, N., & Mahmoud, S. T. (2020). Cellulose-Copper Oxide hybrid nanocomposites membranes for H2S gas detection at low temperatures. *Scientific Reports*, *10*(1), 2940. doi:10.103841598-020-60069-4 PMID:32076095

Hoffmann, M. R., Martin, S. T., Choi, W., & Bahemann, D. W. (1995). Environmental Applications of Semiconductor Photocatalysis. *Chemical Reviews*, *95*(1), 69–96. doi:10.1021/cr00033a004

Hong, Y., Li, C., Yin, B., Li, D., Zhang, Z., Mao, B., Fan, W., Gu, W., & Shi, W. (2018). Promoting visible-light-induced photocatalytic degradation of tetracycline by an efficient and stable beta-Bi_2O_3@g-C_3N_4 core/shell nanocomposite. *Chemical Engineering Journal*, *338*, 137–146. doi:10.1016/j.cej.2017.12.108

Hossain, N., Islam, M. A., & Chowdhury, M. A. (2022). Synthesis and characterization of plant extracted silver nanoparticles and advances in dental implant applications. *Heliyon*, *8*(12), e12313. doi:10.1016/j.heliyon.2022.e12313 PMID:36590472

Hotze, E. M., Phenrat, T., & Lowry, G. V. (2010). Nanoparticle Aggregation: Challenges to Understanding Transport and Reactivity in the Environment. *Journal of Environmental Quality*, *39*(6), 1909–1924. doi:10.2134/jeq2009.0462 PMID:21284288

Hua, M., Zhang, S., Pan, B., Zhang, W., Lv, L., & Zhang, Q. (2012). Heavy metal removal from water/wastewater by nanosized metal oxides: A review. *Journal of Hazardous Materials*, *211-212*, 317–331. doi:10.1016/j.jhazmat.2011.10.016 PMID:22018872

Huang, D., Yang, Z., Li, X., Zhang, L., Hu, J., Su, Y., Hu, N., Yin, G., He, D., & Zhang, Y. (2017). Three-dimensional conductive networks based on stacked SiO 2@ graphene frameworks for enhanced gas sensing. *Nanoscale*, *9*(1), 109–118. doi:10.1039/C6NR06465E PMID:27763653

Huang, F., Yan, A., & Zhao, H. (2016). Influences of doping on photocatalytic properties of TiO_2 photocatalyst. In W. Cao (Ed.), *Semiconductor photocatalysis - materials, mechanisms and applications* (1st ed., p. 51). Intech Open. doi:10.5772/63234

Huang, Z. F., Song, J., Pan, L., Zhang, X., Wang, L., & Zou, J.-J. (2015). Tungsten oxides for photocatalysis, electrochemistry, and phototherapy. *Advanced Materials*, *27*(36), 5309–5327. doi:10.1002/adma.201501217 PMID:26287959

Humayun, M., Ullah, H., Usman, M., Habibi-Yangjeh, A., Tahir, A. A., Wang, C., & Luo, W. (2022). Perovskite-type lanthanum ferrite based photocatalysts: Preparation, properties, and applications. *Journal of Energy Chemistry*, *66*, 314–338. doi:10.1016/j.jechem.2021.08.023

Hunter, G. W., Akbar, S., Bhansali, S., Daniele, M., Erb, P. D., Johnson, K., & Hesketh, P. J. (2020). Editors' choice—Critical review—A critical review of solid state gas sensors. *Journal of the Electrochemical Society*, *167*(3), 037570. doi:10.1149/1945-7111/ab729c

Hurum, D., Gray, K., Rajh, T., & Thurnaer, M. (2004). Photoinitiated Reactions of 2,4,6 TCP on Degussa P25 Formulation TiO2: Wavelength Sensitive Decomposition. *The Journal of Physical Chemistry B, 108*(42), 16483–16487. doi:10.1021/jp047097m

Huy, B. T., Paeng, D. S., Thao, C. T. B., Phuong, N. T. K., & Lee, Y. I. (2020). ZnO-Bi2O3/graphitic carbon nitride photocatalytic system with H_2O_2-assisted enhanced degradation of Indigo carmine under visible light. *Arabian Journal of Chemistry, 13*(2), 3780–3800. doi:10.1016/j.arabjc.2019.01.003

Hwang, L.-L., Chen, J.-C., & Wey, M.-Y. (2013). The properties and filtration efficiency of activated carbon polymer composite membranes for the removal of humic acid. *Desalination, 313*, 166–175. doi:10.1016/j.desal.2012.12.019

Ibrahim, A. A. M., Khan, I., Iqbal, N., & Qurashi, A. (2017). Facile synthesis of tungsten oxide-bismuth vanadate nanoflakes as photoanode material for solar water splitting. *International Journal of Hydrogen Energy, 42*(5), 3423–3430. doi:10.1016/j.ijhydene.2016.09.095

Ibrahim, R. K., Hayyan, M., AlSaadi, M. A., Hayyan, A., & Ibrahim, S. (2016). Environmental application of nanotechnology: Air, soil, and water. *Environmental Science and Pollution Research International, 23*(14), 13754–13788. doi:10.100711356-016-6457-z PMID:27074929

Ihsanullah, A., Abbas, A., Al-Amer, A. M., Laoui, T., Al-Marri, M. J., Nasser, M. S., Khraisheh, M., & Atieh, M. A. (2016). Heavy metal removal from aqueous solution by advanced carbon nanotubes: Critical review of adsorption applications. *Separation and Purification Technology, 157*, 141–161. doi:10.1016/j.seppur.2015.11.039

Ikram, M., Liu, Y., Lv, H., Liu, L., Rehman, A. U., Kan, K., & Wang, R. (2019). 3D-multilayer MoS2 nanosheets vertically grown on highly mesoporous cubic In2O3 for high-performance gas sensing at room temperature. *Applied Surface Science, 466*, 1–11. doi:10.1016/j.apsusc.2018.10.018

Ilgar, M., Baytemir, G., Taşaltın, N., Güllülü, S., Yeşilyurt, İ. S., & Karakuş, S. (2022). Multifunctional maca extract coated CuO nanoparticles with antimicrobial and dopamine sensing activities: A dual electrochemical – Smartphone colorimetric detection system. *Journal of Photochemistry and Photobiology A Chemistry, 431*, 114075. doi:10.1016/j.jphotochem.2022.114075

Iliger, K. S., Sofi, T. A., Bhat, N. A., Ahanger, F. A., Sekhar, J. C., Elhendi, A. Z., Al-Huqail, A. A., & Khan, F. (2021). Copper nanoparticles: Green synthesis and managing fruit rot disease of chilli caused by Colletotrichum capsici. *Saudi Journal of Biological Sciences, 28*(2), 1477–1486. doi:10.1016/j.sjbs.2020.12.003 PMID:33613075

Inglezakis, V. J., Loizidou, M. D., & Grigoropoulou, H. P. (2002). Equilibrium and kinetic ion exchange studies of Pb2+, Cr3+, Fe3+ and Cu2+ on natural clinoptilolite. *Water Research, 36*(11), 2784–2792. doi:10.1016/S0043-1354(01)00504-8 PMID:12146866

Iqbal, N., Khan, I., Yamani, Z. H. A., & Qurashi, A. (2017). A facile one-step strategy for in-situ fabrication of WO_3-$BiVO_4$ nanoarrays for solar-driven photoelectrochemical water splitting applications. *Solar Energy, 144*, 604–611. doi:10.1016/j.solener.2017.01.057

Iravani, S. (2014). Bacteria in Nanoparticle Synthesis: Current Status and Future Prospects. *International Scholarly Research Notices*, *2014*, 1–18. doi:10.1155/2014/359316 PMID:27355054

Isaad, J., & El Achari, A. (2011). Colorimetric sensing of cyanide anions in aqueous media based on functional surface modification of natural cellulose materials. *Tetrahedron*, *67*(26), 4939–4947. doi:10.1016/j.tet.2011.04.061

Ismail, M., Hedhili, M., Anjum, D., Singaravelu, V., & Chung, S. (2021). Synthesis and characterization of iron doped TiO2 nanoparticles using ferrocene from flame spray pyrolysis. *Catalysts*, *11*(4), 438–454. doi:10.3390/catal11040438

Izumi, Y. (2013). Recent advances in the photocatalytic conversion of carbon dioxide to fuels with water and/or hydrogen using solar energy and beyond. *Coordination Chemistry Reviews*, *257*(1), 171–186. doi:10.1016/j.ccr.2012.04.018

Jacobsson, T. J., Fjällström, V., Edoff, M., & Edvinsson, T. (2014). Sustainable solar hydrogen production: From photoelectrochemical cells to PV-electrolyzers and back again. *Energy & Environmental Science*, *7*(7), 2056–2070. doi:10.1039/C4EE00754A

Jafari Kang, A., Baghdadi, M., & Pardakhti, A. (2016). Removal of cadmium and lead from aqueous solutions by magnetic acid-treated activated carbon nanocomposite. *Desalination and Water Treatment*, *57*(40), 18782–18798. doi:10.1080/19443994.2015.1095123

Jahromi, H. D., Kazemi, M., & Sheikhi, M. H. (2021). Room temperature and highly sensitive acetone sensor based on lead sulfide nanosheets. *Materials Science and Engineering B*, *267*, 115082. doi:10.1016/j.mseb.2021.115082

Jamkhande, P. G., Ghule, N. W., Bamer, A. H., & Kalaskar, M. G. (2019). Metal nanoparticles synthesis: An overview on methods of preparation, advantages and disadvantages, and applications. *Journal of Drug Delivery Science and Technology*, *53*, 101174. doi:10.1016/j.jddst.2019.101174

Javadian, H. (2014). Application of kinetic, isotherm and thermodynamic models for the adsorption of Co(II) ions on polyaniline/polypyrrole copolymer nanofibers from aqueous solution. *Journal of Industrial and Engineering Chemistry*, *20*(6), 4233–4241. doi:10.1016/j.jiec.2014.01.026

Javadian, H., Sorkhrodi, F. Z., & Koutenaei, B. B. (2014). Experimental investigation on enhancing aqueous cadmium removal via nanostructure composite of modified hexagonal type mesoporous silica with polyaniline/polypyrrole nanoparticles. *Journal of Industrial and Engineering Chemistry*, *20*(5), 3678–3688. doi:10.1016/j.jiec.2013.12.066

Javanbakht, V., & Mohammadian, M. (2021). Photo assisted advanced oxidation processes form efficient removal of anionic and cationic dyes using Bentonite/TiO2 nanophocatalyst immobilized with silver nanoparticles. *Journal of Molecular Structure*, *1239*, 130496. doi:10.1016/j.molstruc.2021.130496

Jawed, A., Saxena, V., & Pandey, L. M. (2020). Engineered nanomaterials and their surface functionalization for the removal of heavy metals: A review. *Journal of Water Process Engineering*, *33*, 101009. doi:10.1016/j.jwpe.2019.101009

Jawor, A., & Hoek, E. M. V. (2010). Removing Cadmium Ions from Water via Nanoparticle-Enhanced Ultrafiltration. *Environmental Science & Technology*, *44*(7), 2570–2576. doi:10.1021/es902310e PMID:20230021

Jayarambabu, N., Saraswathi, K., Akshaykranth, A., Anitha, N., Venkatappa Rao, T., & Rakesh kumar, R. (2023). Bamboo-mediated silver nanoparticles functionalized with activated carbon and their application for non-enzymatic glucose sensing. *Inorganic Chemistry Communications*, *147*, 110249. doi:10.1016/j.inoche.2022.110249

Jayaweera, H. D. A. C., Siriwardane, I., de Silva, K. M. N., & de Silva, R. M. (2018). Synthesis of multifunctional activated carbon nanocomposite comprising biocompatible flake nano hydroxyapatite and natural turmeric extract for the removal of bacteria and lead ions from aqueous solution. *Chemistry Central Journal*, *12*(1), 18. doi:10.118613065-018-0384-7 PMID:29468333

Jelić, A., Petrović, M., & Barceló, D. (2012). Pharmaceuticals in Drinking Water. In D. Barceló (Ed.), *Emerging Organic Contaminants and Human Health* (pp. 47–70). Springer Berlin Heidelberg. doi:10.1007/698_2011_133

Jewell, J., Vinichenko, V., McCollum, D. (2016). Comparison and interactions between the long-term pursuit of energy independence and climate policies. *Nat Energy*, *1*, 201673.

Jhaveri, J. H., & Murthy, Z. V. P. (2016). A comprehensive review on anti-fouling nanocomposite membranes for pressure driven membrane separation processes. *Desalination*, *379*, 137–154. doi:10.1016/j.desal.2015.11.009

Jiang, C., Moniz, S. J., Wang, A., Zhang, T., & Tang, J. (2017). Photoelectrochemical devices for solar water splitting—Materials and challenges. *Chemical Society Reviews*, *46*(15), 4645–4660. doi:10.1039/C6CS00306K PMID:28644493

Jiao, Y., Zheng, Y., Jaroniec, M., & Qiao, S. Z. (2014). Origin of the electrocatalytic oxygen reduction activity of graphene-based catalysts: A roadmap to achieve the best performance. *Journal of the American Chemical Society*, *136*(11), 4394–4403. doi:10.1021/ja500432h PMID:24580116

Jia, Y., Yu, H., Zhang, Y., Dong, F., & Li, Z. (2016). Cellulose acetate nanofibers coated layer-by-layer with polyethylenimine and graphene oxide on a quartz crystal microbalance for use as a highly sensitive ammonia sensor. *Colloids and Surfaces. B, Biointerfaces*, *148*, 263–269. doi:10.1016/j.colsurfb.2016.09.007 PMID:27619179

Jijun, Z., Alper, B., Jie, H., & Jian Ping, L. (2002). Gas molecule adsorption in carbon nanotubes and nanotube bundles. *Nanotechnology*, *13*(2), 195–200. doi:10.1088/0957-4484/13/2/312

Jin, B. J., Jung, E., Ma, M., Kim, S., Zhang, K., Kim, J. I., Son, Y., & Park, J. H. (2018). Solution-processed yolk-shell-shaped WO_3/$BiVO_4$ heterojunction photoelectrodes for efficient solar water splitting. *Journal of Materials Chemistry. A, Materials for Energy and Sustainability*, *6*(6), 2585–2592. doi:10.1039/C7TA08452H

Jin, T., Xu, D., Diao, P., He, W., Wang, H., & Liao, S. (2016). Tailored preparation of WO_3 nano-grassblades on FTO substrate for photoelectrochemical water splitting. *CrystEngComm*, *18*(36), 6798–6808. doi:10.1039/C6CE01186A

Ju, M.-G., Dai, J., Ma, L., & Zeng, X. C. (2017). Perovskite Chalcogenides with Optimal Bandgap and Desired Optical Absorption for Photovoltaic Devices. *Advanced Energy Materials*, *7*(18), 1700216. doi:10.1002/aenm.201700216

Jung, M. W., Kang, S. M., Nam, K.-H., An, K.-S., & Ku, B.-C. (2018). Highly transparent and flexible NO2 gas sensor film based on MoS2/rGO composites using soft lithographic patterning. *Applied Surface Science*, *456*, 7–12. doi:10.1016/j.apsusc.2018.06.086

Jurado-Sánchez, B., & Wang, J. (2018). Micromotors for environmental applications: A review. *Environmental Science. Nano*, *5*(7), 1530–1544. doi:10.1039/C8EN00299A

Kalantar-zadeh, K., Vijayaraghavan, A., Ham, M. H., Zheng, H., Breedon, M., & Strano, M. S. (2010). Synthesis of atomically thin WO_3 sheets from hydrated tungsten trioxide. *Chemistry of Materials*, *22*(19), 5660–5666. doi:10.1021/cm1019603

Kalanur, S. S., Hwang, Y. J., Chae, S. Y., & Joo, O. S. (2013). Facile growth of aligned WO_3 nanorods on FTO substrate for enhanced photoanodic water oxidation activity. *Journal of Materials Chemistry. A, Materials for Energy and Sustainability*, *1*(10), 3479–3488. doi:10.1039/c3ta01175e

Kalanur, S. S., Yoo, I. H., Eom, K., & Seo, H. (2018). Enhancement of photoelectrochemical water splitting response of WO_3 by means of Bi doping. *Journal of Catalysis*, *357*, 127–137. doi:10.1016/j.jcat.2017.11.012

Kalanur, S. S., Yoo, I. H., Park, J., & Seo, H. (2017). Insights into the electronic bands of WO_3/$BiVO_4$/TiO_2, revealing high solar water splitting efficiency. *Journal of Materials Chemistry. A, Materials for Energy and Sustainability*, *5*(4), 1455–1461. doi:10.1039/C6TA07592D

Kanazawa, E., Sakai, G., Shimanoe, K., Kanmura, Y., Teraoka, Y., Miura, N., & Yamazoe, N. (2001). Metal oxide semiconductor N2O sensor for medical use. *Sensors and Actuators. B, Chemical*, *77*(1-2), 72–77. doi:10.1016/S0925-4005(01)00675-X

Karabulut, G., Beköz Üllen, N., Akyüz, E., & Karakuş, S. (2023). Surface modification of 316L stainless steel with multifunctional locust gum/polyethylene glycol-silver nanoparticles using different coating methods. *Progress in Organic Coatings*, *174*, 107291. doi:10.1016/j.porgcoat.2022.107291

Karabulut, G., Üllen, N. B., & Karakuş, S. (2022). *Nanostructures in Biosensors: Development and Applications*. IntechOpen., doi:10.5772/intechopen.108508

Karakuş, S., Taşaltın, N., Taşaltın, C., & Üllen, N. B. (2021). Synthesis and Characterization of Konjac Gum/Polyethylene Glycol-Silver Nanoparticles and their Potential Application as a Colorimetric Sensor for Hydrogen Peroxide. *Journal of Inorganic and Organometallic Polymers and Materials*, *1*(9), 1–14. doi:10.100710904-021-01984-5

Karelin, A., Baranov, A. M., Akbari, S., Mironov, S., & Karpova, E. (2019). Measurement algorithm for determining unknown flammable gas concentration based on temperature sensitivity of catalytic sensor. *IEEE Sensors Journal*, *19*(11), 4173–4180. doi:10.1109/JSEN.2019.2897626

Katyayan, S., & Agrawal, S. (2018a). Effect of rare earth doping on optical and spectroscopic characteristics of BaZrO3:Eu3+,Tb3+ perovskites. *Methods and Applications in Fluorescence*, *6*(3), 035002. doi:10.1088/2050-6120/aaae0b PMID:29422457

Katyayan, S., & Agrawal, S. (2018b, September 10). Molten salt synthesis and photoluminescent studies of ATiO3:Eu2+, Yb2+ (A = Ca, Ba, Sr) perovskites phosphors. *Journal of Photonics for Energy*, *8*(3), 1. doi:10.1117/1.JPE.8.036001

Katyayan, S., & Agrawal, S. (2018c). Facile molten salt synthesis, structural, morphological and optical studies of ASiO3: Eu2+, Er3+ (A = Ca, Ba, Sr) perovskites. *Journal of Materials Science Materials in Electronics*, *29*(19), 16609–16629. doi:10.100710854-018-9754-0

Katyayan, S., & Agrawal, S. (2019a). Optical Behavior of CaSiO3:Eu2+, Er3+, BaSiO3:Eu2+, Er3+and SrSiO3:Eu2+, Er3+ Phosphors. *Journal of the Minerals Metals & Materials Society*, *71*(8), 2899–2905. doi:10.100711837-019-03545-1

Katyayan, S., & Agrawal, S. (2019b). Optical behavior and TL kinetics of Eu3+ and Tb3+ doped zirconate thermoluminescent phosphors. *Optical and Quantum Electronics*, *51*(8), 277. Advance online publication. doi:10.100711082-019-1993-7

Katyayan, S., & Agrawal, S. (2020). Low temperature salt melt synthesis, thermal and optical studies of AZrO3:Eu3+,Tb3+ perovskite phosphors. *Optik (Stuttgart)*, *204*, 164125. doi:10.1016/j.ijleo.2019.164125

Kaur, N., Singh, M., & Comini, E. (2022). Materials engineering strategies to control metal oxides nanowires sensing properties. *Advanced Materials Interfaces*, *9*(12), 2101629. doi:10.1002/admi.202101629

Kausar, F., Bagheri, A. R., Rasheed, T., Bilal, M., Rizwan, K., Nguyen, T. A., & Iqbal, H. M. N. (2022). Chapter 7 - Nanomaterials for removal of heavy metals from wastewater. In A. Denizli, N. Ali, M. Bilal, A. Khan, & T. A. Nguyen (Eds.), Nano-Biosorbents for Decontamination of Water, Air, and Soil Pollution (pp. 135-161): Elsevier.

Kehl, W. L., Hay, R. G., & Wahl, D. (1952). The structure of tetragonaltungsten trioxide. *Journal of Applied Physics*, *23*(2), 212–215. doi:10.1063/1.1702176

Kenney, M. J., Gong, M., Li, Y., Wu, J. Z., Feng, J., Lanza, M., & Dai, H. (2013). High-performance silicon photoanodes passivated with ultrathin nickel films for water oxidation. *Science*, *342*(6160), 836–840. doi:10.1126cience.1241327 PMID:24233719

Khavar, A., Moussavi, G., Mahjoub, A. M. S., & Abdolmaleki, P. (2018). Synthesis and visible-light photocatalytic activity of In,S-Ti)2@rGO nanocomposite for degradation and detoxification of pesticide atrazine in water. *Chemical Engineering Journal*, *345*, 300–311. doi:10.1016/j.cej.2018.03.095

Khlyustrova, A., Sirotkin, N., Kusova, T., Kraev, A., Titov, V., & Agafonov, A. (2020). Doped TiO2: The effect of doping elements on photocatalytic activity. *Materials Advances*, *1*(5), 1193–1201. doi:10.1039/D0MA00171F

Khoso, W. A., Haleem, N., Baig, M. A., & Jamal, Y. (2021). Synthesis, characterization and heavy metal removal efficiency of nickel ferrite nanoparticles (NFN's). *Scientific Reports*, *11*(1), 3790. doi:10.103841598-021-83363-1 PMID:33589710

Kim, H.-J., & Lee, J.-H. (2014). Highly sensitive and selective gas sensors using p-type oxide semiconductors: Overview. *Sensors and Actuators. B, Chemical*, *192*, 607–627. doi:10.1016/j.snb.2013.11.005

Kim, H.-J., Yoon, J.-W., Choi, K.-I., Jang, H. W., Umar, A., & Lee, J.-H. (2013). Ultraselective and sensitive detection of xylene and toluene for monitoring indoor air pollution using Cr-doped NiO hierarchical nanostructures. *Nanoscale*, *5*(15), 7066–7073. doi:10.1039/c3nr01281f PMID:23807747

Kimouche, A., Ervasti, M. M., Drost, R., Halonen, S., Harju, A., Joensuu, P. M., Sainio, J., & Liljeroth, P. (2015). Ultra-narrow metallic armchair graphene nanoribbons. *Nature Communications*, *6*(1), 10177. doi:10.1038/ncomms10177 PMID:26658960

Kim, Y. K., & Park, H. (2011). Light-harvesting multi-walled carbon nanotubes and CdS hybrids: Application to photocatalytic hydrogen production from water. *Energy & Environmental Science*, *4*(3), 685–694. doi:10.1039/C0EE00330A

Ko, K. Y., Lee, S., Park, K., Kim, Y., Woo, W. J., Kim, D., & Lee, Z. (2018). High-Performance Gas Sensor Using a Large-Area WS2 x Se2–2 x Alloy for Low-Power Operation Wearable Applications. *ACS Applied Materials & Interfaces*, *10*(40), 34163–34171. doi:10.1021/acsami.8b10455 PMID:30222310

Konishi, Y., Ohno, K., Saitoh, N., Nomura, T., Nagamine, S., Hishida, H., Takahashi, Y., & Uruga, T. (2007). Bioreductive deposition of platinum nanoparticles on the bacterium Shewanella algae. *Journal of Biotechnology*, *128*(3), 648–653. doi:10.1016/j.jbiotec.2006.11.014 PMID:17182148

Kotresh, S., Ravikiran, Y., Tiwari, S., & Vijaya Kumari, S. (2017). Polyaniline–cadmium ferrite nanostructured composite for room-temperature liquefied petroleum gas sensing. *Journal of Electronic Materials*, *46*(8), 5240–5247. doi:10.100711664-017-5535-4

Kroke, E., Schwarz, M., Bordon, E. H., Kroll, P., Noll, B., & Norman, A. D. (2002). Tri-s-triazine derivatives. Part I. From trichloro-tri-s-triazine to graphitic C_3N_4 structures. *New Journal of Chemistry*, *26*(5), 508–512. doi:10.1039/b111062b

Kroutil, J., Laposa, A., Voves, J., Davydova, M., Nahlik, J., Kulha, P., & Husak, M. (2018). Performance evaluation of low-cost flexible gas sensor array with nanocomposite polyaniline films. *IEEE Sensors Journal*, *18*(9), 3759–3766. doi:10.1109/JSEN.2018.2811461

Kumar, R., Chauhan, G., & Martinez-Chapa, S. O. (2022). Metal oxides and their composites as flow-through biosensors for biomonitoring. *Metal Oxides for Biomedical and Biosensor Applications*, 291–319. doi:10.1016/B978-0-12-823033-6.00010-7

Kumar, A. S. K., Kalidhasan, S., Rajesh, V., & Rajesh, N. (2012). Application of Cellulose-Clay Composite Biosorbent toward the Effective Adsorption and Removal of Chromium from Industrial Wastewater. *Industrial & Engineering Chemistry Research*, *51*(1), 58–69. doi:10.1021/ie201349h

Kunduru, K. R., Nazarkovsky, M., Farah, S., Pawar, R. P., Basu, A., & Domb, A. J. (2017). 2 - Nanotechnology for water purification: applications of nanotechnology methods in wastewater treatment. In A. M. Grumezescu (Ed.), *Water Purification* (pp. 33–74). Academic Press. doi:10.1016/B978-0-12-804300-4.00002-2

Kuwabara, M. (2012). Innovative nanocrystal-based technologies for ceramic devices with novel electronic functions. *Journal of Advanced Ceramics*, *1*(2), 79–88. doi:10.100740145-012-0016-y

Kwong, W. L., Lee, C. C., & Messinger, J. (2016). Transparent nanoparticulate FeOOH improves the performance of a WO_3 photoanode in a tandem water-splitting device. *The Journal of Physical Chemistry C*, *120*(20), 10941–10950. doi:10.1021/acs.jpcc.6b02432

Kwon, S., Fan, M., Cooper, A., & Yang, H. (2008). Photocatalytic application of micro- and nano TiO2 in environmental engineering. *Critical Reviews in Environmental Science and Technology*, *38*(3), 197–227. doi:10.1080/10643380701628933

Lai, C. W., & Sreekantan, S. (2013). Preparation of hybrid WO_3-TiO_2 nanotube photoelectrodes using anodization and wet impregnation: Improved water-splitting hydrogen generation performance. *International Journal of Hydrogen Energy*, *38*(5), 2156–2166. doi:10.1016/j.ijhydene.2012.12.025

Lai, C.-H., Lu, M.-Y., & Chen, L.-J. (2012). Metal sulfide nanostructures: Synthesis, properties and applications in energy conversion and storage. *Journal of Materials Chemistry*, *22*(1), 19–30. doi:10.1039/C1JM13879K

Lang, X., Chen, X., & Zhao, J. (2013). Heterogeneous visible light photocatalysis for selective organic transformations. *Chemical Society Reviews*, *43*(1), 473–486. doi:10.1039/C3CS60188A PMID:24162830

Le, A. T., Pung, S.-Y., Sreekantan, S., Matsuda, A., & Huynh, D. P. (2019). Mechanisms of removal of heavy metal ions by ZnO particles. *Heliyon*, *5*(4), e01440. doi:10.1016/j.heliyon.2019.e01440 PMID:31008388

Lee, K. J., Nallathamby, P. D., Browning, L. M., Osgood, C. J., & Nancy Xu, X. H. (2007). In vivo imaging of transport and biocompatibility of single silver nanoparticles in early development of zebrafish embryos. *ACS Nano*, *1*(2), 133–143. doi:10.1021/nn700048y PMID:19122772

Lee, K. S., & El-Sayed, M. A. (2006). Gold and silver nanoparticles in sensing and imaging: Sensitivity of plasmon response to size, shape, and metal composition. *The Journal of Physical Chemistry B*, *110*(39), 19220–19225. doi:10.1021/jp062536y PMID:17004772

Lee, K., Gatensby, R., McEvoy, N., Hallam, T., & Duesberg, G. S. (2013). High-performance sensors based on molybdenum disulfide thin films. *Advanced Materials*, *25*(46), 6699–6702. doi:10.1002/adma.201303230 PMID:24027114

Lee, K.-Y., Tammelin, T., Schulfter, K., Kiiskinen, H., Samela, J., & Bismarck, A. (2012). High Performance Cellulose Nanocomposites: Comparing the Reinforcing Ability of Bacterial Cellulose and Nanofibrillated Cellulose. *ACS Applied Materials & Interfaces*, *4*(8), 4078–4086. doi:10.1021/am300852a PMID:22839594

Lee, M. G., Kim, D. H., Sohn, W., Moon, C. W., Park, H., Lee, S., & Jang, H. W. (2016). Conformally coated $BiVO_4$ nanodots on porosity-controlled WO_3 nanorods as highly efficient type II heterojunction photoanodes for water oxidation. *Nano Energy*, *28*, 250–260. doi:10.1016/j.nanoen.2016.08.046

Lee, S., Lintang, H., & Yuliati, L. (2017). HIghly photocatalytic activity of Fe2O3/TiO2 nanocomposites prepared by photodeposition for degradation of 2,4-dicholorophenoxyacetic acid. *Beilstein Journal of Nanotechnology*, *8*, 915–926. doi:10.3762/bjnano.8.93 PMID:28546886

Lee, S.-Y., & Park, S. (2013). TiO2 photocatalysts for water treatment applications. *Journal of Industrial and Engineering Chemistry*, *19*(6), 1761–1769. doi:10.1016/j.jiec.2013.07.012

Lee, W. S., & Choi, J. (2019). Hybrid Integration of Carbon Nanotubes and Transition Metal Dichalcogenides on Cellulose Paper for Highly Sensitive and Extremely Deformable Chemical Sensors. *ACS Applied Materials & Interfaces*, *11*(21), 19363–19371. doi:10.1021/acsami.9b03296 PMID:31062579

Lemański, K., & Dereń, P. J. (2014). Spectroscopic properties of Dy3+ ions in CaTiO3 nano-perovskites. *Journal of Luminescence*, *145*, 661–664. doi:10.1016/j.jlumin.2013.08.048

Leonard, K. C., Nam, K. M., Lee, H. C., Kang, S. H., Park, H. S., & Bard, A. J. (2013). $ZnWO_4$/WO_3 composite for improving photoelectrochemical water oxidation. *The Journal of Physical Chemistry C*, *117*(31), 15901–15910. doi:10.1021/jp403506q

Li, G., Shrotriya, V., Huang, J. (2005). High-efficiency solution processable polymer photovoltaic cells by self-organization of polymer blends. Nat Mater 4:nmat, 1500.

Liang, Q., Li, Z., Yu, X., Huang, Z. H., Kang, F., & Yang, Q. H. (2015). Macroscopic 3D porous graphitic carbon nitride monolith for enhanced photocatalytic hydrogen evolution. *Advanced Materials*, *27*(31), 4634–4639. doi:10.1002/adma.201502057 PMID:26135383

Liew, S. L., Zhang, Z., Goh, T. W. G., Subramanian, G. S., Seng, H. L. D., Hor, T. S. A., Luo, H. K., & Chi, D. Z. (2014). Yb-doped WO_3 photocatalysts for water oxidation with visible light. *International Journal of Hydrogen Energy*, *39*(9), 4291–4298. doi:10.1016/j.ijhydene.2013.12.204

Li, J., Chen, C., Zhang, R., & Wang, X. (2015). Nanoscale Zero-Valent Iron Particles Supported on Reduced Graphene Oxides by Using a Plasma Technique and Their Application for Removal of Heavy-Metal Ions. *Chemistry, an Asian Journal*, *10*(6), 1410–1417. doi:10.1002/asia.201500242 PMID:25917859

Li, J., Li, H., Zhu, Y., Hao, Y., Sun, X., & Wang, L. (2011). Dual roles of amphiphilic triblock copolymer P123 in synthesis of α-Fe nanoparticle/ordered mesoporous silica composites. *Applied Surface Science*, *258*(2), 657–661. doi:10.1016/j.apsusc.2011.07.037

Lim, S., Shi, J., von Gunten, U., & McCurry, D. (2022). Ozonation of organic compounds in water and wastewater: A critical review. *Water Research*, *213*, 118053. doi:10.1016/j.watres.2022.118053 PMID:35196612

Lin, H. S., & Lin, L. Y. (2017). Improving visible-light responses and electric conductivities by Incorporating Sb_2S_3 and reduced grapheme oxide in a WO_3 nanoplate array for photoelectrochemical water oxidation. *Electrochimica Acta*, *252*, 235–244. doi:10.1016/j.electacta.2017.09.007

Lin, X., Xu, D., Zhao, R., Xi, Y., Zhao, L., Song, M., Zhai, H., Che, G., & Chang, L. (2017). Highly efficient photocatalytic activity of g-C3N4 quantum dots (CNQDs)/Ag/Bi2MoO6 nanoheterostructure under visible light. *Separation and Purification Technology*, *178*, 163–168. doi:10.1016/j.seppur.2017.01.020

Lin, Y., Yao, W., Cheng, Y., Qian, H., Wang, X., Ding, Y., Wu, W., & Jiang, X. (2012). Multifold enhanced T2 relaxation of ZnFe2O4 nanoparticles by jamming them inside chitosan nanospheres. *Journal of Materials Chemistry*, *22*(12), 5684–5693. doi:10.1039/c2jm15133b

Li, Q., Gu, Y., & Wang, N.-F. (2017). Application of random forest classifier by means of a QCM-based e-nose in the identification of Chinese liquor flavors. *IEEE Sensors Journal*, *17*(6), 1788–1794. doi:10.1109/JSEN.2017.2657653

Li, R., Jiang, K., Chen, S., Lou, Z., Huang, T., Chen, D., & Shen, G. (2017). SnO 2/SnS 2 nanotubes for flexible room-temperature NH 3 gas sensors. *RSC Advances*, *7*(83), 52503–52509. doi:10.1039/C7RA10537A

Liu, C. J., Li, Y. M., Li, W. Z., Zhu, J., Li, J., Chen, Q. Y., & Yang, Y. H. (2014). CdS quantum dots sensitized plate-like WO_3 photoelectrodes with a TiO_2 buffer-layer. *Materials Letters*, *120*, 170–173. doi:10.1016/j.matlet.2014.01.053

Liu, C. J., Yang, Y. H., Chen, J., Li, S., Li, W. Z., & Tang, X. D. (2017). An in-situ transformation approach for fabrication of $BiVO_4/WO_3$ heterojunction photoanode with high photoelectrochemical activity. *Chemical Engineering Journal*, *326*, 603–611. doi:10.1016/j.cej.2017.05.179

Liu, C. J., Yang, Y. H., Li, W. Z., Li, J., Li, Y. M., & Chen, Q. Y. (2016). In situ synthesis of Bi_2S_3 sensitized WO_3 nanoplate arrays with less interfacial defects and enhanced photoelectrochemical performance. *Scientific Reports*, *6*(1), 23451. doi:10.1038rep23451 PMID:26988275

Liu, F., Chu, X., Dong, Y., Zhang, W., Sun, W., & Shen, L. (2013). Acetone gas sensors based on graphene-ZnFe2O4 composite prepared by solvothermal method. *Sensors and Actuators. B, Chemical*, *188*, 469–474. doi:10.1016/j.snb.2013.06.065

Liu, G., Pan, J., Yin, L., Irvine, J., Li, F., Tan, J., & Cheng, H.-M. (2012). Heteroatom- Modulated Switching of photocatalytic Hydrogen and Oxygen Evolution Preferences of Anatase Another approach that has been applied in the Heteroatom-Modulated Switching of photocatalytic Hydrogen and Oxygen Evolution Preferences of Anatase TiO2. *Advanced Functional Materials*, *22*, 3233–3238. doi:10.1002/adfm.201200414

Liu, J., Bao, S., & Wang, X. (2022). Applications of graphene-based materials in sensors: A review. *Micromachines*, *13*(2), 184. doi:10.3390/mi13020184 PMID:35208308

Liu, S., Zou, Q., Ma, Y., Chi, D., Chen, R., Fang, H., Hu, W., Zhang, K., & Chen, L.-F. (2022). Metal-organic frameworks derived TiO2/carbon nitride heterojunction photocatalyst with efficient catalytic performance under visible light. *Inorganica Chimica Acta*, *536*, 120918. doi:10.1016/j.ica.2022.120918

Liu, X., Cui, J., Sun, J., & Zhang, X. (2014). 3D graphene aerogel-supported SnO 2 nanoparticles for efficient detection of NO 2. *RSC Advances*, *4*(43), 22601–22605. doi:10.1039/c4ra02453b

Liu, X., Hu, Q., Fang, Z., Zhang, X., & Zhang, B. (2009). Magnetic Chitosan Nanocomposites: A Useful Recyclable Tool for Heavy Metal Ion Removal. *Langmuir*, *25*(1), 3–8. doi:10.1021/la802754t PMID:19032060

Liu, X., Ma, T., Pinna, N., & Zhang, J. (2017). Two-dimensional nanostructured materials for gas sensing. *Advanced Functional Materials*, *27*(37), 1702168. doi:10.1002/adfm.201702168

Liu, X., Wang, F., & Wang, Q. (2012). Nanostructure-based WO_3 photoanodes for photoelectrochemical water splitting. *Physical Chemistry Chemical Physics*, *14*(22), 7894–7911. doi:10.1039/c2cp40976c PMID:22534756

Liu, X., Wang, W., Zhang, Y., Pan, Y., Liang, Y., & Li, J. (2018). Enhanced sensitivity of a hydrogen sulfide sensor based on surface acoustic waves at room temperature. *Sensors (Basel)*, *18*(11), 3796. doi:10.339018113796 PMID:30404204

Liu, Y., Zhao, L., Su, J., Li, M., & Guo, L. (2015). Fabrication and properties of a branched $(NH_4)_xWO_3$ nanowire array film and a porous WO_3 nanorod array film. *ACS Applied Materials & Interfaces*, *7*(6), 3532–3538. doi:10.1021/am507230t PMID:25623076

Liu, Z. L., Lin, X. H., Lee, J. Y., Zhang, W., Han, M., & Gan, L. M. (2002). Preparation and characterization of platinum-based electrocatalysts on multi-walled carbon nanotubes for proton exchange membrane fuel cells. *Langmuir*, *18*(10), 4054–4060. doi:10.1021/la0116903

Liu, Z., Luo, F., & Chen, T. (2004). Phenolphthalein immobilized membrane for an optical pH sensor. *Analytica Chimica Acta*, *510*(2), 189–194. doi:10.1016/j.aca.2004.01.019

Li, W., Da, P., Zhang, Y., Wang, Y., Lin, X., Gong, X., & Zheng, G. (2014). WO_3 nanoflakes for enhanced photoelectrochemical conversion. *ACS Nano*, *8*(11), 11770–11777. doi:10.1021/nn5053684 PMID:25347213

Li, W., Ismat Shah, S., Huang, C.-P., Jung, O., & Ni, C. (2002). Metallorganic chemical vapour deposition and characterization of TiO2 nanoparticles. *Materials Science and Engineering B*, *96*(3), 247–253. doi:10.1016/S0921-5107(02)00352-5

Li, W., Liu, C., Yang, Y., Li, J., Chen, Q., & Liu, F. (2012). Plate like WO_3 from hydrothermal RF sputtered tungsten thin films for photoelectrochemical water oxidation. *Materials Letters*, *84*, 41–43. doi:10.1016/j.matlet.2012.06.022

Li, W., Zhan, F., Li, J., Liu, C., Yang, Y., Li, Y., & Chen, Q. (2015). Enhancing photoelectrochemical water splitting by aluminum-doped plate-like WO_3 electrodes. *Electrochimica Acta*, *160*, 57–63. doi:10.1016/j.electacta.2015.01.095

Li, X., Ai, L., & Jiang, J. (2016). Nanoscale zerovalent iron decorated on graphene nanosheets for Cr(VI) removal from aqueous solution: Surface corrosion retard induced the enhanced performance. *Chemical Engineering Journal*, *288*, 789–797. doi:10.1016/j.cej.2015.12.022

Li, X., Li, W., Wang, M., & Liao, Z. (2021). Magnetic nanoparticles for cancer theranostics: Advances and prospects. *Journal of Controlled Release*, *335*, 437–448. doi:10.1016/j.jconrel.2021.05.042 PMID:34081996

Li, X., Li, X., Li, Z., Wang, J., & Zhang, J. (2017). WS2 nanoflakes based selective ammonia sensors at room temperature. *Sensors and Actuators. B, Chemical*, *240*, 273–277. doi:10.1016/j.snb.2016.08.163

Li, Y. G., Feng, J., Li, H. J., Wei, X. L., Wang, R. R., & Zhou, A. N. (2016). Photoelectrochemical splitting of natural sea water with a-Fe_2O_3/WO_3 nanorod arrays. *International Journal of Hydrogen Energy*, *41*(7), 4096–4105. doi:10.1016/j.ijhydene.2016.01.027

Li, Y., Guo, W., Hao, H., Wang, L., Su, Q., Jin, S., Qin, L., Gao, W., Liu, G., & Hu, Z. (2015). Enhancing photoelectrical performance of dye–sensitized solar cell by doping SrTiO 3:Sm 3+ @SiO 2 core–shell nanoparticles in the photoanode. *Electrochimica Acta*, *173*, 656–664. doi:10.1016/j.electacta.2015.05.116

Li, Y., Li, X., Tang, Z., Tang, Z., Yu, J., & Wang, J. (2015). Hydrogen sensing of the mixed-potential-type MnWO4/YSZ/Pt sensor. *Sensors and Actuators. B, Chemical*, *206*, 176–180. doi:10.1016/j.snb.2014.09.055

Li, Y., Luo, N., Sun, G., Zhang, B., Ma, G., Jin, H., Wang, Y., Cao, J., & Zhang, Z. (2018). Facile synthesis of ZnFe2O4/α-Fe2O3 porous microrods with enhanced TEA-sensing performance. *Journal of Alloys and Compounds*, *737*, 255–262. doi:10.1016/j.jallcom.2017.12.068

Li, Y., Wu, S., Huang, L., Xu, H., Zhang, R., Qu, M., Gao, Q., & Li, H. (2015). g-C3N4 modified Bi2O3 composites with enhanced visible-light photocatalytic activity. *Journal of Physics and Chemistry of Solids*, *76*, 112–119. doi:10.1016/j.jpcs.2014.08.012

Li, Z.-J., Wang, L., Yuan, L.-Y., Xiao, C.-L., Mei, L., Zheng, L.-R., Zhang, J., Yang, J.-H., Zhao, Y.-L., Zhu, Z.-T., Chai, Z.-F., & Shi, W.-Q. (2015). Efficient removal of uranium from aqueous solution by zero-valent iron nanoparticle and its graphene composite. *Journal of Hazardous Materials, 290*, 26–33. doi:10.1016/j.jhazmat.2015.02.028 PMID:25734531

Li, Z., Wang, J., Xu, Y., Shen, M., Duan, C., Dai, L., & Ni, Y. (2021). Green and sustainable cellulose-derived humidity sensors: A review. *Carbohydrate Polymers, 270*, 118385. doi:10.1016/j.carbpol.2021.118385 PMID:34364627

Llobet, E. (2013). Gas sensors using carbon nanomaterials: A review. *Sensors and Actuators. B, Chemical, 179*, 32–45. doi:10.1016/j.snb.2012.11.014

Long, L. H., Evans, P. J., & Halliwell, B. (1999). Hydrogen Peroxide in Human Urine: Implications for Antioxidant Defense and Redox Regulation. *Biochemical and Biophysical Research Communications, 262*(3), 605–609. doi:10.1006/bbrc.1999.1263 PMID:10471371

Low, J., Yu, J., Janroniec, M., Wageh, S., & Al-Ghamdi, A. (2017). Heterojunction photocatalysts. *Advanced Materials, 29*(20), 1601694. doi:10.1002/adma.201601694 PMID:28220969

Lu, H., Wang, J., Stoller, M., Wang, T., Bao, Y., & Hao, H. (2016). An Overview of Nanomaterials for Water and Wastewater Treatment. *Advances in Materials Science and Engineering, 4964828*, 1–10. Advance online publication. doi:10.1155/2016/4964828

Luo, B., Song, R., Geng, J., Jing, D., & Zhang, Y. (2018). Facile preparation with high yield of a 3D porous graphitic carbon nitride for dramatically enhanced photocatalytic H_2 evolution under visible light. *Applied Catalysis B: Environmental, 238*, 294–301. doi:10.1016/j.apcatb.2018.07.039

Lü, T., Qi, D., Zhang, D., Lü, Y., & Zhao, H. (2018). A facile method for emulsified oil-water separation by using polyethylenimine-coated magnetic nanoparticles. *Journal of Nanoparticle Research, 20*(4), 88. doi:10.100711051-018-4193-7

Lv, R., Terrones, H., Elias, A. L., Perea-Lopez, N., Gutierrez, H. R., Cruz-Silva, E., Rajukumar, L. P., Dresselhaus, M. S., & Terrones, M. (2015). Two-dimensional transition metal dichalcogenides: Clusters, ribbons, sheets and more. *Nano Today, 10*(5), 559–592. doi:10.1016/j.nantod.2015.07.004

Mahadeva, S. K., Yun, S., & Kim, J. (2011). Flexible humidity and temperature sensor based on cellulose–polypyrrole nanocomposite. *Sensors and Actuators. A, Physical, 165*(2), 194–199. doi:10.1016/j.sna.2010.10.018

Mahdavi, K., Zinatloo-Ajabshir, S., Yousif, Q. A., & Salavati-Niasari, M. (2022). Enhanced photocatalytic degradation of toxic contaminants using Dy2O3-SiO2 ceramic nanostructured materials fabricated by a new, simple and rapid sonochemical approach. *Ultrasonics Sonochemistry, 82*, 105892. doi:10.1016/j.ultsonch.2021.105892 PMID:34959201

Majeed, M., Hakeem, K. R., & Rehman, R. U. (2022). Synergistic effect of plant extract coupled silver nanoparticles in various therapeutic applications- present insights and bottlenecks. *Chemosphere, 288*, 132527. doi:10.1016/j.chemosphere.2021.132527 PMID:34637861

Maji, S. K., Dutta, A. K., Bhadu, G. R., Paul, P., Mondal, A., & Adhikary, B. (2013). A novel amperometric biosensor for hydrogen peroxide and glucose based on cuprous sulfide nanoplates. *Journal of Materials Chemistry. B, Materials for Biology and Medicine*, *1*(33), 4127–4134. doi:10.1039/c3tb20846j PMID:32260965

Maji, S. K., Dutta, A. K., Biswas, P., Srivastava, D. N., Paul, P., Mondal, A., & Adhikary, B. (2012). Synthesis and characterization of FeS nanoparticles and their photocatalytic, electrocatalytic and biomimic peroxidase behavior. *Applied Catalysis A, General*, *419*, 170–177. doi:10.1016/j.apcata.2012.01.025

Maji, S. K., Dutta, A. K., Srivastava, D. N., Paul, P., Mondal, A., & Adhikary, B. (2011). Effective photocatalytic degradation of organic pollutant by ZnS nanocrystals synthesized via thermal decomposition of single-source precursor. *Polyhedron*, *30*(15), 2493–2498. doi:10.1016/j.poly.2011.06.029

Maji, S. K., Dutta, A. K., Srivastava, D. N., Paul, P., Mondal, A., & Adhikary, B. (2012). Peroxidase-like behavior, amperometric biosensing of hydrogen peroxide and photocatalytic activity by cadmium sulfide nanoparticles. *Journal of Molecular Catalysis A Chemical*, *358*, 1–9. doi:10.1016/j.molcata.2012.03.007

Maji, S. K., Sreejith, S., Mandal, A. K., Dutta, A. K., & Zhao, Y. (2014). Synthesis of Ag_2S quantum dots by a single-source precursor: An efficient electrode material for rapid detection of phenol. *Analytical Methods*, *6*(7), 2059–2064. doi:10.1039/c3ay42264j

Malakootian, M., Olama, N., Malakootian, M., & Nasisri, A. (2019). Photocatalytic degradation of metrodinazole from aquatic systems by TiO2- doped Fe3+ nanophotocatalyst. *International Journal of Environmental Science and Technology*, *16*(8), 4275–4284. doi:10.100713762-018-1836-2

Malhotra, S., Verma, A., Tyagi, N., & Kumar, V. (2017). *BIOSENSORS: PRINCIPLE, TYPES AND APPLICATIONS*, *3*(2).

Mali, M. G., Yoon, H., Kim, M., Swihart, M. T., Al-Deyab, S. S., & Yoon, S. S. (2015). Electrosprayed heterojunction $WO_3/BiVO_4$ films with nano textured pillar structure for enhanced photoelectrochemical water splitting. *Applied Physics Letters*, *106*(15), 151603. doi:10.1063/1.4918583

Manikandan, S., Karmegam, N., Subbaiya, R., Karthiga Devi, G., Arulvel, R., Ravindran, B., & Kumar Awasthi, M. (2021). Emerging nano-structured innovative materials as adsorbents in wastewater treatment. *Bioresource Technology*, *320*, 124394. doi:10.1016/j.biortech.2020.124394 PMID:33220545

Ma, Q., Kutilike, B., Kari, N., Abliz, S., & Yimit, A. (2021). Study on surface sensitization of g-C3N4 by functioned different aggregation behavior porphyrin and its optical properties. *Materials Science in Semiconductor Processing*, *121*, 105316. doi:10.1016/j.mssp.2020.105316

Marcus, R. A. (1964). Chemical and electrochemical electron-transfer theory. *Annual Review of Physical Chemistry*, *15*(1), 155–196. doi:10.1146/annurev.pc.15.100164.001103

Marí, B., Singh, K. C., Cembrero-Coca, P., Singh, I., Singh, D., & Chand, S. (2013). Red emitting MTiO3 (M=Ca or Sr) phosphors doped with Eu3+ or Pr3+ with some cations as co-dopants. *Displays*, *34*(4), 346–351. doi:10.1016/j.displa.2013.07.003

Marimuthu, S., Antonisamy, A. J., Malayandi, S., Rajendran, K., Tsai, P. C., Pugazhendhi, A., & Ponnusamy, V. K. (2020). Silver nanoparticles in dye effluent treatment: A review on synthesis, treatment methods, mechanisms, photocatalytic degradation, toxic effects and mitigation of toxicity. *Journal of Photochemistry and Photobiology. B, Biology*, *205*, 111823. doi:10.1016/j.jphotobiol.2020.111823 PMID:32120184

Martinez, S. C., Hernández, S., & Russo, N. (2015). $BiVO_4$ as photocatalyst for solar fuels production through water splitting: A short review. *Applied Catalysis A, General*, *504*, 158–170. doi:10.1016/j.apcata.2014.11.044

Masar, M., Ali, H., Guler, A. C., Urbanek, M., Urbanek, P., Hanulikova, B., Pistekova, H., Annusova, A., Machovsky, M., & Kuritka, I. (2023). Multifunctional bandgap-reduced ZnO nanocrystals for photocatalysis, self-cleaning, and antibacterial glass surfaces. *Colloids and Surfaces. A, Physicochemical and Engineering Aspects*, *656*, 130447. doi:10.1016/j.colsurfa.2022.130447

Matindoust, S., Baghaei-Nejad, M., Abadi, M. H. S., Zou, Z., & Zheng, L.-R. (2016). Food quality and safety monitoring using gas sensor array in intelligent packaging. *Sensor Review*, *36*(2), 169–183. doi:10.1108/SR-07-2015-0115

Ma, X., Zhou, X., Gong, Y., Han, N., Liu, H., & Chen, Y. (2017). MOF-derived hierarchical ZnO/ZnFe 2 O 4 hollow cubes for enhanced acetone gas-sensing performance. *RSC Advances*, *7*(55), 34609–34617. doi:10.1039/C7RA04437B

Ma, Y., Jia, Y. L., Wang, L., Yang, M., Bi, Y. P., & Qia, Y. X. (2016). Exfoliated thin Bi_2MoO_6 nanosheets supported on WO_3 electrode for enhanced photoelectrochemical water splitting. *Applied Surface Science*, *390*, 399–405. doi:10.1016/j.apsusc.2016.08.116

Mehmood, F., Iqbal, J., Jan, T., Gul, A., Mansoor, Q., & Faryal, R. (2017). Structural, photoluminescence, electrical, anti-cancer and visible light driven photocatalytic characteristics of Co doped WO_3 nanoplates. *Vibrational Spectroscopy*, *93*, 78–89. doi:10.1016/j.vibspec.2017.09.005

Mekaru, H., Lu, J., & Tamanoi, F. (2015). Development of mesoporous silica-based nanoparticles with controlled release capability for cancer therapy. *Advanced Drug Delivery Reviews*, *95*, 40–49. doi:10.1016/j.addr.2015.09.009 PMID:26434537

Méndez, E. G., Posada, C. M., & Ocampo, J. M. J. (2022). Statistical analysis of Sr substituted NiFe2O4 thin films for liquefied petroleum gas sensor applications. *Materials Science and Engineering B*, *278*, 115614. doi:10.1016/j.mseb.2022.115614

Meng, S., Sun, W., Zhang, S., Zheng, X., Fu, X., & Chen, S. (2018). Insight in the transfer mechanism of photogenerated carriers for WO3/TiO2 heterojunction photocatalysts is it transfer of Band-Band or Z-scheme? Why? *Journal of Physical Chemistry C*, *122*, 26326-26336.

Meshram, P., & Abhilash. (2022). Strategies for Recycling of Primary and Secondary Resources for Germanium Extraction. *Mining, Metallurgy & Exploration*, *39*(2), 1–19. doi:10.100742461-022-00549-5

Midhuna, A. SD, B. S., & Sreenidhi, P. (2021). *Optimizing Design of Surface Acoustic Wave Gas Sensor.* Paper presented at the 2021 5th International Conference on Trends in Electronics and Informatics (ICOEI).

Migas, D. B., Shaposhnikov, V. L., Rodin, V. N., & Borisenko, V. E. (2010). Tungsten oxides. I: Effects of oxygen vacancies and doping on electronic and optical properties of different phases of WO_3. *Journal of Applied Physics*, *108*(9), 093713. doi:10.1063/1.3505688

Millis, A., & Le Hunte, S. (1997). An overview of semiconductor photocatalysis. *Journal of Photochemistry and Photobiology A Chemistry*, *108*(1), 1–35. doi:10.1016/S1010-6030(97)00118-4

Mi, Q., Zhanaidarova, A., Brunschwig, B. S., Gray, H. B., & Lewis, N. S. (2012). A quantitative assessment of the competition between water and anion oxidation at WO_3 photoanodes in acidic aqueous electrolytes. *Energy & Environmental Science*, *5*(2), 5694–5700. doi:10.1039/c2ee02929d

Mirica, K. A., Azzarelli, J. M., Weis, J. G., Schnorr, J. M., & Swager, T. M. (2013). Rapid prototyping of carbon-based chemiresistive gas sensors on paper. *Proceedings of the National Academy of Sciences of the United States of America*, *110*(35), E3265–E3270. doi:10.1073/pnas.1307251110 PMID:23942132

Mohamed, A. M., Shaban, S. A., El Sayed, H. A., Alanadouli, B. E., & Allam, N. K. (2016). Morphology photoactivity relationship: WO_3 nanostructured films for solar hydrogen production. *International Journal of Hydrogen Energy*, *41*(2), 866–872. doi:10.1016/j.ijhydene.2015.09.108

Mohammed, M. K. (2020). Sol-gel synthesis of Au-doped TiO2 supported SWCNT nanohybrid with visible-light driven photocatalytic for high degradation performance toward methylene blue dye. *Optik (Stuttgart)*, *223*, 165607. doi:10.1016/j.ijleo.2020.165607

Mohd Faheem, M., Bhagat, M., Sharma, P., & Anand, R. (2022). Induction of p53 mediated mitochondrial apoptosis and cell cycle arrest in human breast cancer cells by plant mediated synthesis of silver nanoparticles from Bergenia ligulata (Whole plant). *International Journal of Pharmaceutics*, *619*, 121710. doi:10.1016/j.ijpharm.2022.121710 PMID:35367334

Mokammel, M. A., Islam, M. J., Hasanuzzaman, M., & Hashmi, M. S. J. (2022). Nanoscale Materials for Self-Cleaning and Antibacterial Applications. Encyclopedia of Smart Materials, 315–324. doi:10.1016/B978-0-12-815732-9.00015-2

Molina, G. A., Esparza, R., López-Miranda, J. L., Hernández-Martínez, A. R., España-Sánchez, B. L., Elizalde-Peña, E. A., & Estevez, M. (2019). Green synthesis of Ag nanoflowers using Kalanchoe Daigremontiana extract for enhanced photocatalytic and antibacterial activities. *Colloids and Surfaces. B, Biointerfaces*, *180*, 141–149. doi:10.1016/j.colsurfb.2019.04.044 PMID:31039515

Moo, J. G. S., & Pumera, M. (2015). Chemical Energy Powered Nano/Micro/Macromotors and the Environment. *Chemistry (Weinheim an der Bergstrasse, Germany)*, *21*(1), 58–72. doi:10.1002/chem.201405011 PMID:25410790

Mukhopadhyay, R., Bhaduri, D., Sarkar, B., Rusmin, R., Hou, D., Khanam, R., Sarkar, S., Kumar Biswas, J., Vithanage, M., Bhatnagar, A., & Ok, Y. S. (2020). Clay–polymer nanocomposites: Progress and challenges for use in sustainable water treatment. *Journal of Hazardous Materials*, *383*, 121125. doi:10.1016/j.jhazmat.2019.121125 PMID:31541959

Muthuvel, A., Said, N., Jothibas, M., Gurushankar, K., & Mohana, V. (2021). Microwave assisted green synthesis of nanoscaled: Photocatalyst, antibacterial and antioxidant properties. *Journal of Materials Science Materials in Electronics*, *32*(18), 23522–23539. doi:10.100710854-021-06840-3

Mu, W., Xie, X., Li, X., Zhang, R., Yu, Q., Lv, K., Wei, H., & Jian, Y. (2014). Characterizations of Nb-doped WO_3 nanomaterials and their enhanced photocatalytic performance. *RSC Advances*, *4*(68), 36064. doi:10.1039/C4RA04080E

Nabi, G., Khalid, N., Tahir, M., Rafique, M., Rizwan, M., Hussain, S., ... Majid, A. (2018). A Review on Novel Eco-friendly Green Approach to Synthesis TiO2 nanoparticles using different extracts. *Journal of Inorganic and Organometallic Polymers and Materials*, *28*(4), 15521564. doi:10.100710904-018-0812-0

Naje, A. N., Ibraheem, R. R., & Ibrahim, F. T. (2016). Parametric analysis of NO 2 gas sensor based on carbon nanotubes. *Photonic Sensors*, *6*(2), 153–157. doi:10.100713320-016-0304-1

Nakata, K., & Fujishima, A. (2012). TiO_2 photocatalysis: Design and applications. J. Photochem. Photobiol. C Photochem. *Journal of Photochemistry and Photobiology C, Photochemistry Reviews*, *13*(3), 169–189. doi:10.1016/j.jphotochemrev.2012.06.001

Nasrollahzadeh, M., Shafiei, N., & Nezafat, Z. (2021). Synthesis of biopolymer-based metal nanoparticles. *Biopolymer-Based Metal Nanoparticle Chemistry for Sustainable Applications*, 255–316. doi:10.1016/B978-0-12-822108-2.00018-1

Navarro, Y. R. M. (2009). Water splitting on semiconductor catalysts under visible-light irradiation. *ChemSusChem*, *2*(6), 471–485. doi:10.1002/cssc.200900018 PMID:19536754

Ng, K. H., Minggua, L. J., Mark-Lee, W. F., Arifin, K., Jumali, M. H. H., & Kassim, M. B. (2018). A new method for the fabrication of a bilayer WO_3/Fe_2O_3 photoelectrode for enhanced photoelectrochemical performance. *Materials Research Bulletin*, *98*, 47–52. doi:10.1016/j.materresbull.2017.04.019

Ngo, Y. L. T., Choi, W. M., Chung, J. S., & Hur, S. H. (2019). Highly biocompatible phenylboronic acid-functionalized graphitic carbon nitride quantum dots for the selective glucose sensor. *Sensors and Actuators. B, Chemical*, *282*, 36–44. doi:10.1016/j.snb.2018.11.031

Nguyen, C., & Juang, R.-S. (2019). Efficient removal of methylene blue dye by a hybrid adsorption–photocatalysis process using reduced graphene oxide/titanate nanotube composites for water reuse. *Journal of Industrial and Engineering Chemistry, 76*, 296–309. doi:10.1016/j.jiec.2019.03.054

Nirmala, N., Shriniti, V., Aasresha, K., Arun, J., Gopinath, K. P., Dawn, S. S., Sheeladevi, A., Priyadharsini, P., Birindhadevi, K., Chi, N. T. L., & Pugazhendhi, A. (2022). Removal of toxic metals from wastewater environment by graphene-based composites: A review on isotherm and kinetic models, recent trends, challenges and future directions. *The Science of the Total Environment, 840*, 156564. doi:10.1016/j.scitotenv.2022.156564 PMID:35690214

Noto, L. L., Pitale, S. S., Gusowki, M. A., Terblans, J. J., Ntwaeaborwa, O. M., & Swart, H. C. (2013). Afterglow enhancement with In3+ codoping in CaTiO3:Pr3+ red phosphor. *Powder Technology, 237*, 141–146. doi:10.1016/j.powtec.2013.01.029

Novoselov, K. S., Fal′ko, V. I., Colombo, L., Gellert, P. R., Schwab, M. G., & Kim, K. (. (2012). A roadmap for graphene. *Nature, 490*(7419), 192–200. doi:10.1038/nature11458 PMID:23060189

Nozik, A. J. (1978). Photoelectrochemistry: Applications to solar energy conversion. *Annual Review of Physical Chemistry, 29*(1), 189–222. doi:10.1146/annurev.pc.29.100178.001201

Nujić, M., & Habuda-Stanić, M. (2019). Toxic Metal Ions in Drinking Water and Effective Removal Using Graphene Oxide Nanocomposite. In M. Naushad (Ed.), *A New Generation Material Graphene: Applications in Water Technology* (pp. 373–395). Springer International Publishing. doi:10.1007/978-3-319-75484-0_15

Nukui, Y., Srinivasan, N., Shoji, S., Atarashi, D., Sakai, E., & Miyauchi, M. (2015). Vertically aligned hexagonal WO_3 nanotree electrode for photoelectrochemical water oxidation. *Chemical Physics Letters, 635*, 306–311. doi:10.1016/j.cplett.2015.07.006

Oh, T., Kim, M., Choi, J., & Kim, J. (2018). Design of graphitic carbon nitride nanowires with captured mesoporous carbon spheres for EDLC electrode materials. *Ionics, 24*(12), 3957–3965. doi:10.100711581-018-2544-0

Olagunju, M. O., Zahran, E. M., Reed, J. M., Zeynaloo, E., Shukla, D., Cohn, J. L., Surnar, B., Dhar, S., Bachas, L. G., & Knecht, M. R. (2021). Halide Effects in BiVO4/BiOX Heterostructures Decorated with Pd Nanoparticles for Photocatalytic Degradation of Rhodamine B as a Model Organic Pollutant. *ACS Applied Nano Materials, 4*(3), 3262–3272. doi:10.1021/acsanm.1c00481

Oluwafemi, O. S., Anyik, J. L., Zikalala, N. E., & Sakho, E. H. M. (2019). Biosynthesis of silver nanoparticles from water hyacinth plant leaves extract for colourimetric sensing of heavy metals. *Nano-Structures & Nano-Objects, 20*, 100387. doi:10.1016/j.nanoso.2019.100387

Ong, C. B., Ng, L. Y., & Mohammad, A. W. (2018). A review of ZnO nanoparticles as solar photocatalysts: Synthesis, mechanisms and applications. *Renewable & Sustainable Energy Reviews, 81*, 536–551. doi:10.1016/j.rser.2017.08.020

Ossoliński, K., Ruman, T., Ossoliński, T., Ossolińska, A., Arendowski, A., Kołodziej, A., Płaza-Altamer, A., & Nizioł, J. (2023). Monoisotopic silver nanoparticles-based mass spectrometry imaging of human bladder cancer tissue: Biomarker discovery. *Advances in Medical Sciences*, *68*(1), 38–45. doi:10.1016/j.advms.2022.12.002 PMID:36566601

Osterloh, F. E. (2013). Inorganic nanostructures for photoelectrochemical and photocatalytic water splitting. *Chemical Society Reviews*, *42*(6), 2294–2320. doi:10.1039/C2CS35266D PMID:23072874

Oun, A. A., Shankar, S., & Rhim, J.-W. (2020). Multifunctional nanocellulose/metal and metal oxide nanoparticle hybrid nanomaterials. *Critical Reviews in Food Science and Nutrition*, *60*(3), 435–460. doi:10.1080/10408398.2018.1536966 PMID:31131614

Palmstrom, A. F., Santra, P. K., & Bent, S. F. (2015). Atomic layer deposition in nanostructured photovoltaics: Tuning optical, electronic and surface properties. *Nanoscale*, *7*(29), 12266–12283. doi:10.1039/C5NR02080H PMID:26147328

Pan, G., & Kurumada, K.-I. (2008). Hybrid gel reinforced with coating layer for removal of phenol from aqueous solution. *Chemical Engineering Journal*, *138*(1), 194–199. doi:10.1016/j.cej.2007.06.025

Parlayici, S., Eskizeybek, V., Avcı, A., & Pehlivan, E. (2015). Removal of chromium (VI) using activated carbon-supported-functionalized carbon nanotubes. *Journal of Nanostructure in Chemistry*, *5*(3), 255–263. doi:10.100740097-015-0156-z

Pei, Y., Zhang, X., Hui, Z., Zhou, J., Huang, X., Sun, G., & Huang, W. (2021). Ti3C2TX MXene for sensing applications: Recent progress, design principles, and future perspectives. *ACS Nano*, *15*(3), 3996–4017. doi:10.1021/acsnano.1c00248 PMID:33705113

Pena-Pereira, F., Lavilla, I., de la Calle, I., Romero, V., & Bendicho, C. (2023). Detection of gases and organic vapors by cellulose-based sensors. *Analytical and Bioanalytical Chemistry*, *415*(18), 4039–4060. Advance online publication. doi:10.100700216-023-04649-z PMID:37000210

Peng, F., Xu, T., Wu, F., Ma, C., Liu, Y., Li, J., Zhao, B., & Mao, C. (2017). Novel biomimetic enzyme for sensitive detection of superoxide anions. *Talanta*, *174*, 82–91. doi:10.1016/j.talanta.2017.05.028 PMID:28738660

Persaud, K., & Dodd, G. (1982). Analysis of discrimination mechanisms in the mammalian olfactory system using a model nose. *Nature*, *299*(5881), 352–355. doi:10.1038/299352a0 PMID:7110356

Phattepur, H., Nagaraju, G., Shruthi, R., Sandhya, R., & Banupriya, C. (2023). Synthesis of Zn doped black TiO2 nanoparticles for degradation of 2, 4, 6 tri-chloro-phenol under visible light. *Journal of the Indian Chemical Society*, *100*, 100881. doi:10.1016/j.jics.2023.100881

Philip, D., Unni, C., Aromal, S. A., & Vidhu, V. K. (2011). Murraya Koenigii leaf-assisted rapid green synthesis of silver and gold nanoparticles. *Spectrochimica Acta. Part A: Molecular and Biomolecular Spectroscopy*, *78*(2), 899–904. doi:10.1016/j.saa.2010.12.060 PMID:21215687

Pleskov, Y. V., & Gurevich, Y. Y. (1986). *Semiconductor Photoelectrochemistry*. Consultants Bureau. doi:10.1007/978-1-4684-9078-7

Poli, J.-P., Mer-Calfati, C., Scorsone, E., & Saada, S. (2017). SAW sensor's frequency shift characterization for odor recognition and concentration estimation. *IEEE Sensors Journal*, *17*(21), 7011–7018. doi:10.1109/JSEN.2017.2751666

Poole, A. (2004). Treatment of biorefactory organic compounds in wool scour effluents by hydroxyl radical oxidation. *Water Research*, *38*(14-15), 3458–3464. doi:10.1016/j.watres.2004.06.001 PMID:15276763

Pooseekheaw, P., Thongpan, W., Panthawan, A., Kantarak, E., Sroila, W., & Singjai, P. (2020). Porous V2O5/TiO2 Nanoheterostructure Films with Enhanced Visible-Light Photocatalytic Performance Prepared by the Sparking Method. *Molecules (Basel, Switzerland)*, *25*(15), 3327. Advance online publication. doi:10.3390/molecules25153327 PMID:32707984

Popov, V. N. (2004). Carbon nanotubes: Properties and application. *Materials Science and Engineering R Reports*, *43*(3), 61–102. doi:10.1016/j.mser.2003.10.001

Porosoff, M. D., Yan, B., & Chen, J. G. (2016). Catalytic reduction of CO_2 by H_2 for synthesis of CO, methanol and hydrocarbons: Challenges and opportunities. *Energy & Environmental Science*, *9*(1), 62–73. doi:10.1039/C5EE02657A

Potdar, S., Huang, C.-M., Praveen, B., Manickam, S., & Sonawane, S. (2022). Highly photoactive titanium dioxide supported platinum catalyst: Synthesis using cleaner ultrasound approach. *Catalysts*, *12*(1), 78. doi:10.3390/catal12010078

Power, A. C., Gorey, B., Chandra, S., & Chapman, J. (2018). Carbon nanomaterials and their application to electrochemical sensors: a review. *7*(1), 19-41. 10.1515/ntrev-2017-0160

Pragathiswaran, C., Violetmary, J., Faritha, A., Selvarani, K., & Nawas, P. M. A. (2021). Photocatalytic degradation, sensing of Cd2+ using silver nanoparticles synthesised from plant extract of cissus quadrangularis and their microbial activity. *Materials Today: Proceedings*, *45*, 3348–3356. doi:10.1016/j.matpr.2020.12.656

Prasannalakshmi, P., & Shanmugam, N. (2017). Fabrication of TiO2/ZnO nanocomposites for solar energy driven photocatalysis. *Materials Science in Semiconductor Processing*, *61*, 114–124. doi:10.1016/j.mssp.2017.01.008

Purohit, B., Vernekar, P. R., Shetti, N. P., & Chandra, P. (2020). Biosensor nanoengineering: Design, operation, and implementation for biomolecular analysis. *Sensors International*, *1*, 100040. doi:10.1016/j.sintl.2020.100040

Qin, D. D., Tao, C. L., Friesen, S. A., Wang, T.-H., Varghese, O. K., Bao, N.-Z., Yang, Z.-Y., Mallouk, T. E., & Grimes, C. A. (2011). Dense layers of vertically oriented WO_3 crystals as anodes for photoelectrochemical water oxidation. *Chemical Communications*, *48*(5), 729–731. doi:10.1039/C1CC15691H PMID:22117192

Qin, R., Li, F., Chen, M., & Jiang, W. (2009). Preparation of chitosan–ethylenediaminetetraacetate-enwrapped magnetic CoFe2O4 nanoparticles via zero-length emulsion crosslinking method. *Applied Surface Science*, *256*(1), 27–32. doi:10.1016/j.apsusc.2009.07.032

Qi, X. P., She, G. W., Liu, Y. Y., Mua, L. X., & Shi, W. S. (2012). Electrochemical synthesis of CdS/ZnO nanotube arrays with excellent photoelectrochemical properties. *Chemical Communications*, *48*(2), 242–244. doi:10.1039/C1CC15674H PMID:22083259

Rafiq, A., Ikram, M., Ali, S., Niaz, F., Khan, M., Khan, Q., & Maqbool, M. (2021). Photocatalytic degradation of dyes using semiconductor photocatalysts to clean industrial water pollution. *Journal of Industrial and Engineering Chemistry*, *97*, 111–128. doi:10.1016/j.jiec.2021.02.017

Rahman, M. Z., Davey, K., & Mullins, C. B. (2018). Tuning the intrinsic properties of carbon nitride for high quantum yield photocatalytic hydrogen production. *Advancement of Science*, *5*(10), 1800820. doi:10.1002/advs.201800820 PMID:30356987

Rajapaksha, P. P., Power, A., Chandra, S., & Chapman, J. (2018). Graphene, electrospun membranes and granular activated carbon for eliminating heavy metals, pesticides and bacteria in water and wastewater treatment processes. *Analyst (London)*, *143*(23), 5629–5645. doi:10.1039/C8AN00922H PMID:30357153

Raj, V. B., Singh, H., Nimal, A., Tomar, M., Sharma, M., & Gupta, V. (2013). Effect of metal oxide sensing layers on the distinct detection of ammonia using surface acoustic wave (SAW) sensors. *Sensors and Actuators. B, Chemical*, *187*, 563–573. doi:10.1016/j.snb.2013.04.063

Ramachandiran, D., & Rajesh, K. (2022). Selective colorimetric and fluorimertic sensor of Hg (II) ion from silver nanoparticles using Acacia chundra leaves extract. *Materials Chemistry and Physics*, *287*, 126284. doi:10.1016/j.matchemphys.2022.126284

Ramu, A., Umar, A., Gopi, S., Algadi, H., Albargi, H., Ibrahim, A. A., Alsaiari, M. A., Wang, Y., & Choi, D. (2021). Tetracyanonickelate (II)/KOH/reduced graphene oxide fabricated carbon felt for mediated electron transfer type electrochemical sensor for efficient detection of N2O gas at room temperature. *Environmental Research*, *201*, 111591. doi:10.1016/j.envres.2021.111591 PMID:34186081

Rana, A., Yadav, K., & Jagadevan, S. (2020). A comprehensive review on green synthesis of nature-inspired metal nanoparticles: Mechanism, application and toxicity. *Journal of Cleaner Production*, *272*, 122880. doi:10.1016/j.jclepro.2020.122880

Rao, P. M., Cai, L. L., Liu, C., Cho, I. S., Lee, C. H., Weisse, J. M., Yang, P. D., & Zheng, X. L. (2014). Simultaneously efficient light absorption and charge separation in WO_3/$BiVO_4$ core/shell nanowire photoanode for photoelectrochemical water oxidation. *Nano Letters*, *14*(2), 1099–1105. doi:10.1021/nl500022z PMID:24437363

Rao, P. M., Cho, I. S., & Zheng, X. (2013). Flame synthesis of WO_3 nanotubes and nanowires for efficient photoelectrochemical water-splitting. *Proceedings of the Combustion Institute*, *34*(2), 2187–2195. doi:10.1016/j.proci.2012.06.122

Rasheed, T., Li, C., Fu, L., Nabeel, F., Yu, C., Gong, L., & Zhou, Y. (2018). Development and characterization of newly engineered chemosensor with intracellular monitoring potentialities and lowest detection of toxic elements. *Journal of Molecular Liquids*, *272*, 440–449. doi:10.1016/j.molliq.2018.09.112

Rasheed, T., Li, C., Nabeel, F., Huang, W., & Zhou, Y. (2019). Self-assembly of alternating copolymer vesicles for the highly selective, sensitive and visual detection and quantification of aqueous Hg2+. *Chemical Engineering Journal*, *358*, 101–109. doi:10.1016/j.cej.2018.09.216

Rasheed, T., Li, C., Zhang, Y., Nabeel, F., Peng, J., Qi, J., Gong, L., & Yu, C. (2018). Rhodamine-based multianalyte colorimetric probe with potentialities as on-site assay kit and in biological systems. *Sensors and Actuators. B, Chemical*, *258*, 115–124. doi:10.1016/j.snb.2017.11.100

Rasheed, T., Nabeel, F., Adeel, M., Rizwan, K., Bilal, M., & Iqbal, H. M. N. (2019). Carbon nanotubes-based cues: A pathway to future sensing and detection of hazardous pollutants. *Journal of Molecular Liquids*, *292*, 111425. doi:10.1016/j.molliq.2019.111425

Rasheed, T., Nabeel, F., Li, C., & Bilal, M. (2019). Rhodamine-assisted fluorescent strategy for the sensitive and selective in-field mapping of environmental pollutant Hg(II) with potential bioimaging. *Journal of Luminescence*, *208*, 519–526. doi:10.1016/j.jlumin.2019.01.032

Rasheed, T., Nabeel, F., Li, C., & Zhang, Y. (2019). Rhodol assisted alternating copolymer based chromogenic vesicles for the aqueous detection and quantification of hydrazine via switch-on strategy. *Journal of Molecular Liquids*, *274*, 461–469. doi:10.1016/j.molliq.2018.11.014

Rasheed, T., Nabeel, F., & Shafi, S. (2019). Chromogenic vesicles for aqueous detection and quantification of Hg2+/Cu2+ in real water samples. *Journal of Molecular Liquids*, *282*, 489–498. doi:10.1016/j.molliq.2019.03.048

Reddy, D. A., Choi, J., Lee, S., Ma, R., & Kim, T. K. (2015). Self-assembled macro porous ZnS–graphene aerogels for photocatalytic degradation of contaminants in water. *RSC Advances*, *5*(24), 18342–18351. doi:10.1039/C4RA16494F

Reddy, D. H. K., & Lee, S.-M. (2013). Application of magnetic chitosan composites for the removal of toxic metal and dyes from aqueous solutions. *Advances in Colloid and Interface Science*, *201-202*, 68–93. doi:10.1016/j.cis.2013.10.002 PMID:24182685

Reddy, I. N., Jang, W. Y., Manjunath, V., & Shim, J. (2021). Fe_3O_4-Bi_2O_3 nanostructures for efficient energy generation application-water oxidation under visible light irradiation. *Journal of Electroanalytical Chemistry (Lausanne, Switzerland)*, *895*, 115484. doi:10.1016/j.jelechem.2021.115484

Rivas, G. A., Rubianes, M. D., Rodriguez, M. C., Ferreyra, N. F., Luque, G. L., Pedano, M. L., Miscoria, S. A., & Parrado, C. (2007). Carbon nanotubes for electrochemical biosensing. *Talanta*, *74*(3), 291–307. doi:10.1016/j.talanta.2007.10.013 PMID:18371643

Rizzi, V., Gubitosa, T., Signovile, R., Fini, P., Cecone, P., Matencio, A., & Cosma, P. (2021). Cyclodextrin nanosponges as adsorbent material to remove hazardous pollutants from water. *Chemical Engineering Journal, 411*, 128514. doi:10.1016/j.cej.2021.128514

Rodríguez-Pérez, M., Chacón, C., Palacios-González, E., Rodríguez-Gattorno, G., & Oskam, G. (2014). Photoelectrochemical water oxidation at electrophoretically deposited WO_3 films as a function of crystal structure and morphology. *Electrochimica Acta, 140*, 320–331. doi:10.1016/j.electacta.2014.03.022

Rono, N., Kibet, J. K., Martincigh, B. S., & Nyamori, V. O. (2020). A review of the current status of graphitic carbon nitride. *Critical Reviews in Solid State and Material Sciences, 46*(3), 189–217. doi:10.1080/10408436.2019.1709414

Rust, C., Li, H., Gordeev, G., Spari, M., Guttmann, M., Jin, Q., Reich, S., & Flavel, B. S. (2022). Global Alignment of Carbon Nanotubes via High Precision Microfluidic Dead-End Filtration. *Advanced Functional Materials, 32*(10), 2107411. doi:10.1002/adfm.202107411

Sahebi, S., Sheikhi, M., & Ramavandi, B. (2019). A new biomimetic aquaporin thin film composite membrane for forward osmosis: Characterization and performance assessment. *Desalination and Water Treatment, 148*, 42–50. doi:10.5004/dwt.2019.23748

Sahnesarayi, M. K., Rastegari, S., & Sarpoolaky, H. (2022). Enhanced photoelectrochemical water splitting performance of vertically aligned Bi_2O_3 nanosheet arrays derived from chemical bath deposition method by controlling chemical bath temperature and complexing agent concentration. *Surfaces and Interfaces, 30*, 101819. doi:10.1016/j.surfin.2022.101819

Said, K., Ismail, A. F., Karim, Z., Abdullah, M., & Hafeez, A. (2021). A review of technologies for the phenolic compounds recovery and phenol removal from wastewater. *Process Safety and Environmental Protection, 151*, 257–289. doi:10.1016/j.psep.2021.05.015

Saito, R., Miseki, Y., & Sayama, K. (2012). Highly efficient photoelectrochemical water splitting using a thin film photoanode of $BiVO_4/SnO_2/WO_3$ multi-composite in a carbonate electrolyte. *Chemical Communications (Cambridge), 48*(32), 3833–3835. doi:10.1039/c2cc30713h PMID:22437791

Sakhuja, N., Jha, R. K., Chaurasiya, R., Dixit, A., & Bhat, N. (2020). 1T-phase titanium disulfide nanosheets for sensing H2S and O2. *ACS Applied Nano Materials, 3*(4), 3382–3394. doi:10.1021/acsanm.0c00127

Sakthivel, S., Janczarek, M., & Kisch, H. (2004). Visible light activity and photoelectrochemical properties of nitrogen-doped TiO2. *The Journal of Physical Chemistry B, 108*(50), 19384–19387. doi:10.1021/jp046857q

Salam, M. A., Makki, M. S. I., & Abdelaal, M. Y. A. (2011). Preparation and characterization of multi-walled carbon nanotubes/chitosan nanocomposite and its application for the removal of heavy metals from aqueous solution. *Journal of Alloys and Compounds, 509*(5), 2582–2587. doi:10.1016/j.jallcom.2010.11.094

Salim, W., & Ho, W. S. W. (2015). Recent developments on nanostructured polymer-based membranes. *Current Opinion in Chemical Engineering*, *8*, 76–82. doi:10.1016/j.coche.2015.03.003

Salje, E. (1976). Structural phase transitions in the system WO_3-$NaWO_3$. *Ferroelectrics*, *12*(1), 215–217. doi:10.1080/00150197608244166

Salje, E. (1977). The orthorhombic phase of WO_3. *Acta Crystallographica. Section B, Structural Crystallography and Crystal Chemistry*, *33*(2), 574–577. doi:10.1107/S0567740877004130

Sankar Sana, S., Haldhar, R., Parameswaranpillai, J., Chavali, M., & Kim, S. C. (2022). Silver nanoparticles-based composite for dye removal: A comprehensive review. *Cleaner Materials*, *6*, 100161. doi:10.1016/j.clema.2022.100161

Sankareswari, M., Amutha, C., Vasantha, V. S., Arunpandian, M., & Nagarajan, E. R. (2022). Biosynthesized silver nanoparticles using Rosary Pea seed Extract: Evaluation of Antibacterial, cytotoxic and photocatalytic activity. *Inorganic Chemistry Communications*, *145*, 109977. doi:10.1016/j.inoche.2022.109977

Santhosh, A., Theertha, V., Prakash, P., & Smitha Chandran, S. (2021). From waste to a value added product: Green synthesis of silver nanoparticles from onion peels together with its diverse applications. *Materials Today: Proceedings*, *46*, 4460–4463. doi:10.1016/j.matpr.2020.09.680

Sarkar, A., Ghosh, A. B., Saha, N., Dutta, A. K., Srivastava, D. N., Paul, P., & Adhikary, B. (2015). Enhanced photocatalytic activity of Eu-doped Bi_2S_3 nanoflowers for degradation of organic pollutants under visible light illumination. *Catalysis Science & Technology*, *5*(8), 4055–4063. doi:10.1039/C5CY00473J

Schiff, H. (1859). Zur Nachweisung der Harnsäure. *Justus Liebigs Annalen der Chemie*, *109*(1), 65–71. doi:10.1002/jlac.18591090107

Seabold, J. A., & Choi, K. S. (2011). Effect of a cobalt-based oxygen evolution catalyst on the stability and the selectivity of photo-oxidation reactions of a WO_3 photoanode. *Chemistry of Materials*, *23*(5), 1105–1112. doi:10.1021/cm1019469

Seo, H., Park, G., Oh, K. H., Kang, S. H., Lee, H. C., Cho, S. K., & Nam, K. M. (2017). Analysis of charge separation processes in WO_3-$BiVO_4$ composite for efficient photoelectrochemical water oxidation. *Journal of Electroanalytical Chemistry (Lausanne, Switzerland)*, *789*, 17–23. doi:10.1016/j.jelechem.2017.02.021

Shaddad, M. N., Arunachalam, P., Hezam, M. A. L., Saeedan, N. M., Gimenez, S., Bisquert, J., & Al Mayouf, A. M. (2021). Unprecedented solar water splitting of dendritic nanostructured Bi_2O_3 films by combined oxygen vacancy formation and Na_2MoO_4 doping. *International Journal of Hydrogen Energy*, *46*(46), 23702–23714. doi:10.1016/j.ijhydene.2021.04.184

Shannon, M. A., Bohn, P. W., Elimelech, M., Georgiadis, J. G., Mariñas, B. J., & Mayes, A. M. (2009). Science and technology for water purification in the coming decades. In *Nanoscience and Technology* (pp. 337–346). Co-Published with Macmillan Publishers Ltd. doi:10.1142/9789814287005_0035

Shao, D., Hu, J., & Wang, X. (2010). Plasma Induced Grafting Multiwalled Carbon Nanotube with Chitosan and Its Application for Removal of UO, Cu2+, and Pb2+ from Aqueous Solutions. *Plasma Processes and Polymers*, *7*(12), 977–985. doi:10.1002/ppap.201000062

Sharma, S., Mehta, S. K., Ibhadon, A. O., & Kansal, S. K. (2019). Fabrication of novel carbon quantum dots modified bismuth oxide (α-BO/C-dots): Material properties and catalytic applications. *Journal of Colloid and Interface Science*, *533*, 227–237. doi:10.1016/j.jcis.2018.08.056 PMID:30165300

Shekhirev, M., Lipatov, A., Torres, A., Vorobeva, N. S., Harkleroad, A., Lashkov, A., Sysoev, V., & Sinitskii, A. (2020). Highly selective gas sensors based on graphene nanoribbons grown by chemical vapor deposition. *ACS Applied Materials & Interfaces*, *12*(6), 7392–7402. doi:10.1021/acsami.9b13946 PMID:32011111

Shen, Y., Saboe, P. O., Sines, I. T., Erbakan, M., & Kumar, M. (2014). Biomimetic membranes: A review. *Journal of Membrane Science*, *454*, 359–381. doi:10.1016/j.memsci.2013.12.019

Shin, S., Han, H. S., Kim, J. S., Park, I. J., Lee, M. H., Hong, K. S., & Cho, I. S. (2015). A tree-like nanoporous WO_3 photoanode with enhanced charge transport efficiency for photoelectrochemical water oxidation. *Journal of Materials Chemistry. A, Materials for Energy and Sustainability*, *3*(24), 12920–12926. doi:10.1039/C5TA00823A

Shi, Y., Wang, C.-H., & Jing, X. P. (2008).. . *Journal of the Electrochemical Society*, *155*(6), J148–J151. doi:10.1149/1.2898897

Shukla, S., Masih, A., Aryan, & Mehata, M. S. (2022). Catalytic activity of silver nanoparticles synthesized using Crinum asiaticum (Sudarshan) leaf extract. *Materials Today: Proceedings*, *56*, 3714–3720. doi:10.1016/j.matpr.2021.12.468

Singh, J., Srivastava, M., Dutta, J., & Dutta, P. K. (2011). Preparation and properties of hybrid monodispersed magnetic α-Fe2O3 based chitosan nanocomposite film for industrial and biomedical applications. *International Journal of Biological Macromolecules*, *48*(1), 170–176. doi:10.1016/j.ijbiomac.2010.10.016 PMID:21056054

Singh, J., Srivastava, M., Kalita, P., & Malhotra, B. D. (2012). A novel ternary NiFe2O4/CuO/FeO-chitosan nanocomposite as a cholesterol biosensor. *Process Biochemistry (Barking, London, England)*, *47*(12), 2189–2198. doi:10.1016/j.procbio.2012.08.012

Siqueira, G., Bras, J., & Dufresne, A. (2010). Cellulosic Bionanocomposites: A Review of Preparation, Properties and Applications. *Polymers*, *2*(4), 728–765. doi:10.3390/polym2040728

Smith, S. C., & Rodrigues, D. F. (2015). Carbon-based nanomaterials for removal of chemical and biological contaminants from water: A review of mechanisms and applications. *Carbon*, *91*, 122–143. doi:10.1016/j.carbon.2015.04.043

Soltaninejad, V., Ahghari, M. R., Taheri-Ledari, R., & Maleki, A. (2021). Bifunctional PVA/ZnO/AgI/Chlorophyll Nanocomposite Film: Enhanced Photocatalytic Activity for Degradation of Pollutants and Antimicrobial Property under Visible-Light Irradiation. *Langmuir*, *37*(15), 4700–4713. doi:10.1021/acs.langmuir.1c00501 PMID:33821643

Som, S., Kunti, A. K., Kumar, V., Kumar, V., Dutta, S., Chowdhury, M., Sharma, S. K., Terblans, J. J., & Swart, H. C. (2014). Defect correlated fluorescent quenching and electron phonon coupling in the spectral transition of Eu3+ in CaTiO3 for red emission in display application. *Journal of Applied Physics, 115*(19), 193101. doi:10.1063/1.4876316

Song, H., Li, Y., Lou, Z., Xiao, M., Hu, L., Ye, Z., & Zhu, L. (2015). Synthesis of Fe-doped WO_3 nanostructures with high visible-light-driven photocatalytic activities. *Applied Catalysis B: Environmental*, *166–167*, 112–120. doi:10.1016/j.apcatb.2014.11.020

Song, X., Hu, R., Xu, S., Liu, Z., Wang, J., Shi, Y., Xu, J., Chen, K., & Yu, L. (2021). Highly Sensitive Ammonia Gas Detection at Room Temperature by Integratable Silicon Nanowire Field-Effect Sensors. *ACS Applied Materials & Interfaces*, *13*(12), 14377–14384. doi:10.1021/acsami.1c00585 PMID:33750109

Srivastava, R., & Yadav, B. C. (2012). Ferrite materials: Introduction, synthesis techniques, and applications as sensors. *International Journal of Green Nanotechnology*, *4*(2), 141–154. doi:10.1080/19430892.2012.676918

Su, J. Z., Guo, L. J., Bao, N. Z., & Grimes, C. A. (2011). Nanostructured WO_3/$BiVO_4$ heterojunction films for efficient photoelectrochemical water splitting. *Nano Letters*, *11*(5), 1928–1933. doi:10.1021/nl2000743 PMID:21513345

Su, J. Z., Zhang, T., & Wang, L. (2017). Engineered WO_3 nanorods for conformal growth of WO_3/$BiVO_4$ core-shell heterojunction towards efficient photoelectrochemical water oxidation. *Journal of Materials Science Materials in Electronics*, *28*(5), 4481–4491. doi:10.100710854-016-6082-0

Sumi, M. B., Devadiga, A., Shetty, K. V., & Saidutta, M. B. (2016). Solar photocatalytically active, engineered silver nanoparticle synthesis using aqueous extract of mesocarp of Cocos nucifera (Red Spicata Dwarf), *12*(1), 14–32. doi:10.1080/17458080.2016.1251622

Sun, M., Li, D., Li, W., Chen, Y., Chen, Z., He, Y., & Fu, X. (2008). New Photocatalyst, Sb_2S_3, for Degradation of Methyl Orange under Visible-Light Irradiation. *The Journal of Physical Chemistry C*, *112*(46), 18076–18081. doi:10.1021/jp806496d

Sun, S. P., Gu, S., Sun, J. H., Xia, F. F., & Chen, G. H. (2018). First principles investigation of the electronic properties of graphitic carbon nitride with different building block and sheet staggered arrangement. *Journal of Alloys and Compounds*, *735*, 131–139. doi:10.1016/j.jallcom.2017.11.061

Sun, T., Li, Z., Liu, X., Ma, L., Wang, J., & Yang, S. (2016). Facile construction of 3D graphene/MoS2 composites as advanced electrode materials for supercapacitors. *Journal of Power Sources*, *331*, 180–188. doi:10.1016/j.jpowsour.2016.09.036

Suzuki, T. M., Kobayashi, S., Pacheco Tanaka, D. A., Llosa Tanco, M. A., Nagase, T., & Onodera, Y. (2004). Separation and concentration of trace Pb(II) by the porous resin loaded with α-zirconium phosphate crystals. *Reactive & Functional Polymers*, *58*(2), 131–138. doi:10.1016/j.reactfunctpolym.2003.11.012

Tahir, M. B., Sohaib, M., Sagir, M., & Rafique, M. (2022). Role of Nanotechnology in Photocatalysis. Encyclopedia of Smart Materials, 578–589. doi:10.1016/B978-0-12-815732-9.00006-1

Tai, H., Duan, Z., Wang, Y., Wang, S., & Jiang, Y. (2020). Paper-Based Sensors for Gas, Humidity, and Strain Detections: A Review. *ACS Applied Materials & Interfaces*, *12*(28), 31037–31053. doi:10.1021/acsami.0c06435 PMID:32584534

Takafuji, M., Ide, S., Ihara, H., & Xu, Z. (2004). Preparation of Poly(1-vinylimidazole)-Grafted Magnetic Nanoparticles and Their Application for Removal of Metal Ions. *Chemistry of Materials*, *16*(10), 1977–1983. doi:10.1021/cm030334y

Tan, E., Kahyaoğlu, İ. M., & Karakuş, S. (2021). A sensitive and smartphone colorimetric assay for the detection of hydrogen peroxide based on antibacterial and antifungal matcha extract silver nanoparticles enriched with polyphenol. *Polymer Bulletin*, 1–27. doi:10.1007/S00289-021-03857-W/TABLES/5 PMID:34413556

Tang, W.-W., Zeng, G.-M., Gong, J.-L., Liu, Y., Wang, X.-Y., Liu, Y.-Y., Liu, Z.-F., Chen, L., Zhang, X.-R., & Tu, D.-Z. (2012). Simultaneous adsorption of atrazine and Cu (II) from wastewater by magnetic multi-walled carbon nanotube. *Chemical Engineering Journal*, *211-212*, 470–478. doi:10.1016/j.cej.2012.09.102

Tang, Z., Zhang, X., Duan, L., Wu, A., & Lu, W. (2019). Three-Dimensional Carbon Nitride Nanowire Scaffold for Flexible Supercapacitors. *Nanoscale Research Letters*, *14*(1), 98. doi:10.118611671-019-2932-z PMID:30874966

Tanisaki, S. (1960). Crystal structure of monoclinic tungsten trioxideat room temperature. *Journal of the Physical Society of Japan*, *15*(4), 573–581. doi:10.1143/JPSJ.15.573

Tee, S. Y., Win, K. Y., & Teo, W. S. (2017). Recent progress in energy-driven water splitting. *Advancement of Science*, 1600337. PMID:28546906

Teodoro, K. B. R., Sanfelice, R. C., Migliorini, F. L., Pavinatto, A., Facure, M. H. M., & Correa, D. S. (2021). A Review on the Role and Performance of Cellulose Nanomaterials in Sensors. *ACS Sensors*, *6*(7), 2473–2496. doi:10.1021/acssensors.1c00473 PMID:34182751

Terashi, Y., Purwanto, A., Wang, W., Iskandar, F., & Okuyama, K. (2008). Role of urea addition in the preparation of tetragonal BaTiO3 nanoparticles using flame-assisted spray pyrolysis. *Journal of the European Ceramic Society*, *28*(13), 2573–2580. doi:10.1016/j.jeurceramsoc.2008.03.002

Tetyana, P., Shumbula, P. M., & Njengele-Tetyana, Z. (2021). *Biosensors: Design, Development and Applications*. Nanopores. doi:10.5772/intechopen.97576

Thambiliyagodage, C. (2022). Efficient photocatalysis of carbon coupled TiO2 to degrade pollutants in wastewater – A review. *Environmental Nanotechnology, Monitoring & Management, 18*, 100737. doi:10.1016/j.enmm.2022.100737

Thomas, S., Saji, K. J., & Jayaraj, M. K. (2022). An introduction to biosensors. *Nanomaterials for Sensing and Optoelectronic Applications*, 91–107. doi:10.1016/B978-0-12-824008-3.00013-8

Tortella, G., Rubilar, O., Diez, M. C., Cuozzo, S., Pieretti, J. C., & Seabra, A. B. (2022). Role of bacteria and actinobacteria in the biosynthesis of silver nanoparticles. *Green Synthesis of Silver Nanomaterials*, 493–512. doi:10.1016/B978-0-12-824508-8.00003-4

Tounsadi, H., Khalidi, A., Abdennouri, M., & Barka, N. (2016). Activated carbon from Diplotaxis Harra biomass: Optimization of preparation conditions and heavy metal removal. *Journal of the Taiwan Institute of Chemical Engineers, 59*, 348–358. doi:10.1016/j.jtice.2015.08.014

Travis, W., Glover, E. N. K., Bronstein, H., Scanlon, D. O., & Palgrave, R. G. (2016). On the application of the tolerance factor to inorganic and hybrid halide perovskites: A revised system. *Chemical Science (Cambridge), 7*(7), 4548–4556. doi:10.1039/C5SC04845A PMID:30155101

Tyagi, S., Kumar, A., Kumar, A., Gautam, Y. K., Kumar, V., Kumar, Y., & Singh, B. P. (2022). Enhancement in the sensitivity and selectivity of Cu functionalized MoS2 nanoworm thin films for nitrogen dioxide gas sensor. *Materials Research Bulletin, 150*, 111784. doi:10.1016/j.materresbull.2022.111784

Uddin, I., Ahmad, K., Khan, A. A., & Kazmi, M. A. (2017). Synthesis of silver nanoparticles using Matricaria recutita (Babunah) plant extract and its study as mercury ions sensor. *Sensing and Bio-Sensing Research, 16*, 62–67. doi:10.1016/j.sbsr.2017.11.005

Vaid, P., Raizada, P., Saini, A. K., & Saini, R. V. (2020). Biogenic silver, gold and copper nanoparticles - A sustainable green chemistry approach for cancer therapy. *Sustainable Chemistry and Pharmacy, 16*(August 2019). doi:10.1016/j.scp.2020.100247

Valdés, Á., & Kroes, G. J. (2009). First principles study of the photo oxidation of water on tungsten trioxide (WO_3). *The Journal of Chemical Physics, 130*(11), 114701. doi:10.1063/1.3088845 PMID:19317549

Van der Schueren, L., De Clerck, K., Brancatelli, G., Rosace, G., Van Damme, E., & De Vos, W. (2012). Novel cellulose and polyamide halochromic textile sensors based on the encapsulation of Methyl Red into a sol–gel matrix. *Sensors and Actuators. B, Chemical, 162*(1), 27–34. doi:10.1016/j.snb.2011.11.077

Velasco-Garcia, M. N., & Mottram, T. (2003). Biosensor Technology addressing Agricultural Problems. *Biosystems Engineering, 84*(1), 1–12. doi:10.1016/S1537-5110(02)00236-2

Venkatesan, A., Al-onazi, W. A., Elshikh, M., Pham, T. H., Suganya, S., Boobas, S., & Priyadharsan, A. (2022). Study of synergistic effect of cobalt and carbon codoped with TiO2 photocatalyst for visible light induced degradation of phenol. *Chemosphere, 305*, 135333. doi:10.1016/j.chemosphere.2022.135333 PMID:35709834

Verma, R., Asthana, A., Singh, A. K., Prasad, S., & Susan, M. A. B. H. (2017). Novel glycine-functionalized magnetic nanoparticles entrapped calcium alginate beads for effective removal of lead. *Microchemical Journal, 130*, 168–178. doi:10.1016/j.microc.2016.08.006

Vigneshvar, S., Sudhakumari, C. C., Senthilkumaran, B., & Prakash, H. (2016). Recent advances in biosensor technology for potential applications - an overview. *Frontiers in Bioengineering and Biotechnology, 4*(FEB), 11. doi:10.3389/fbioe.2016.00011 PMID:26909346

Vishali, D., Manikandan, B., John, R., & Murali, K. (2021). Degradation of Methylene Blue Dye by TiO2 Nanoparticles BIologically Synthesized Using Leaf Extract of Citrus Aurantifolia. *Transactions on Electrical and Electronic Materials, 22*(5), 622–629. doi:10.100742341-020-00270-4

Voloshina, E., Mollenhauer, D., Chiappisi, L., & Paulus, B. (2011). Theoretical study on the adsorption of pyridine derivatives on graphene. *Chemical Physics Letters, 510*(4-6), 220–223. doi:10.1016/j.cplett.2011.05.025

Waich, K., Mayr, T., & Klimant, I. (2008). Fluorescence sensors for trace monitoring of dissolved ammonia. *Talanta, 77*(1), 66–72. doi:10.1016/j.talanta.2008.05.058 PMID:18804600

Waisman, H., Rozenberg, J., Sassi, O., & Hourcade, J. C. (2012). Peakoil profiles through the lens of a general equilibrium assessment. *Energy Policy, 48*, 744–753. doi:10.1016/j.enpol.2012.06.005

Walter, M. G., Warren, E. L., McKone, J. R., Boettcher, S. W., Mi, Q., Santori, E. A., & Lewis, N. S. (2010). Solar water splitting cells. *Chemical Reviews, 110*(11), 6446–6473. doi:10.1021/cr1002326 PMID:21062097

Wang, D. P., Bassi, P. S., Qi, H., Zhao, X., Gurudayal, Wong, L., Xu, R., Sritharan, T., & Chen, Z. (2016). Improved charge separation in $WO_3/CuWO_4$ composite photoanodes for photoelectrochemical water oxidation. *Materials (Basel), 9*(5), 348. doi:10.3390/ma9050348 PMID:28773473

Wang, F., Di Valentin, C., & Pacchioni, G. (2012). Doping of WO_3 for photocatalytic water splitting: Hints from density functional theory. *The Journal of Physical Chemistry C, 116*(16), 8901–8909. doi:10.1021/jp300867j

Wang, G. (2007). Hydrothermal synthesis and photocatalytic activity of nanocrystalline TiO2 powders in ethanol-water mixed solutions. *Journal of Molecular Catalysis A Chemical, 274*(1-2), 185–191. doi:10.1016/j.molcata.2007.05.009

Wang, G., Ling, Y., Wang, H., Yang, X., Wang, C., Zhang, J. Z., & Li, Y. (2012). Hydrogen-treated WO_3 nanoflakes show enhanced photostability. *Energy & Environmental Science, 5*(3), 6180–6187. doi:10.1039/c2ee03158b

Wang, H., Liang, D., Xu, Y., Liang, X., Qiu, X., & Lin, Z. (2021). A highly efficient photoelectrochemical sensor for detection of chlorpyrifos based on 2D/2D β-Bi_2O_3/ g-C_3N_4 heterojunctions. *Environmental Science. Nano, 8*(3), 773–783. doi:10.1039/D0EN01243B

Wang, H., Yuan, X., Wu, Y., Chen, X., Leng, L., Wang, H., Li, H., & Zeng, G. (2015). Facile synthesis of polypyrrole decorated reduced graphene oxide–Fe3O4 magnetic composites and its application for the Cr(VI) removal. *Chemical Engineering Journal, 262*, 597–606. doi:10.1016/j.cej.2014.10.020

Wang, J., Pan, K., He, Q., & Cao, B. (2013). Polyacrylonitrile/polypyrrole core/shell nanofiber mat for the removal of hexavalent chromium from aqueous solution. *Journal of Hazardous Materials, 244-245*, 121–129. doi:10.1016/j.jhazmat.2012.11.020 PMID:23246947

Wang, M., Wang, T., Luo, Y., He, K., Pan, L., Li, Z., Cui, Z., Liu, Z., Tu, J., & Chen, X. (2021). Fusing Stretchable Sensing Technology with Machine Learning for Human–Machine Interfaces. *Advanced Functional Materials, 31*(39), 2008807. doi:10.1002/adfm.202008807

Wang, N., Zhu, J., Zheng, X., Xiong, F., Huang, B., Shi, J., & Li, C. (2015). A facile two-step method for fabrication of plate-like WO_3 photoanode under mild conditions. *Faraday Discussions, 176*, 185–197. doi:10.1039/C4FD00139G PMID:25406337

Wang, S. C., Chen, P., Bai, Y., Yun, J. H., Liu, G., & Wang, L. (2018). New $BiVO_4$ dual photoanodes with enriched oxygen vacancies for efficient solar-driven water splitting. *Advanced Materials, 20*(20), 1800486. doi:10.1002/adma.201800486 PMID:29602201

Wang, S. C., Chen, P., Yun, J. H., Hu, Y. X., & Wang, L. Z. (2017). An electrochemically treated $BiVO_4$ photoanode for efficient photoelectrochemical water splitting. *Angewandte Chemie International Edition, 56*(29), 8500–8504. doi:10.1002/anie.201703491 PMID:28516511

Wang, S., Chen, H., Gao, G., Butburee, T., Lyu, M., Thaweesak, S., Yun, J.-H., Du, A., Liu, G., & Wang, L. (2016). Synergistic crystal facet engineering and structural control of WO_3 films exhibiting unprecedented photoelectrochemical performance. *Nano Energy, 24*, 94–102. doi:10.1016/j.nanoen.2016.04.010

Wang, S.-F., Shen, L., Zhang, W.-D., & Tong, Y.-J. (2005). Preparation and Mechanical Properties of Chitosan/Carbon Nanotubes Composites. *Biomacromolecules, 6*(6), 3067–3072. doi:10.1021/bm050378v PMID:16283728

Wang, W., Yu, J. C., Shen, Z., Chan, D. K. L., & Gu, T. (2014). g-C_3N_4 quantum dots: Direct synthesis, upconversion properties and photocatalytic application. *Chemical Communications (Cambridge), 50*(70), 10148–10150. doi:10.1039/C4CC02543A PMID:25052166

Wang, X., Maeda, K., Chen, X., Takanabe, K., Domen, K., Hou, Y., Fu, X., & Antonietti, M. (2009). Polymer semiconductors for artificial photosynthesis: Hydrogen evolution by mesoporousgraphitic carbon nitride with visible light. *Journal of the American Chemical Society, 131*(5), 1680–1681. doi:10.1021/ja809307s PMID:19191697

Wang, X., Tang, Y., Leiw, M.-Y., & Lim, T.-T. (2011). Solvothermal synthesis of Fe-C codoped TiO2 nanoparticles for visible light photocatalytic removal of emerging organic contaminants in water. *Applied Catalysis A, General, 409-410*, 257–266. doi:10.1016/j.apcata.2011.10.011

Wang, X., Xie, Z., Huang, H., Liu, Z., Chen, D., & Shen, G. (2012). Gas sensors, thermistor and photodetector based on ZnS nanowires. *Journal of Materials Chemistry*, *22*(14), 6845–6850. doi:10.1039/c2jm16523f

Wang, Y. D., Tian, W., Chen, L., Cao, F. R., Guo, J., & Li, L. (2017). Three-Dimensional WO_3 nanoplate/Bi_2S_3 nanorod heterojunction as a highly efficient photoanode for improved photoelectrochemical water splitting. *ACS Applied Materials & Interfaces*, *46*(46), 40235–40243. doi:10.1021/acsami.7b11510 PMID:29067799

Wang, Y., Chinnathambi, A., Nasif, O., & Alharbi, S. A. (2021). Green synthesis and chemical characterization of a novel anti-human pancreatic cancer supplement by silver nanoparticles containing Zingiber officinale leaf aqueous extract. *Arabian Journal of Chemistry*, *14*(4), 103081. doi:10.1016/j.arabjc.2021.103081

Wang, Y., Chou, S., & Zhang, Z. (2019). Nanomaterials Innovation. *Small*, *15*(32), 15–17. doi:10.1002mll.201902246 PMID:31397544

Wang, Y., Jin, Y., Li, S., Han, J., Ju, Z., & Jia, M. (2018). Flower-like MoS 2 supported on three-dimensional graphene aerogels as high-performance anode materials for sodium-ion batteries. *Ionics*, *24*(11), 3431–3437. doi:10.100711581-018-2528-0

Wang, Z., Zhu, L., Sun, S., Wang, J., & Yan, W. (2021). One-dimensional nanomaterials in resistive gas sensor: From material design to application. *Chemosensors (Basel, Switzerland)*, *9*(8), 198. doi:10.3390/chemosensors9080198

Weng, X., Yang, K., Owens, G., & Chen, Z. (2022). Biosynthesis of silver nanoparticles using three different fruit extracts: Characterization, formation mechanism and estrogen removal. *Journal of Environmental Management*, *316*, 115224. doi:10.1016/j.jenvman.2022.115224 PMID:35550961

Wen, X., Wang, Q., Dai, T., Shao, J., Wu, X., Jiang, Z., Jacob, J. A., & Jiang, C. (2020). Identification of possible reductants in the aqueous leaf extract of mangrove plant Rhizophora apiculata for the fabrication and cytotoxicity of silver nanoparticles against human osteosarcoma MG-63 cells. *Materials Science and Engineering C*, *116*, 111252. doi:10.1016/j.msec.2020.111252 PMID:32806252

Wenzhang, L., Jie, L., Xuan, W., & Qiyuan, C. (2012). Preparation and water-splitting photocatalytic behavior of S-doped WO_3. *Applied Surface Science*, *263*, 157–162. doi:10.1016/j.apsusc.2012.09.021

Wijs, G. A., & de. (1999). Anomalous behavior of the semiconducting gap in WO_3 from first principles calculations. *Physical Review B: Condensed Matter*, *59*(4), 2684–2693. doi:10.1103/PhysRevB.59.2684

Wilson, A. D., & Baietto, M. (2009). Applications and advances in electronic-nose technologies. *Sensors (Basel)*, *9*(7), 5099–5148. doi:10.339090705099 PMID:22346690

Woan, K., Pyrgiotakis, G., & Sigmund, W. (2009). Photocatalytic Carbon-Nanotube-TiO_2 Composites. *Advanced Materials*, *21*(21), 2233–2239. doi:10.1002/adma.200802738

Woodward, P. M., Sleight, A. W., & Vogt, T. (1995). Structure refinement of triclinic tungsten trioxide. *Journal of Physics and Chemistry of Solids*, *56*(10), 1305–1315. doi:10.1016/0022-3697(95)00063-1

Wu, F.-C., Tseng, R.-L., & Juang, R.-S. (2010). A review and experimental verification of using chitosan and its derivatives as adsorbents for selected heavy metals. *Journal of Environmental Management*, *91*(4), 798–806. doi:10.1016/j.jenvman.2009.10.018 PMID:19917518

Wu, J., Tao, K., Zhang, J., Guo, Y., Miao, J., & Norford, L. K. (2016). Chemically functionalized 3D graphene hydrogel for high performance gas sensing. *Journal of Materials Chemistry. A, Materials for Energy and Sustainability*, *4*(21), 8130–8140. doi:10.1039/C6TA01426G

Wu, K., Li, J., & Zhang, C. (2019). Zinc ferrite based gas sensors: A review. *Ceramics International*, *45*(9), 11143–11157. doi:10.1016/j.ceramint.2019.03.086

Wu, M.-T., Tsai, Y.-L., Chiu, C.-W., & Cheng, C.-C. (2016). Synthesis, characterization, and highly acid-resistant properties of crosslinking β-chitosan with polyamines for heavy metal ion adsorption. *RSC Advances*, *6*(106), 104754–104762. doi:10.1039/C6RA21993D

Wu, M., Zhang, J., Liu, C., Gong, Y., Wang, R., He, B., & Wang, H. (2018). Rational design and fabrication of noble-metal-free NixPco-catalyst embedded 3D N-TiO_2/g-C_3N_4 heterojunctions with enhanced photocatalytic hydrogen evolution. *ChemCatChem*, *10*(14), 3069–3077. doi:10.1002/cctc.201800197

Wu, P. D., Liu, Z. F., Chen, D., Zhou, M., & Wei, J. D. (2018). Flake-like NiO/WO_3 p-n heterojunction photocathode for photoelectrochemical water splitting. *Applied Surface Science*, *440*, 1101–1106. doi:10.1016/j.apsusc.2018.01.292

Wu, S.-J., Liou, T.-H., & Mi, F.-L. (2009). Synthesis of zero-valent copper-chitosan nanocomposites and their application for treatment of hexavalent chromium. *Bioresource Technology*, *100*(19), 4348–4353. doi:10.1016/j.biortech.2009.04.013 PMID:19414251

Xiang, Q., Yu, J., & Wong, P. (2011). Quantitative characterization of hydroxyl radicals produced by various photocatalysts. *Journal of Colloid and Interface Science*, *357*(1), 163–167. doi:10.1016/j.jcis.2011.01.093 PMID:21349533

Xiao, Q., & Gao, L. (2013). One-step hydrothermal synthesis of C, W-codoped mesoporous TiO_2 with enhanced visible light photocatalytic activity. *Journal of Alloys and Compounds*, *551*, 286–292. doi:10.1016/j.jallcom.2012.10.040

Xiao, Z., Zhang, J., Jin, L., Xia, Y., Lei, L., Wang, H., Zhang, J., & Xu, S. (2017). Enhanced up-conversion luminescence intensity in single-crystal SrTiO3: Er3+ nanocubes by codoping with Yb3+ ions. *Journal of Alloys and Compounds*, *724*, 139–145. doi:10.1016/j.jallcom.2017.07.024

Xie, M., Wei, W., Jiang, Z., Xu, Y., & Xie, J. (2016). Carbon nitride nanowires/nanofibers: A novel template-free synthesis from a cyanuric chloride-melamine precursor towards enhanced adsorption and visible-light photocatalytic performance. *Ceramics International*, *42*(3), 4158–4170. doi:10.1016/j.ceramint.2015.11.089

Xie, Y. P., Liu, G., Yin, L., & Cheng, H. M. (2012). Crystal facet dependent photocatalytic oxidation and reduction reactivity of monoclinic WO_3 for solar energy conversion. *Journal of Materials Chemistry*, *22*(14), 6746–6751. doi:10.1039/c2jm16178h

Xu, C., Ravi Anusuyadevi, P., Aymonier, C., Luque, R., & Marre, S. (2019). Nanostructured materials for photocatalysis. *Chemical Society Reviews*, *48*(14), 3868–3902. doi:10.1039/C9CS00102F PMID:31173018

Xue, S., Hou, X., Xie, W., Wei, X., & He, D. (2015). Dramatic improvement of photocatalytic activity for N-doped Bi2O3/g-C3N4 composites. *Materials Letters*, *161*, 640–643. doi:10.1016/j.matlet.2015.09.067

Xu, T., Pei, Y., Liu, Y., Wu, D., Shi, Z., Xu, J., Tian, Y., & Li, X. (2017). High-response NO2 resistive gas sensor based on bilayer MoS2 grown by a new two-step chemical vapor deposition method. *Journal of Alloys and Compounds*, *725*, 253–259. doi:10.1016/j.jallcom.2017.06.105

Xu, Y., Sheng, K., Li, C., & Shi, G. (2010). Self-assembled graphene hydrogel via a one-step hydrothermal process. *ACS Nano*, *4*(7), 4324–4330. doi:10.1021/nn101187z PMID:20590149

Yadav, D. K., & Srivastava, S. (2017). Carbon nanotubes as adsorbent to remove heavy metal ion (Mn+7) in wastewater treatment. *Materials Today: Proceedings*, *4*(2, 2, Part A), 4089–4094. doi:10.1016/j.matpr.2017.02.312

Yamazoe, N., & Miura, N. (1994). Environmental gas sensing. *Sensors and Actuators. B, Chemical*, *20*(2-3), 95–102. doi:10.1016/0925-4005(93)01183-5

Yan, A., Xie, C., Huang, F., Zhang, S. S., & Zhang, S. S. (2014). An efficient method to modulate the structure, morphology and properties of WO_3 through niobium doping. *Journal of Alloys and Compounds*, *610*, 132–137. doi:10.1016/j.jallcom.2014.04.188

Yang, D., Gopal, R. A., Lkhagvaa, T., & Choi, D. (2021). Metal-oxide gas sensors for exhaled-breath analysis: A review. *Measurement Science & Technology*, *32*(10), 102004. doi:10.1088/1361-6501/ac03e3

Yang, F., Wang, T., Deng, X., Dang, J., Huang, Z., Hu, S., Li, Y., & Ouyang, M. (2021). Review on hydrogen safety issues: Incident statistics, hydrogen diffusion, and detonation process. *International Journal of Hydrogen Energy*, *46*(61), 31467–31488. doi:10.1016/j.ijhydene.2021.07.005

Yang, G., Lee, C., Kim, J., Ren, F., & Pearton, S. J. (2013). Flexible graphene-based chemical sensors on paper substrates. *Physical Chemistry Chemical Physics*, *15*(6), 1798–1801. doi:10.1039/C2CP43717A PMID:23262787

Yang, J., Geng, B., Ye, Y., & Yu, X. (2012). Stick-like titania precursor route to MTiO3 (M = Sr, Ba, and Ca) polyhedra. *CrystEngComm*, *14*(8), 2959–2965. doi:10.1039/c2ce06532k

Yang, J., Hou, B., Wang, J., Tian, B., Bi, J., Wang, N., Li, X., & Huang, X. (2019). Nanomaterials for the Removal of Heavy Metals from Wastewater. *Nanomaterials (Basel, Switzerland)*, *9*(3), 424. Advance online publication. doi:10.3390/nano9030424 PMID:30871096

Yang, J., Li, W., Li, J., Sun, D., & Chen, Q. (2012). Hydrothermal synthesis and photoelectrochemical properties of vertically aligned tungsten trioxide (hydrate) plate-like arrays fabricated directly on FTO substrates. *Journal of Materials Chemistry*, *22*(34), 17744–17752. doi:10.1039/c2jm33199c

Yang, L., Park, C.-H., Son, Y.-W., Cohen, M. L., & Louie, S. G. (2007). Quasiparticle energies and band gaps in graphene nanoribbons. *Physical Review Letters*, *99*(18), 186801. doi:10.1103/PhysRevLett.99.186801 PMID:17995426

Yang, M., He, J., Hu, X., Yan, C., & Cheng, Z. (2011). CuO nanostructures as quartz crystal microbalance sensing layers for detection of trace hydrogen cyanide gas. *Environmental Science & Technology*, *45*(14), 6088–6094. doi:10.1021/es201121w PMID:21699255

Yang, X., Biswas, S. K., Han, J., Tanpichai, S., Li, M.-C., Chen, C., Zhu, S., Das, A. K., & Yano, H. (2021). Surface and Interface Engineering for Nanocellulosic Advanced Materials. *Advanced Materials*, *33*(28), 2002264. doi:10.1002/adma.202002264 PMID:32902018

Yang, Y., Xie, Y., Pang, L., Li, M., Song, X., Wen, J., & Zhao, H. (2013). Preparation of Reduced Graphene Oxide/Poly(acrylamide) Nanocomposite and Its Adsorption of Pb(II) and Methylene Blue. *Langmuir*, *29*(34), 10727–10736. doi:10.1021/la401940z PMID:23895359

Yan, H., Yang, L., Yang, Z., Yang, H., Li, A., & Cheng, R. (2012). Preparation of chitosan/poly(acrylic acid) magnetic composite microspheres and applications in the removal of copper(II) ions from aqueous solutions. *Journal of Hazardous Materials*, *229-230*, 371–380. doi:10.1016/j.jhazmat.2012.06.014 PMID:22749139

Yan, W., Worsley, M. A., Pham, T., Zettl, A., Carraro, C., & Maboudian, R. (2018). Effects of ambient humidity and temperature on the NO2 sensing characteristics of WS2/graphene aerogel. *Applied Surface Science*, *450*, 372–379. doi:10.1016/j.apsusc.2018.04.185

Yan, X., Pu, R., Xie, R., Zhang, B., Shi, Y., Liu, W., Ma, G., & Yang, N. (2021). Design and fabrication of $Bi_2O_3/BiFeO_3$ heterojunction film with improved photoelectrochemical performance. *Applied Surface Science*, *552*, 149442. doi:10.1016/j.apsusc.2021.149442

Yao, J., & Yang, G. (2022). 2D layered material alloys: Synthesis and application in electronic and optoelectronic devices. *Advancement of Science*, *9*(1), 2103036. doi:10.1002/advs.202103036 PMID:34719873

Yaqoob, A. A., Parveen, T., Umar, K., & Mohamad Ibrahim, M. N. (2020). Role of Nanomaterials in the Treatment of Wastewater: A Review. *Water (Basel)*, *12*(2), 495. Advance online publication. doi:10.3390/w12020495

Yasir, M., Asabuwa Ngwabebhoh, F., Šopík, T., Ali, H., & Sedlařík, V. (2022). Electrospun polyurethane nanofibers coated with polyaniline/polyvinyl alcohol as ultrafiltration membranes for the removal of ethinylestradiol hormone micropollutant from aqueous phase. *Journal of Environmental Chemical Engineering*, *10*(3), 107811. doi:10.1016/j.jece.2022.107811

Yasir, M., Masar, M., Sopik, T., Ali, H., Urbanek, M., Antos, J., Machovsky, M., & Kuritka, I. (2022). ZnO nanowires and nanorods based ZnO/WO3/Pt heterojunction for efficient photocatalytic degradation of estriol (E3) hormone. *Materials Letters*, *319*, 132291. doi:10.1016/j.matlet.2022.132291

Yuan, W., Yang, K., Peng, H., Li, F., & Yin, F. (2018). A flexible VOCs sensor based on a 3D Mxene framework with a high sensing performance. *Journal of Materials Chemistry. A, Materials for Energy and Sustainability*, *6*(37), 18116–18124. doi:10.1039/C8TA06928J

Yu, G., Lu, Y., Guo, J., Patel, M., Bafana, A., Wang, X., Qiu, B., Jeffryes, C., Wei, S., Guo, Z., & Wujcik, E. K. (2018). Carbon nanotubes, graphene, and their derivatives for heavy metal removal. *Advanced Composites and Hybrid Materials*, *1*(1), 56–78. doi:10.100742114-017-0004-3

Yu, L., Ruan, S., Xu, X., Zou, R., & Hu, J. (2017). One-dimensional nanomaterial-assembled macroscopic membranes for water treatment. *Nano Today*, *17*, 79–95. doi:10.1016/j.nantod.2017.10.012

Yuwono, A. S., & Lammers, P. S. (2004). Odor pollution in the environment and the detection instrumentation. *Agricultural Engineering International: CIGR Journal.*

Zaleska, A. (2008). Doped TiO2: A review. *Recent Patents on Engineering*, *2*(3), 157–164. doi:10.2174/187221208786306289

Zamarchi, F., & Vieira, I. C. (2021). Determination of paracetamol using a sensor based on green synthesis of silver nanoparticles in plant extract. *Journal of Pharmaceutical and Biomedical Analysis*, *196*, 113912. doi:10.1016/j.jpba.2021.113912 PMID:33581590

Zeebaree, A., Zeebaree, S., Rashid, R., Zebari, O., Albarwry, A., Ali, A., & Zebari, A. Y. (2022). Sustainable engineering of plant-synthesized TiO2 nanocatalysts: Diagnosis, properties and their photocatalytic performance in removing of methylene blue dye from effluent. A review. *Current Research in Green and Sustainable Chemistry*, *5*, 100312. doi:10.1016/j.crgsc.2022.100312

Zeng, Q. Y., Li, J. H., Li, L. S., Bai, J., Xia, L. G., & Zhou, B. X. (2017). Synthesis of $WO_3/BiVO_4$ photoanode using a reaction of bismuth nitrate with peroxovanadate on WO_3 film for efficient photoelectrocatalytic water splitting and organic pollutant degradation. *Applied Catalysis B: Environmental*, *217*, 21–29. doi:10.1016/j.apcatb.2017.05.072

Zhan, F. Q., Li, J., Li, W. Z., Liu, Y. S., Xie, R. R., Yang, Y. H., Li, Y. M., & Chen, Q. Y. (2015). In situ formation of $CuWO_4/WO_3$ heterojunction plates array films with enhanced photoelectrochemical properties. *International Journal of Hydrogen Energy*, *40*(20), 6512–6520. doi:10.1016/j.ijhydene.2015.03.131

Zhang, D., Wei, A., Zhang, J., & Qiu, R. (2015). The photocatalytic interaction of Cr(VI) ions and phenol on polymer-modified TiO2 under visible light irradiation. *Kinetics and Catalysis*, *56*(5), 569–573. doi:10.1134/S0023158415050195

Zhang, F., Zhang, K., Chen, T., & Oh, W. (2012). Photocatalytic degradation of methyl orange on platinum and palladium co-doped TiO2 nanoparticles. *Synthesis and Reactivity in Inorganic, Metal-Organic, and Nano-Metal Chemistry, 42*(5), 685–691. doi:10.1080/15533174.2011.615040

Zhang, H. F., Zhou, W. W., Yang, Y. P., & Cheng, C. W. (2017). 3D WO_3/$BiVO_4$/cobalt phosphate composites inverse opal photoanode for efficient photoelectrochemical water splitting. *Small, 13*(16), 1603840. doi:10.1002mll.201603840 PMID:28165199

Zhang, J., Hu, Y., Jiang, X., Chen, S., Meng, S., & Fu, X. (2014). Design of a direct Z-scheme photocatalyst: Preparation and characterization of Bi_2O_3/g-C_3N_4 with high visible light activity. *Journal of Hazardous Materials, 280*, 713–722. doi:10.1016/j.jhazmat.2014.08.055 PMID:25232654

Zhang, J., Ling, Y., Gao, W., Wang, S., & Li, J. (2013). Enhanced photoelectrochemical water splitting on novel nanoflake WO_3 electrodes by dealloying of amorphous Fe–W alloys. *Journal of Materials Chemistry. A, Materials for Energy and Sustainability, 1*(36), 10677–10685. doi:10.1039/c3ta12273e

Zhang, J., Liu, Z. H., & Liu, Z. F. (2016). Novel WO_3/Sb_2S_3 heterojunction photocatalyst based on WO_3 of different morphologies for enhanced efficiency in photoelectrochemical water splitting. *ACS Applied Materials & Interfaces, 8*(15), 9684–9691. doi:10.1021/acsami.6b00429 PMID:27032422

Zhang, J., Ma, H. P., & Liu, Z. F. (2017). Highly efficient photocatalyst based on all oxides WO_3/Cu_2O heterojunction for photoelectrochemical water splitting. *Applied Catalysis B: Environmental, 201*, 84–91. doi:10.1016/j.apcatb.2016.08.025

Zhang, J., Qin, Z., Zeng, D., & Xie, C. (2017). Metal-oxide-semiconductor based gas sensors: Screening, preparation, and integration. *Physical Chemistry Chemical Physics, 19*(9), 6313–6329. doi:10.1039/C6CP07799D PMID:28198897

Zhang, K., Shi, X. J., Kima, J. K., & Park, J. H. (2012). Photoelectrochemical cells with tungsten trioxide/Mo-doped $BiVO_4$ bilayers. *Physical Chemistry Chemical Physics, 14*(31), 11119–11124. doi:10.1039/c2cp40991g PMID:22772604

Zhang, L., Wang, G., Xiong, Z., Tang, H., & Jiang, C. (2018). Fabrication of flower-like direct Z-scheme β-Bi_2O_3/g-C_3N_4 photocatalyst with enhanced visible light photoactivity for Rhodamine B degradation. *Applied Surface Science, 436*, 162–171. doi:10.1016/j.apsusc.2017.11.280

Zhang, L., Yu, Y., Zheng, S., Zhong, L., & Xue, J. (2021). Preparation and properties of conductive bacterial cellulose-based graphene oxide-silver nanoparticles antibacterial dressing. *Carbohydrate Polymers, 257*, 117671. doi:10.1016/j.carbpol.2021.117671 PMID:33541624

Zhang, Q., Pan, B., Zhang, S., Wang, J., Zhang, W., & Lv, L. (2011). New insights into nanocomposite adsorbents for water treatment: A case study of polystyrene-supported zirconium phosphate nanoparticles for lead removal. *Journal of Nanoparticle Research, 13*(10), 5355–5364. doi:10.100711051-011-0521-x

Zhang, Q., Pan, B., Zhang, W., Pan, B., Lv, L., Wang, X., Wu, J., & Tao, X. (2009). Selective removal of Pb(II), Cd(II), and Zn(II) ions from waters by an inorganic exchanger Zr(HPO3S)2. *Journal of Hazardous Materials*, *170*(2), 824–828. doi:10.1016/j.jhazmat.2009.05.038 PMID:19493618

Zhang, T., Wang, L., Su, J., & Guo, L. (2016). Branched tungsten oxide nanorod arrays synthesized by controlled phase transformation for solar water oxidation. *ChemCatChem*, *8*(12), 2119–2127. doi:10.1002/cctc.201600267

Zhang, X., Zhang, Y., Jia, X., Zhang, N., Xia, R., Zhang, X., Wang, Z., & Yu, M. (2021). In situ fabrication of a novel S-scheme heterojunctionphotocatalyts Bi2O3/P-C3N4 to enhance levofloxacin removal from water. *Separation and Purification Technology*, *268*, 118691. doi:10.1016/j.seppur.2021.118691

Zhang, Y., Hawboldt, K., Zhang, L., Lu, J., Chang, L., & Dwyer, A. (2022). Carbonaceous nanomaterial TiO2 heterojunctions for visible light driven photocatalytic degradation of aqueous organic pollutants. *Applied Catalysis A, General*, *630*, 118460. doi:10.1016/j.apcata.2021.118460

Zhang, Y., Wu, B., Xu, H., Liu, H., Wang, M., He, Y., & Pan, B. (2016). Nanomaterials-enabled water and wastewater treatment. *NanoImpact*, *3-4*, 22–39. doi:10.1016/j.impact.2016.09.004

Zhao, C., Zhu, D. C., Cheng, X., & Cao, S. (2017). Highly ordered Ag-TiO2 nanocomposited arrays with high visible photocatalytic activity. *Frontiers of Materials Science*, *11*(3), 241–249. doi:10.100711706-017-0386-8

Zhao, G., Huang, X., Tang, Z., Huang, Q., Niu, F., & Wang, X. (2018). Polymer-based nanocomposites for heavy metal ions removal from aqueous solution: A review. *Polymer Chemistry*, *9*(26), 3562–3582. doi:10.1039/C8PY00484F

Zhao, H., Zhang, J., & Zheng, H. (2020). Facile preparation of dual Z-scheme Bi_2O_3/g-C_3N_4/ $Ag_6Si_2O_7$ photocatalyst for highly efficient visible-light photocatalytic degradation of organic pollutants. *Applied Surface Science*, *527*, 146901. doi:10.1016/j.apsusc.2020.146901

Zhao, W., Wang, Z., Shen, X., Li, J., Xu, C., & Gan, Z. (2012). Hydrogen generation via photoelectrocatalytic water splitting using a tungsten tri oxide catalyst under visible light irradiation. *International Journal of Hydrogen Energy*, *37*(1), 908–915. doi:10.1016/j.ijhydene.2011.03.161

Zhao, X., Dou, X., Mohan, D., Pittman, C. U. Jr, Ok, Y. S., & Jin, X. (2014). Antimonate and antimonite adsorption by a polyvinyl alcohol-stabilized granular adsorbent containing nanoscale zero-valent iron. *Chemical Engineering Journal*, *247*, 250–257. doi:10.1016/j.cej.2014.02.096

Zhao, Y., Song, J.-G., Ryu, G. H., Ko, K. Y., Woo, W. J., Kim, Y., ... Lee, Z. (2018). Low-temperature synthesis of 2D MoS 2 on a plastic substrate for a flexible gas sensor. *Nanoscale*, *10*(19), 9338–9345. doi:10.1039/C8NR00108A PMID:29737989

Zhao, Z., Butburee, T., Lyv, M., Peerakiatkhajohn, P., Wang, S., Wang, L., & Zheng, H. (2016). Etching treatment of vertical WO_3 nanoplates as a photoanode for enhanced photoelectrochemical performance. *RSC Advances*, *6*(72), 68204–68210. doi:10.1039/C6RA11750C

Zhou, C., Shi, N., Jiang, X., Chen, M., Jiang, J., Zheng, Y., Wu, W., Cui, D., Haick, H., & Tang, N. (2022). Techniques for wearable gas sensors fabrication. *Sensors and Actuators. B, Chemical*, *353*, 131133. doi:10.1016/j.snb.2021.131133

Zhou, L., Wang, Y., Liu, Z., & Huang, Q. (2009). Characteristics of equilibrium, kinetics studies for adsorption of Hg(II), Cu(II), and Ni(II) ions by thiourea-modified magnetic chitosan microspheres. *Journal of Hazardous Materials*, *161*(2), 995–1002. doi:10.1016/j.jhazmat.2008.04.078 PMID:18538924

Zhou, Y. G., Zhang, L. Y., Lin, L. H., Wygant, B. R., Liu, Y., Zhu, Y., Zheng, Y. B., Mullins, C. B., Zhao, Y., Zhang, X. H., & Yu, G. H. (2017). Highly efficient photoelectrochemical water splitting from hierarchical WO_3/$BiVO_4$ nanoporous sphere arrays. *Nano Letters*, *17*(12), 8012–8017. doi:10.1021/acs.nanolett.7b04626 PMID:29185764

Zhu, J., Cho, M., Li, Y., Cho, I., Suh, J.-H., Orbe, D. D., Jeong, Y., Ren, T.-L., & Park, I. (2019). Biomimetic turbinate-like artificial nose for hydrogen detection based on 3D porous laser-induced graphene. *ACS Applied Materials & Interfaces*, *11*(27), 24386–24394. doi:10.1021/acsami.9b04495 PMID:31192578

Zhu, J., Li, W., Li, J., Li, Y., Hu, H., & Yang, Y. (2013). Photoelectrochemical activity of $NiWO_4$/WO_3 heterojunction photoanode under visible light irradiation. *Electrochimica Acta*, *112*, 191–198. doi:10.1016/j.electacta.2013.08.146

Zhu, M., Li, X., Ge, L., Zi, Y., Qi, M., Li, Y., Li, D., & Mu, C. (2020). Green synthesis of κ-carrageenan@Ag submicron-particles with high aqueous stability, robust antibacterial activity and low cytotoxicity. *Materials Science and Engineering C*, *106*, 110185. doi:10.1016/j.msec.2019.110185 PMID:31753345

Zhu, Z. Z., Wang, Z., & Li, H. L. (2009). Self-assembly of palladium nanoparticles on functional multi-walled carbon nanotubes for formaldehyde oxidation. *Journal of Power Sources*, *186*(2), 339–343. doi:10.1016/j.jpowsour.2008.10.006

Zu, M., Zhou, X., Zhang, S., Qian, S., Li, D.-S., Liu, X., & Zhang, S. (2021). Sustainable engineering of TiO2-based advanced oxidation technologies: From photocatalysis to application devices. *Journal of Materials Science and Technology*, *78*, 202–222. doi:10.1016/j.jmst.2020.10.061

Related References

To continue our tradition of advancing academic research, we have compiled a list of recommended IGI Global readings. These references will provide additional information and guidance to further enrich your knowledge and assist you with your own research and future publications.

Abbasnejad, B., Moeinzadeh, S., Ahankoob, A., & Wong, P. S. (2021). The Role of Collaboration in the Implementation of BIM-Enabled Projects. In J. Underwood & M. Shelbourn (Eds.), *Handbook of Research on Driving Transformational Change in the Digital Built Environment* (pp. 27–62). IGI Global. https://doi.org/10.4018/978-1-7998-6600-8.ch002

Abdulrahman, K. O., Mahamood, R. M., & Akinlabi, E. T. (2022). Additive Manufacturing (AM): Processing Technique for Lightweight Alloys and Composite Material. In K. Kumar, B. Babu, & J. Davim (Ed.), *Handbook of Research on Advancements in the Processing, Characterization, and Application of Lightweight Materials* (pp. 27-48). IGI Global. https://doi.org/10.4018/978-1-7998-7864-3.ch002

Agrawal, R., Sharma, P., & Saxena, A. (2021). A Diamond Cut Leather Substrate Antenna for BAN (Body Area Network) Application. In V. Singh, V. Dubey, A. Saxena, R. Tiwari, & H. Sharma (Eds.), *Emerging Materials and Advanced Designs for Wearable Antennas* (pp. 54–59). IGI Global. https://doi.org/10.4018/978-1-7998-7611-3.ch004

Ahmad, F., Al-Ammar, E. A., & Alsaidan, I. (2022). Battery Swapping Station: A Potential Solution to Address the Limitations of EV Charging Infrastructure. In M. Alam, R. Pillai, & N. Murugesan (Eds.), *Developing Charging Infrastructure and Technologies for Electric Vehicles* (pp. 195–207). IGI Global. doi:10.4018/978-1-7998-6858-3.ch010

Aikhuele, D. (2018). A Study of Product Development Engineering and Design Reliability Concerns. *International Journal of Applied Industrial Engineering*, *5*(1), 79–89. doi:10.4018/IJAIE.2018010105

Al-Khatri, H., & Al-Atrash, F. (2021). Occupants' Habits and Natural Ventilation in a Hot Arid Climate. In R. González-Lezcano (Ed.), *Advancements in Sustainable Architecture and Energy Efficiency* (pp. 146–168). IGI Global. https://doi.org/10.4018/978-1-7998-7023-4.ch007

Al-Shebeeb, O. A., Rangaswamy, S., Gopalakrishan, B., & Devaru, D. G. (2017). Evaluation and Indexing of Process Plans Based on Electrical Demand and Energy Consumption. *International Journal of Manufacturing, Materials, and Mechanical Engineering*, *7*(3), 1–19. doi:10.4018/IJMMME.2017070101

Amuda, M. O., Lawal, T. F., & Akinlabi, E. T. (2017). Research Progress on Rheological Behavior of AA7075 Aluminum Alloy During Hot Deformation. *International Journal of Materials Forming and Machining Processes*, *4*(1), 53–96. doi:10.4018/IJMFMP.2017010104

Amuda, M. O., Lawal, T. F., & Mridha, S. (2021). Microstructure and Mechanical Properties of Silicon Carbide-Treated Ferritic Stainless Steel Welds. In L. Burstein (Ed.), *Handbook of Research on Advancements in Manufacturing, Materials, and Mechanical Engineering* (pp. 395–411). IGI Global. https://doi.org/10.4018/978-1-7998-4939-1.ch019

Anikeev, V., Gasem, K. A., & Fan, M. (2021). Application of Supercritical Technologies in Clean Energy Production: A Review. In L. Chen (Ed.), *Handbook of Research on Advancements in Supercritical Fluids Applications for Sustainable Energy Systems* (pp. 792–821). IGI Global. https://doi.org/10.4018/978-1-7998-5796-9.ch022

Arafat, M. Y., Saleem, I., & Devi, T. P. (2022). Drivers of EV Charging Infrastructure Entrepreneurship in India. In M. Alam, R. Pillai, & N. Murugesan (Eds.), *Developing Charging Infrastructure and Technologies for Electric Vehicles* (pp. 208–219). IGI Global. https://doi.org/10.4018/978-1-7998-6858-3.ch011

Araujo, A., & Manninen, H. (2022). Contribution of Project-Based Learning on Social Skills Development: An Industrial Engineer Perspective. In A. Alves & N. van Hattum-Janssen (Eds.), *Training Engineering Students for Modern Technological Advancement* (pp. 119–145). IGI Global. https://doi.org/10.4018/978-1-7998-8816-1.ch006

Armutlu, H. (2018). Intelligent Biomedical Engineering Operations by Cloud Computing Technologies. In U. Kose, G. Guraksin, & O. Deperlioglu (Eds.), *Nature-Inspired Intelligent Techniques for Solving Biomedical Engineering Problems* (pp. 297–317). Hershey, PA: IGI Global. doi:10.4018/978-1-5225-4769-3.ch015

Atik, M., Sadek, M., & Shahrour, I. (2017). Single-Run Adaptive Pushover Procedure for Shear Wall Structures. In V. Plevris, G. Kremmyda, & Y. Fahjan (Eds.), *Performance-Based Seismic Design of Concrete Structures and Infrastructures* (pp. 59–83). Hershey, PA: IGI Global. doi:10.4018/978-1-5225-2089-4.ch003

Attia, H. (2021). Smart Power Microgrid Impact on Sustainable Building. In R. González-Lezcano (Ed.), *Advancements in Sustainable Architecture and Energy Efficiency* (pp. 169–194). IGI Global. https://doi.org/10.4018/978-1-7998-7023-4.ch008

Aydin, A., Akyol, E., Gungor, M., Kaya, A., & Tasdelen, S. (2018). Geophysical Surveys in Engineering Geology Investigations With Field Examples. In N. Ceryan (Ed.), *Handbook of Research on Trends and Digital Advances in Engineering Geology* (pp. 257–280). Hershey, PA: IGI Global. doi:10.4018/978-1-5225-2709-1.ch007

Ayoobkhan, M. U. D., Y., A., J., Easwaran, B., & R., T. (2021). Smart Connected Digital Products and IoT Platform With the Digital Twin. In P. Vasant, G. Weber, & W. Punurai (Ed.), Research Advancements in Smart Technology, Optimization, and Renewable Energy (pp. 330-350). IGI Global. https://doi.org/ doi:10.4018/978-1-7998-3970-5.ch016

Baeza Moyano, D., & González Lezcano, R. A. (2021). The Importance of Light in Our Lives: Towards New Lighting in Schools. In R. González-Lezcano (Ed.), *Advancements in Sustainable Architecture and Energy Efficiency* (pp. 239–256). IGI Global. https://doi.org/10.4018/978-1-7998-7023-4.ch011

Bagdadee, A. H. (2021). A Brief Assessment of the Energy Sector of Bangladesh. *International Journal of Energy Optimization and Engineering*, *10*(1), 36–55. doi:10.4018/IJEOE.2021010103

Baklezos, A. T., & Hadjigeorgiou, N. G. (2021). Magnetic Sensors for Space Applications and Magnetic Cleanliness Considerations. In C. Nikolopoulos (Ed.), *Recent Trends on Electromagnetic Environmental Effects for Aeronautics and Space Applications* (pp. 147–185). IGI Global. https://doi.org/10.4018/978-1-7998-4879-0.ch006

Bas, T. G. (2017). Nutraceutical Industry with the Collaboration of Biotechnology and Nutrigenomics Engineering: The Significance of Intellectual Property in the Entrepreneurship and Scientific Research Ecosystems. In T. Bas & J. Zhao (Eds.), *Comparative Approaches to Biotechnology Development and Use in Developed and Emerging Nations* (pp. 1–17). Hershey, PA: IGI Global. doi:10.4018/978-1-5225-1040-6.ch001

Bazeer Ahamed, B., & Periakaruppan, S. (2021). Taxonomy of Influence Maximization Techniques in Unknown Social Networks. In P. Vasant, G. Weber, & W. Punurai (Eds.), *Research Advancements in Smart Technology, Optimization, and Renewable Energy* (pp. 351-363). IGI Global. https://doi.org/10.4018/978-1-7998-3970-5.ch017

Beale, R., & André, J. (2017). *Design Solutions and Innovations in Temporary Structures*. Hershey, PA: IGI Global. doi:10.4018/978-1-5225-2199-0

Behnam, B. (2017). Simulating Post-Earthquake Fire Loading in Conventional RC Structures. In P. Samui, S. Chakraborty, & D. Kim (Eds.), *Modeling and Simulation Techniques in Structural Engineering* (pp. 425–444). Hershey, PA: IGI Global. doi:10.4018/978-1-5225-0588-4.ch015

Ben Hamida, I., Salah, S. B., Msahli, F., & Mimouni, M. F. (2018). Distribution Network Reconfiguration Using SPEA2 for Power Loss Minimization and Reliability Improvement. *International Journal of Energy Optimization and Engineering*, *7*(1), 50–65. doi:10.4018/IJEOE.2018010103

Bentarzi, H. (2021). Fault Tree-Based Root Cause Analysis Used to Study Mal-Operation of a Protective Relay in a Smart Grid. In A. Recioui & H. Bentarzi (Eds.), *Optimizing and Measuring Smart Grid Operation and Control* (pp. 289–308). IGI Global. https://doi.org/10.4018/978-1-7998-4027-5.ch012

Beysens, D. A., Garrabos, Y., & Zappoli, B. (2021). Thermal Effects in Near-Critical Fluids: Piston Effect and Related Phenomena. In L. Chen (Ed.), *Handbook of Research on Advancements in Supercritical Fluids Applications for Sustainable Energy Systems* (pp. 1–31). IGI Global. https://doi.org/10.4018/978-1-7998-5796-9.ch001

Bhaskar, S. V., & Kudal, H. N. (2017). Effect of TiCN and AlCrN Coating on Tribological Behaviour of Plasma-nitrided AISI 4140 Steel. *International Journal of Surface Engineering and Interdisciplinary Materials Science*, *5*(2), 1–17. doi:10.4018/IJSEIMS.2017070101

Bhuyan, D. (2018). Designing of a Twin Tube Shock Absorber: A Study in Reverse Engineering. In K. Kumar & J. Davim (Eds.), *Design and Optimization of Mechanical Engineering Products* (pp. 83–104). Hershey, PA: IGI Global. doi:10.4018/978-1-5225-3401-3.ch005

Blumberg, G. (2021). Blockchains for Use in Construction and Engineering Projects. In J. Underwood & M. Shelbourn (Eds.), *Handbook of Research on Driving Transformational Change in the Digital Built Environment* (pp. 179–208). IGI Global. https://doi.org/10.4018/978-1-7998-6600-8.ch008

Burstein, L. (2021). Simulation Tool for Cable Design. In L. Burstein (Ed.), *Handbook of Research on Advancements in Manufacturing, Materials, and Mechanical Engineering* (pp. 54–74). IGI Global. https://doi.org/10.4018/978-1-7998-4939-1.ch003

Calderon, F. A., Giolo, E. G., Frau, C. D., Rengel, M. G., Rodriguez, H., Tornello, M., ... Gallucci, R. (2018). Seismic Microzonation and Site Effects Detection Through Microtremors Measures: A Review. In N. Ceryan (Ed.), *Handbook of Research on Trends and Digital Advances in Engineering Geology* (pp. 326–349). Hershey, PA: IGI Global. doi:10.4018/978-1-5225-2709-1.ch009

Ceryan, N., & Can, N. K. (2018). Prediction of The Uniaxial Compressive Strength of Rocks Materials. In N. Ceryan (Ed.), *Handbook of Research on Trends and Digital Advances in Engineering Geology* (pp. 31–96). Hershey, PA: IGI Global. doi:10.4018/978-1-5225-2709-1.ch002

Ceryan, S. (2018). Weathering Indices Used in Evaluation of the Weathering State of Rock Material. In N. Ceryan (Ed.), *Handbook of Research on Trends and Digital Advances in Engineering Geology* (pp. 132–186). Hershey, PA: IGI Global. doi:10.4018/978-1-5225-2709-1.ch004

Chen, H., Padilla, R. V., & Besarati, S. (2017). Supercritical Fluids and Their Applications in Power Generation. In L. Chen & Y. Iwamoto (Eds.), *Advanced Applications of Supercritical Fluids in Energy Systems* (pp. 369–402). Hershey, PA: IGI Global. doi:10.4018/978-1-5225-2047-4.ch012

Chen, H., Padilla, R. V., & Besarati, S. (2021). Supercritical Fluids and Their Applications in Power Generation. In L. Chen (Ed.), *Handbook of Research on Advancements in Supercritical Fluids Applications for Sustainable Energy Systems* (pp. 566–599). IGI Global. https://doi.org/10.4018/978-1-7998-5796-9.ch016

Chen, L. (2017). Principles, Experiments, and Numerical Studies of Supercritical Fluid Natural Circulation System. In L. Chen & Y. Iwamoto (Eds.), *Advanced Applications of Supercritical Fluids in Energy Systems* (pp. 136–187). Hershey, PA: IGI Global. doi:10.4018/978-1-5225-2047-4.ch005

Chen, L. (2021). Principles, Experiments, and Numerical Studies of Supercritical Fluid Natural Circulation System. In L. Chen (Ed.), *Handbook of Research on Advancements in Supercritical Fluids Applications for Sustainable Energy Systems* (pp. 219–269). IGI Global. https://doi.org/10.4018/978-1-7998-5796-9.ch007

Chiba, Y., Marif, Y., Henini, N., & Tlemcani, A. (2021). Modeling of Magnetic Refrigeration Device by Using Artificial Neural Networks Approach. *International Journal of Energy Optimization and Engineering*, *10*(4), 68–76. https://doi.org/10.4018/IJEOE.2021100105

Clementi, F., Di Sciascio, G., Di Sciascio, S., & Lenci, S. (2017). Influence of the Shear-Bending Interaction on the Global Capacity of Reinforced Concrete Frames: A Brief Overview of the New Perspectives. In V. Plevris, G. Kremmyda, & Y. Fahjan (Eds.), *Performance-Based Seismic Design of Concrete Structures and Infrastructures* (pp. 84–111). Hershey, PA: IGI Global. doi:10.4018/978-1-5225-2089-4.ch004

Codinhoto, R., Fialho, B. C., Pinti, L., & Fabricio, M. M. (2021). BIM and IoT for Facilities Management: Understanding Key Maintenance Issues. In J. Underwood & M. Shelbourn (Eds.), *Handbook of Research on Driving Transformational Change in the Digital Built Environment* (pp. 209–231). IGI Global. doi:10.4018/978-1-7998-6600-8.ch009

Cortés-Polo, D., Calle-Cancho, J., Carmona-Murillo, J., & González-Sánchez, J. (2017). Future Trends in Mobile-Fixed Integration for Next Generation Networks: Classification and Analysis. *International Journal of Vehicular Telematics and Infotainment Systems*, *1*(1), 33–53. doi:10.4018/IJVTIS.2017010103

Costa, H. G., Sheremetieff, F. H., & Araújo, E. A. (2022). Influence of Game-Based Methods in Developing Engineering Competences. In A. Alves & N. van Hattum-Janssen (Eds.), *Training Engineering Students for Modern Technological Advancement* (pp. 69–88). IGI Global. https://doi.org/10.4018/978-1-7998-8816-1.ch004

Cui, X., Zeng, S., Li, Z., Zheng, Q., Yu, X., & Han, B. (2018). Advanced Composites for Civil Engineering Infrastructures. In K. Kumar & J. Davim (Eds.), *Composites and Advanced Materials for Industrial Applications* (pp. 212–248). Hershey, PA: IGI Global. doi:10.4018/978-1-5225-5216-1.ch010

Dalgıç, S., & Kuşku, İ. (2018). Geological and Geotechnical Investigations in Tunneling. In N. Ceryan (Ed.), *Handbook of Research on Trends and Digital Advances in Engineering Geology* (pp. 482–529). Hershey, PA: IGI Global. doi:10.4018/978-1-5225-2709-1.ch014

Dang, C., & Hihara, E. (2021). Study on Cooling Heat Transfer of Supercritical Carbon Dioxide Applied to Transcritical Carbon Dioxide Heat Pump. In L. Chen (Ed.), *Handbook of Research on Advancements in Supercritical Fluids Applications for Sustainable Energy Systems* (pp. 451–493). IGI Global. https://doi.org/10.4018/978-1-7998-5796-9.ch013

Daus, Y., Kharchenko, V., & Yudaev, I. (2021). Research of Solar Energy Potential of Photovoltaic Installations on Enclosing Structures of Buildings. *International Journal of Energy Optimization and Engineering, 10*(4), 18–34. https://doi.org/10.4018/IJEOE.2021100102

Daus, Y., Kharchenko, V., & Yudaev, I. (2021). Optimizing Layout of Distributed Generation Sources of Power Supply System of Agricultural Object. *International Journal of Energy Optimization and Engineering, 10*(3), 70–84. https://doi.org/10.4018/IJEOE.2021070104

de la Varga, D., Soto, M., Arias, C. A., van Oirschot, D., Kilian, R., Pascual, A., & Álvarez, J. A. (2017). Constructed Wetlands for Industrial Wastewater Treatment and Removal of Nutrients. In Á. Val del Río, J. Campos Gómez, & A. Mosquera Corral (Eds.), *Technologies for the Treatment and Recovery of Nutrients from Industrial Wastewater* (pp. 202–230). Hershey, PA: IGI Global. doi:10.4018/978-1-5225-1037-6.ch008

Deb, S., Ammar, E. A., AlRajhi, H., Alsaidan, I., & Shariff, S. M. (2022). V2G Pilot Projects: Review and Lessons Learnt. In M. Alam, R. Pillai, & N. Murugesan (Eds.), *Developing Charging Infrastructure and Technologies for Electric Vehicles* (pp. 252–267). IGI Global. https://doi.org/10.4018/978-1-7998-6858-3.ch014

Dekhandji, F. Z., & Rais, M. C. (2021). A Comparative Study of Power Quality Monitoring Using Various Techniques. In A. Recioui & H. Bentarzi (Eds.), *Optimizing and Measuring Smart Grid Operation and Control* (pp. 259–288). IGI Global. https://doi.org/10.4018/978-1-7998-4027-5.ch011

Deperlioglu, O. (2018). Intelligent Techniques Inspired by Nature and Used in Biomedical Engineering. In U. Kose, G. Guraksin, & O. Deperlioglu (Eds.), *Nature-Inspired Intelligent Techniques for Solving Biomedical Engineering Problems* (pp. 51–77). Hershey, PA: IGI Global. doi:10.4018/978-1-5225-4769-3.ch003

Dhurpate, P. R., & Tang, H. (2021). Quantitative Analysis of the Impact of Inter-Line Conveyor Capacity for Throughput of Manufacturing Systems. *International Journal of Manufacturing, Materials, and Mechanical Engineering*, *11*(1), 1–17. https://doi.org/10.4018/IJMMME.2021010101

Dinkar, S., & Deep, K. (2021). A Survey of Recent Variants and Applications of Antlion Optimizer. *International Journal of Energy Optimization and Engineering*, *10*(2), 48–73. doi:10.4018/IJEOE.2021040103

Dixit, A. (2018). Application of Silica-Gel-Reinforced Aluminium Composite on the Piston of Internal Combustion Engine: Comparative Study of Silica-Gel-Reinforced Aluminium Composite Piston With Aluminium Alloy Piston. In K. Kumar & J. Davim (Eds.), *Composites and Advanced Materials for Industrial Applications* (pp. 63–98). Hershey, PA: IGI Global. doi:10.4018/978-1-5225-5216-1.ch004

Drabecki, M. P., & Kułak, K. B. (2021). Global Pandemics on European Electrical Energy Markets: Lessons Learned From the COVID-19 Outbreak. *International Journal of Energy Optimization and Engineering*, *10*(3), 24–46. https://doi.org/10.4018/IJEOE.2021070102

Dutta, M. M. (2021). Nanomaterials for Food and Agriculture. In M. Bhat, I. Wani, & S. Ashraf (Eds.), *Applications of Nanomaterials in Agriculture, Food Science, and Medicine* (pp. 75–97). IGI Global. doi:10.4018/978-1-7998-5563-7.ch004

Dutta, M. M., & Goswami, M. (2021). Coating Materials: Nano-Materials. In S. Roy & G. Bose (Eds.), *Advanced Surface Coating Techniques for Modern Industrial Applications* (pp. 1–30). IGI Global. doi:10.4018/978-1-7998-4870-7.ch001

Elsayed, A. M., Dakkama, H. J., Mahmoud, S., Al-Dadah, R., & Kaialy, W. (2017). Sustainable Cooling Research Using Activated Carbon Adsorbents and Their Environmental Impact. In T. Kobayashi (Ed.), *Applied Environmental Materials Science for Sustainability* (pp. 186–221). Hershey, PA: IGI Global. doi:10.4018/978-1-5225-1971-3.ch009

Ercanoglu, M., & Sonmez, H. (2018). General Trends and New Perspectives on Landslide Mapping and Assessment Methods. In N. Ceryan (Ed.), *Handbook of Research on Trends and Digital Advances in Engineering Geology* (pp. 350–379). Hershey, PA: IGI Global. doi:10.4018/978-1-5225-2709-1.ch010

Faroz, S. A., Pujari, N. N., Rastogi, R., & Ghosh, S. (2017). Risk Analysis of Structural Engineering Systems Using Bayesian Inference. In P. Samui, S. Chakraborty, & D. Kim (Eds.), *Modeling and Simulation Techniques in Structural Engineering* (pp. 390–424). Hershey, PA: IGI Global. doi:10.4018/978-1-5225-0588-4.ch014

Fekik, A., Hamida, M. L., Denoun, H., Azar, A. T., Kamal, N. A., Vaidyanathan, S., Bousbaine, A., & Benamrouche, N. (2022). Multilevel Inverter for Hybrid Fuel Cell/PV Energy Conversion System. In A. Fekik & N. Benamrouche (Eds.), *Modeling and Control of Static Converters for Hybrid Storage Systems* (pp. 233–270). IGI Global. https://doi.org/10.4018/978-1-7998-7447-8.ch009

Fekik, A., Hamida, M. L., Houassine, H., Azar, A. T., Kamal, N. A., Denoun, H., Vaidyanathan, S., & Sambas, A. (2022). Power Quality Improvement for Grid-Connected Photovoltaic Panels Using Direct Power Control. In A. Fekik & N. Benamrouche (Eds.), *Modeling and Control of Static Converters for Hybrid Storage Systems* (pp. 107–142). IGI Global. https://doi.org/10.4018/978-1-7998-7447-8.ch005

Fernando, P. R., Hamigah, T., Disne, S., Wickramasingha, G. G., & Sutharshan, A. (2018). The Evaluation of Engineering Properties of Low Cost Concrete Blocks by Partial Doping of Sand with Sawdust: Low Cost Sawdust Concrete Block. *International Journal of Strategic Engineering*, *1*(2), 26–42. doi:10.4018/IJoSE.2018070103

Ferro, G., Minciardi, R., Parodi, L., & Robba, M. (2022). Optimal Charging Management of Microgrid-Integrated Electric Vehicles. In M. Alam, R. Pillai, & N. Murugesan (Eds.), *Developing Charging Infrastructure and Technologies for Electric Vehicles* (pp. 133–155). IGI Global. https://doi.org/10.4018/978-1-7998-6858-3.ch007

Flumerfelt, S., & Green, C. (2022). Graduate Lean Leadership Education: A Case Study of a Program. In A. Alves & N. van Hattum-Janssen (Eds.), *Training Engineering Students for Modern Technological Advancement* (pp. 202–224). IGI Global. https://doi.org/10.4018/978-1-7998-8816-1.ch010

Galli, B. J. (2021). Implications of Economic Decision Making to the Project Manager. *International Journal of Strategic Engineering*, *4*(1), 19–32. https://doi.org/10.4018/IJoSE.2021010102

Gento, A. M., Pimentel, C., & Pascual, J. A. (2022). Teaching Circular Economy and Lean Management in a Learning Factory. In A. Alves & N. van Hattum-Janssen (Eds.), *Training Engineering Students for Modern Technological Advancement* (pp. 183–201). IGI Global. https://doi.org/10.4018/978-1-7998-8816-1.ch009

Ghosh, S., Mitra, S., Ghosh, S., & Chakraborty, S. (2017). Seismic Reliability Analysis in the Framework of Metamodelling Based Monte Carlo Simulation. In P. Samui, S. Chakraborty, & D. Kim (Eds.), *Modeling and Simulation Techniques in Structural Engineering* (pp. 192–208). Hershey, PA: IGI Global. doi:10.4018/978-1-5225-0588-4.ch006

Gil, M., & Otero, B. (2017). Learning Engineering Skills through Creativity and Collaboration: A Game-Based Proposal. In R. Alexandre Peixoto de Queirós & M. Pinto (Eds.), *Gamification-Based E-Learning Strategies for Computer Programming Education* (pp. 14–29). Hershey, PA: IGI Global. doi:10.4018/978-1-5225-1034-5.ch002

Gill, J., Ayre, M., & Mills, J. (2017). Revisioning the Engineering Profession: How to Make It Happen! In M. Gray & K. Thomas (Eds.), *Strategies for Increasing Diversity in Engineering Majors and Careers* (pp. 156–175). Hershey, PA: IGI Global. doi:10.4018/978-1-5225-2212-6.ch008

Godzhaev, Z., Senkevich, S., Kuzmin, V., & Melikov, I. (2021). Use of the Neural Network Controller of Sprung Mass to Reduce Vibrations From Road Irregularities. In P. Vasant, G. Weber, & W. Punurai (Ed.), *Research Advancements in Smart Technology, Optimization, and Renewable Energy* (pp. 69-87). IGI Global. https://doi.org/10.4018/978-1-7998-3970-5.ch005

Gomes de Gusmão, C. M. (2022). Digital Competencies and Transformation in Higher Education: Upskilling With Extension Actions. In A. Alves & N. van Hattum-Janssen (Eds.), *Training Engineering Students for Modern Technological Advancement* (pp. 313–328). IGI Global. https://doi.org/10.4018/978-1-7998-8816-1.ch015A

Goyal, N., Ram, M., & Kumar, P. (2017). Welding Process under Fault Coverage Approach for Reliability and MTTF. In M. Ram & J. Davim (Eds.), *Mathematical Concepts and Applications in Mechanical Engineering and Mechatronics* (pp. 222–245). Hershey, PA: IGI Global. doi:10.4018/978-1-5225-1639-2.ch011

Gray, M., & Lundy, C. (2017). Engineering Study Abroad: High Impact Strategy for Increasing Access. In M. Gray & K. Thomas (Eds.), *Strategies for Increasing Diversity in Engineering Majors and Careers* (pp. 42–59). Hershey, PA: IGI Global. doi:10.4018/978-1-5225-2212-6.ch003

Güler, O., & Varol, T. (2021). Fabrication of Functionally Graded Metal and Ceramic Powders Synthesized by Electroless Deposition. In S. Roy & G. Bose (Eds.), *Advanced Surface Coating Techniques for Modern Industrial Applications* (pp. 150–187). IGI Global. https://doi.org/10.4018/978-1-7998-4870-7.ch007

Guraksin, G. E. (2018). Internet of Things and Nature-Inspired Intelligent Techniques for the Future of Biomedical Engineering. In U. Kose, G. Guraksin, & O. Deperlioglu (Eds.), *Nature-Inspired Intelligent Techniques for Solving Biomedical Engineering Problems* (pp. 263–282). Hershey, PA: IGI Global. doi:10.4018/978-1-5225-4769-3.ch013

Hamida, M. L., Fekik, A., Denoun, H., Ardjal, A., & Bokhtache, A. A. (2022). Flying Capacitor Inverter Integration in a Renewable Energy System. In A. Fekik & N. Benamrouche (Eds.), *Modeling and Control of Static Converters for Hybrid Storage Systems* (pp. 287–306). IGI Global. https://doi.org/10.4018/978-1-7998-7447-8.ch011

Hasegawa, N., & Takahashi, Y. (2021). Control of Soap Bubble Ejection Robot Using Facial Expressions. *International Journal of Manufacturing, Materials, and Mechanical Engineering, 11*(2), 1–16. https://doi.org/10.4018/IJMMME.2021040101

Hejazi, T., & Akbari, L. (2017). A Multiresponse Optimization Model for Statistical Design of Processes with Discrete Variables. In M. Ram & J. Davim (Eds.), *Mathematical Concepts and Applications in Mechanical Engineering and Mechatronics* (pp. 17–37). Hershey, PA: IGI Global. doi:10.4018/978-1-5225-1639-2.ch002

Hejazi, T., & Hejazi, A. (2017). Monte Carlo Simulation for Reliability-Based Design of Automotive Complex Subsystems. In M. Ram & J. Davim (Eds.), *Mathematical Concepts and Applications in Mechanical Engineering and Mechatronics* (pp. 177–200). Hershey, PA: IGI Global. doi:10.4018/978-1-5225-1639-2.ch009

Hejazi, T., & Poursabbagh, H. (2017). Reliability Analysis of Engineering Systems: An Accelerated Life Testing for Boiler Tubes. In M. Ram & J. Davim (Eds.), *Mathematical Concepts and Applications in Mechanical Engineering and Mechatronics* (pp. 154–176). Hershey, PA: IGI Global. doi:10.4018/978-1-5225-1639-2.ch008

Henao, J., Poblano-Salas, C. A., Vargas, F., Giraldo-Betancur, A. L., Corona-Castuera, J., & Sotelo-Mazón, O. (2021). Principles and Applications of Thermal Spray Coatings. In S. Roy & G. Bose (Eds.), *Advanced Surface Coating Techniques for Modern Industrial Applications* (pp. 31–70). IGI Global. https://doi.org/10.4018/978-1-7998-4870-7.ch002

Henao, J., & Sotelo, O. (2018). Surface Engineering at High Temperature: Thermal Cycling and Corrosion Resistance. In A. Pakseresht (Ed.), *Production, Properties, and Applications of High Temperature Coatings* (pp. 131–159). Hershey, PA: IGI Global. doi:10.4018/978-1-5225-4194-3.ch006

Hrnčič, M. K., Cör, D., & Knez, Ž. (2021). Supercritical Fluids as a Tool for Green Energy and Chemicals. In L. Chen (Ed.), *Handbook of Research on Advancements in Supercritical Fluids Applications for Sustainable Energy Systems* (pp. 761–791). IGI Global. doi:10.4018/978-1-7998-5796-9.ch021

Ibrahim, O., Erdem, S., & Gurbuz, E. (2021). Studying Physical and Chemical Properties of Graphene Oxide and Reduced Graphene Oxide and Their Applications in Sustainable Building Materials. In R. González-Lezcano (Ed.), *Advancements in Sustainable Architecture and Energy Efficiency* (pp. 221–238). IGI Global. https://doi.org/10.4018/978-1-7998-7023-4.ch010

Ihianle, I. K., Islam, S., Naeem, U., & Ebenuwa, S. H. (2021). Exploiting Patterns of Object Use for Human Activity Recognition. In A. Nwajana & I. Ihianle (Eds.), *Handbook of Research on 5G Networks and Advancements in Computing, Electronics, and Electrical Engineering* (pp. 382–401). IGI Global. https://doi.org/10.4018/978-1-7998-6992-4.ch015

Ijemaru, G. K., Ngharamike, E. T., Oleka, E. U., & Nwajana, A. O. (2021). An Energy-Efficient Model for Opportunistic Data Collection in IoV-Enabled SC Waste Management. In A. Nwajana & I. Ihianle (Eds.), *Handbook of Research on 5G Networks and Advancements in Computing, Electronics, and Electrical Engineering* (pp. 1–19). IGI Global. https://doi.org/10.4018/978-1-7998-6992-4.ch001

Ilori, O. O., Adetan, D. A., & Umoru, L. E. (2017). Effect of Cutting Parameters on the Surface Residual Stress of Face-Milled Pearlitic Ductile Iron. *International Journal of Materials Forming and Machining Processes*, *4*(1), 38–52. doi:10.4018/IJMFMP.2017010103

Imam, M. H., Tasadduq, I. A., Ahmad, A., Aldosari, F., & Khan, H. (2017). Automated Generation of Course Improvement Plans Using Expert System. *International Journal of Quality Assurance in Engineering and Technology Education*, *6*(1), 1–12. doi:10.4018/IJQAETE.2017010101

Injeti, S. K., & Kumar, T. V. (2018). A WDO Framework for Optimal Deployment of DGs and DSCs in a Radial Distribution System Under Daily Load Pattern to Improve Techno-Economic Benefits. *International Journal of Energy Optimization and Engineering*, *7*(2), 1–38. doi:10.4018/IJEOE.2018040101

Ishii, N., Anami, K., & Knisely, C. W. (2018). *Dynamic Stability of Hydraulic Gates and Engineering for Flood Prevention*. Hershey, PA: IGI Global. doi:10.4018/978-1-5225-3079-4

Iwamoto, Y., & Yamaguchi, H. (2021). Application of Supercritical Carbon Dioxide for Solar Water Heater. In L. Chen (Ed.), *Handbook of Research on Advancements in Supercritical Fluids Applications for Sustainable Energy Systems* (pp. 370–387). IGI Global. https://doi.org/10.4018/978-1-7998-5796-9.ch010

Jayapalan, S. (2018). A Review of Chemical Treatments on Natural Fibers-Based Hybrid Composites for Engineering Applications. In K. Kumar & J. Davim (Eds.), *Composites and Advanced Materials for Industrial Applications* (pp. 16–37). Hershey, PA: IGI Global. doi:10.4018/978-1-5225-5216-1.ch002

Kapetanakis, T. N., Vardiambasis, I. O., Ioannidou, M. P., & Konstantaras, A. I. (2021). Modeling Antenna Radiation Using Artificial Intelligence Techniques: The Case of a Circular Loop Antenna. In C. Nikolopoulos (Ed.), *Recent Trends on Electromagnetic Environmental Effects for Aeronautics and Space Applications* (pp. 186–225). IGI Global. https://doi.org/10.4018/978-1-7998-4879-0.ch007

Karkalos, N. E., Markopoulos, A. P., & Dossis, M. F. (2017). Optimal Model Parameters of Inverse Kinematics Solution of a 3R Robotic Manipulator Using ANN Models. *International Journal of Manufacturing, Materials, and Mechanical Engineering*, *7*(3), 20–40. doi:10.4018/IJMMME.2017070102

Kelly, M., Costello, M., Nicholson, G., & O'Connor, J. (2021). The Evolving Integration of BIM Into Built Environment Programmes in a Higher Education Institute. In J. Underwood & M. Shelbourn (Eds.), *Handbook of Research on Driving Transformational Change in the Digital Built Environment* (pp. 294–326). IGI Global. https://doi.org/10.4018/978-1-7998-6600-8.ch012

Kesimal, A., Karaman, K., Cihangir, F., & Ercikdi, B. (2018). Excavatability Assessment of Rock Masses for Geotechnical Studies. In N. Ceryan (Ed.), *Handbook of Research on Trends and Digital Advances in Engineering Geology* (pp. 231–256). Hershey, PA: IGI Global. doi:10.4018/978-1-5225-2709-1.ch006

Knoflacher, H. (2017). The Role of Engineers and Their Tools in the Transport Sector after Paradigm Change: From Assumptions and Extrapolations to Science. In H. Knoflacher & E. Ocalir-Akunal (Eds.), *Engineering Tools and Solutions for Sustainable Transportation Planning* (pp. 1–29). Hershey, PA: IGI Global. doi:10.4018/978-1-5225-2116-7.ch001

Kose, U. (2018). Towards an Intelligent Biomedical Engineering With Nature-Inspired Artificial Intelligence Techniques. In U. Kose, G. Guraksin, & O. Deperlioglu (Eds.), *Nature-Inspired Intelligent Techniques for Solving Biomedical Engineering Problems* (pp. 1–26). Hershey, PA: IGI Global. doi:10.4018/978-1-5225-4769-3.ch001

Kostić, S. (2018). A Review on Enhanced Stability Analyses of Soil Slopes Using Statistical Design. In N. Ceryan (Ed.), *Handbook of Research on Trends and Digital Advances in Engineering Geology* (pp. 446–481). Hershey, PA: IGI Global. doi:10.4018/978-1-5225-2709-1.ch013

Kumar, A., Patil, P. P., & Prajapati, Y. K. (2018). *Advanced Numerical Simulations in Mechanical Engineering*. Hershey, PA: IGI Global. doi:10.4018/978-1-5225-3722-9

Kumar, G. R., Rajyalakshmi, G., & Manupati, V. K. (2017). Surface Micro Patterning of Aluminium Reinforced Composite through Laser Peening. *International Journal of Manufacturing, Materials, and Mechanical Engineering*, 7(4), 15–27. doi:10.4018/IJMMME.2017100102

Kumar, N., Basu, D. N., & Chen, L. (2021). Effect of Flow Acceleration and Buoyancy on Thermalhydraulics of sCO2 in Mini/Micro-Channel. In L. Chen (Ed.), *Handbook of Research on Advancements in Supercritical Fluids Applications for Sustainable Energy Systems* (pp. 161–182). IGI Global. doi:10.4018/978-1-7998-5796-9.ch005

Kumari, N., & Kumar, K. (2018). Fabrication of Orthotic Calipers With Epoxy-Based Green Composite. In K. Kumar & J. Davim (Eds.), *Composites and Advanced Materials for Industrial Applications* (pp. 157–176). Hershey, PA: IGI Global. doi:10.4018/978-1-5225-5216-1.ch008

Kuppusamy, R. R. (2018). Development of Aerospace Composite Structures Through Vacuum-Enhanced Resin Transfer Moulding Technology (VERTMTy): Vacuum-Enhanced Resin Transfer Moulding. In K. Kumar & J. Davim (Eds.), *Composites and Advanced Materials for Industrial Applications* (pp. 99–111). Hershey, PA: IGI Global. doi:10.4018/978-1-5225-5216-1.ch005

Kurganov, V. A., Zeigarnik, Y. A., & Maslakova, I. V. (2021). Normal and Deteriorated Heat Transfer Under Heating Turbulent Supercritical Pressure Coolants Flows in Round Tubes. In L. Chen (Ed.), *Handbook of Research on Advancements in Supercritical Fluids Applications for Sustainable Energy Systems* (pp. 494–532). IGI Global. https://doi.org/10.4018/978-1-7998-5796-9.ch014

Li, H., & Zhang, Y. (2021). Heat Transfer and Fluid Flow Modeling for Supercritical Fluids in Advanced Energy Systems. In L. Chen (Ed.), *Handbook of Research on Advancements in Supercritical Fluids Applications for Sustainable Energy Systems* (pp. 388–422). IGI Global. https://doi.org/10.4018/978-1-7998-5796-9.ch011

Loy, J., Howell, S., & Cooper, R. (2017). Engineering Teams: Supporting Diversity in Engineering Education. In M. Gray & K. Thomas (Eds.), *Strategies for Increasing Diversity in Engineering Majors and Careers* (pp. 106–129). Hershey, PA: IGI Global. doi:10.4018/978-1-5225-2212-6.ch006

Macher, G., Armengaud, E., Kreiner, C., Brenner, E., Schmittner, C., Ma, Z., ... Krammer, M. (2018). Integration of Security in the Development Lifecycle of Dependable Automotive CPS. In N. Druml, A. Genser, A. Krieg, M. Menghin, & A. Hoeller (Eds.), *Solutions for Cyber-Physical Systems Ubiquity* (pp. 383–423). Hershey, PA: IGI Global. doi:10.4018/978-1-5225-2845-6.ch015

Madhu, M. N., Singh, J. G., Mohan, V., & Ongsakul, W. (2021). Transmission Risk Optimization in Interconnected Systems: Risk-Adjusted Available Transfer Capability. In P. Vasant, G. Weber, & W. Punurai (Ed.), *Research Advancements in Smart Technology, Optimization, and Renewable Energy* (pp. 183-199). IGI Global. https://doi.org/10.4018/978-1-7998-3970-5.ch010

Mahendramani, G., & Lakshmana Swamy, N. (2018). Effect of Weld Groove Area on Distortion of Butt Welded Joints in Submerged Arc Welding. *International Journal of Manufacturing, Materials, and Mechanical Engineering*, *8*(2), 33–44. doi:10.4018/IJMMME.2018040103

Makropoulos, G., Koumaras, H., Setaki, F., Filis, K., Lutz, T., Montowtt, P., Tomaszewski, L., Dybiec, P., & Järvet, T. (2021). 5G and Unmanned Aerial Vehicles (UAVs) Use Cases: Analysis of the Ecosystem, Architecture, and Applications. In A. Nwajana & I. Ihianle (Eds.), *Handbook of Research on 5G Networks and Advancements in Computing, Electronics, and Electrical Engineering* (pp. 36–69). IGI Global. https://doi.org/10.4018/978-1-7998-6992-4.ch003

Meric, E. M., Erdem, S., & Gurbuz, E. (2021). Application of Phase Change Materials in Construction Materials for Thermal Energy Storage Systems in Buildings. In R. González-Lezcano (Ed.), *Advancements in Sustainable Architecture and Energy Efficiency* (pp. 1–20). IGI Global. https://doi.org/10.4018/978-1-7998-7023-4.ch001

Mihret, E. T., & Yitayih, K. A. (2021). Operation of VANET Communications: The Convergence of UAV System With LTE/4G and WAVE Technologies. *International Journal of Smart Vehicles and Smart Transportation*, *4*(1), 29–51. https://doi.org/10.4018/IJSVST.2021010103

Mir, M. A., Bhat, B. A., Sheikh, B. A., Rather, G. A., Mehraj, S., & Mir, W. R. (2021). Nanomedicine in Human Health Therapeutics and Drug Delivery: Nanobiotechnology and Nanobiomedicine. In M. Bhat, I. Wani, & S. Ashraf (Eds.), *Applications of Nanomaterials in Agriculture, Food Science, and Medicine* (pp. 229–251). IGI Global. doi:10.4018/978-1-7998-5563-7.ch013

Mohammadzadeh, S., & Kim, Y. (2017). Nonlinear System Identification of Smart Buildings. In P. Samui, S. Chakraborty, & D. Kim (Eds.), *Modeling and Simulation Techniques in Structural Engineering* (pp. 328–347). Hershey, PA: IGI Global. doi:10.4018/978-1-5225-0588-4.ch011

Molina, G. J., Aktaruzzaman, F., Soloiu, V., & Rahman, M. (2017). Design and Testing of a Jet-Impingement Instrument to Study Surface-Modification Effects by Nanofluids. *International Journal of Surface Engineering and Interdisciplinary Materials Science*, *5*(2), 43–61. doi:10.4018/IJSEIMS.2017070104

Moreno-Rangel, A., & Carrillo, G. (2021). Energy-Efficient Homes: A Heaven for Respiratory Illnesses. In R. González-Lezcano (Ed.), *Advancements in Sustainable Architecture and Energy Efficiency* (pp. 49–71). IGI Global. https://doi.org/10.4018/978-1-7998-7023-4.ch003

Msomi, V., & Jantjies, B. T. (2021). Correlative Analysis Between Tensile Properties and Tool Rotational Speeds of Friction Stir Welded Similar Aluminium Alloy Joints. *International Journal of Surface Engineering and Interdisciplinary Materials Science*, *9*(2), 58–78. https://doi.org/10.4018/IJSEIMS.2021070104

Muigai, M. N., Mwema, F. M., Akinlabi, E. T., & Obiko, J. O. (2021). Surface Engineering of Materials Through Weld-Based Technologies: An Overview. In S. Roy & G. Bose (Eds.), *Advanced Surface Coating Techniques for Modern Industrial Applications* (pp. 247–260). IGI Global. doi:10.4018/978-1-7998-4870-7.ch011

Mukherjee, A., Saeed, R. A., Dutta, S., & Naskar, M. K. (2017). Fault Tracking Framework for Software-Defined Networking (SDN). In C. Singhal & S. De (Eds.), *Resource Allocation in Next-Generation Broadband Wireless Access Networks* (pp. 247–272). Hershey, PA: IGI Global. doi:10.4018/978-1-5225-2023-8.ch011

Mukhopadhyay, A., Barman, T. K., & Sahoo, P. (2018). Electroless Nickel Coatings for High Temperature Applications. In K. Kumar & J. Davim (Eds.), *Composites and Advanced Materials for Industrial Applications* (pp. 297–331). Hershey, PA: IGI Global. doi:10.4018/978-1-5225-5216-1.ch013

Mwema, F. M., & Wambua, J. M. (2022). Machining of Poly Methyl Methacrylate (PMMA) and Other Olymeric Materials: A Review. In K. Kumar, B. Babu, & J. Davim (Eds.), *Handbook of Research on Advancements in the Processing, Characterization, and Application of Lightweight Materials* (pp. 363–379). IGI Global. https://doi.org/10.4018/978-1-7998-7864-3.ch016

Mykhailyshyn, R., Savkiv, V., Boyko, I., Prada, E., & Virgala, I. (2021). Substantiation of Parameters of Friction Elements of Bernoulli Grippers With a Cylindrical Nozzle. *International Journal of Manufacturing, Materials, and Mechanical Engineering, 11*(2), 17–39. https://doi.org/10.4018/IJMMME.2021040102

Náprstek, J., & Fischer, C. (2017). Dynamic Stability and Post-Critical Processes of Slender Auto-Parametric Systems. In V. Plevris, G. Kremmyda, & Y. Fahjan (Eds.), *Performance-Based Seismic Design of Concrete Structures and Infrastructures* (pp. 128–171). Hershey, PA: IGI Global. doi:10.4018/978-1-5225-2089-4.ch006

Nautiyal, L., Shivach, P., & Ram, M. (2018). Optimal Designs by Means of Genetic Algorithms. In M. Ram & J. Davim (Eds.), *Soft Computing Techniques and Applications in Mechanical Engineering* (pp. 151–161). Hershey, PA: IGI Global. doi:10.4018/978-1-5225-3035-0.ch007

Nazir, R. (2017). Advanced Nanomaterials for Water Engineering and Treatment: Nano-Metal Oxides and Their Nanocomposites. In T. Saleh (Ed.), *Advanced Nanomaterials for Water Engineering, Treatment, and Hydraulics* (pp. 84–126). Hershey, PA: IGI Global. doi:10.4018/978-1-5225-2136-5.ch005

Nikolopoulos, C. D. (2021). Recent Advances on Measuring and Modeling ELF-Radiated Emissions for Space Applications. In C. Nikolopoulos (Ed.), *Recent Trends on Electromagnetic Environmental Effects for Aeronautics and Space Applications* (pp. 1–38). IGI Global. https://doi.org/10.4018/978-1-7998-4879-0.ch001

Nogueira, A. F., Ribeiro, J. C., Fernández de Vega, F., & Zenha-Rela, M. A. (2018). Evolutionary Approaches to Test Data Generation for Object-Oriented Software: Overview of Techniques and Tools. In M. Khosrow-Pour, D.B.A. (Ed.), Incorporating Nature-Inspired Paradigms in Computational Applications (pp. 162-194). Hershey, PA: IGI Global. https://doi.org/ doi:10.4018/978-1-5225-5020-4.ch006

Nwajana, A. O., Obi, E. R., Ijemaru, G. K., Oleka, E. U., & Anthony, D. C. (2021). Fundamentals of RF/Microwave Bandpass Filter Design. In A. Nwajana & I. Ihianle (Eds.), *Handbook of Research on 5G Networks and Advancements in Computing, Electronics, and Electrical Engineering* (pp. 149–164). IGI Global. https://doi.org/10.4018/978-1-7998-6992-4.ch005

Ogbodo, E. A. (2021). Comparative Study of Transmission Line Junction vs. Asynchronously Coupled Junction Diplexers. In A. Nwajana & I. Ihianle (Eds.), *Handbook of Research on 5G Networks and Advancements in Computing, Electronics, and Electrical Engineering* (pp. 326–336). IGI Global. https://doi.org/10.4018/978-1-7998-6992-4.ch013

Orosa, J. A., Vergara, D., Fraguela, F., & Masdías-Bonome, A. (2021). Statistical Understanding and Optimization of Building Energy Consumption and Climate Change Consequences. In R. González-Lezcano (Ed.), *Advancements in Sustainable Architecture and Energy Efficiency* (pp. 195–220). IGI Global. https://doi.org/10.4018/978-1-7998-7023-4.ch009

Osho, M. B. (2018). Industrial Enzyme Technology: Potential Applications. In S. Bharati & P. Chaurasia (Eds.), *Research Advancements in Pharmaceutical, Nutritional, and Industrial Enzymology* (pp. 375–394). Hershey, PA: IGI Global. doi:10.4018/978-1-5225-5237-6.ch017

Ouadi, A., & Zitouni, A. (2021). Phasor Measurement Improvement Using Digital Filter in a Smart Grid. In A. Recioui & H. Bentarzi (Eds.), *Optimizing and Measuring Smart Grid Operation and Control* (pp. 100–117). IGI Global. https://doi.org/10.4018/978-1-7998-4027-5.ch005

Padmaja, P., & Marutheswar, G. (2017). Certain Investigation on Secured Data Transmission in Wireless Sensor Networks. *International Journal of Mobile Computing and Multimedia Communications*, *8*(1), 48–61. doi:10.4018/IJMCMC.2017010104

Palmer, S., & Hall, W. (2017). An Evaluation of Group Work in First-Year Engineering Design Education. In R. Tucker (Ed.), *Collaboration and Student Engagement in Design Education* (pp. 145–168). Hershey, PA: IGI Global. doi:10.4018/978-1-5225-0726-0.ch007

Panchenko, V. (2021). Prospects for Energy Supply of the Arctic Zone Objects of Russia Using Frost-Resistant Solar Modules. In P. Vasant, G. Weber, & W. Punurai (Eds.), *Research Advancements in Smart Technology, Optimization, and Renewable Energy* (pp. 149-169). IGI Global. https://doi.org/10.4018/978-1-7998-3970-5.ch008

Panchenko, V. (2021). Photovoltaic Thermal Module With Paraboloid Type Solar Concentrators. *International Journal of Energy Optimization and Engineering*, *10*(2), 1–23. https://doi.org/10.4018/IJEOE.2021040101

Pandey, K., & Datta, S. (2021). Dry Machining of Inconel 825 Superalloys: Performance of Tool Inserts (Carbide, Cermet, and SiAlON). *International Journal of Manufacturing, Materials, and Mechanical Engineering*, *11*(4), 26–39. doi:10.4018/IJMMME.2021100102

Panneer, R. (2017). Effect of Composition of Fibers on Properties of Hybrid Composites. *International Journal of Manufacturing, Materials, and Mechanical Engineering*, *7*(4), 28–43. doi:10.4018/IJMMME.2017100103

Pany, C. (2021). Estimation of Correct Long-Seam Mismatch Using FEA to Compare the Measured Strain in a Non-Destructive Testing of a Pressurant Tank: A Reverse Problem. *International Journal of Smart Vehicles and Smart Transportation*, *4*(1), 16–28. doi:10.4018/IJSVST.2021010102

Paul, S., & Roy, P. (2018). Optimal Design of Power System Stabilizer Using a Novel Evolutionary Algorithm. *International Journal of Energy Optimization and Engineering*, *7*(3), 24–46. doi:10.4018/IJEOE.2018070102

Paul, S., & Roy, P. K. (2021). Oppositional Differential Search Algorithm for the Optimal Tuning of Both Single Input and Dual Input Power System Stabilizer. In P. Vasant, G. Weber, & W. Punurai (Eds.), *Research Advancements in Smart Technology, Optimization, and Renewable Energy* (pp. 256-282). IGI Global. https://doi.org/10.4018/978-1-7998-3970-5.ch013

Pavaloiu, A. (2018). Artificial Intelligence Ethics in Biomedical-Engineering-Oriented Problems. In U. Kose, G. Guraksin, & O. Deperlioglu (Eds.), *Nature-Inspired Intelligent Techniques for Solving Biomedical Engineering Problems* (pp. 219–231). Hershey, PA: IGI Global. doi:10.4018/978-1-5225-4769-3.ch010

Pioro, I., Mahdi, M., & Popov, R. (2017). Application of Supercritical Pressures in Power Engineering. In L. Chen & Y. Iwamoto (Eds.), *Advanced Applications of Supercritical Fluids in Energy Systems* (pp. 404–457). Hershey, PA: IGI Global. doi:10.4018/978-1-5225-2047-4.ch013

Plaksina, T., & Gildin, E. (2017). Rigorous Integrated Evolutionary Workflow for Optimal Exploitation of Unconventional Gas Assets. *International Journal of Energy Optimization and Engineering*, *6*(1), 101–122. doi:10.4018/IJEOE.2017010106

Popat, J., Kakadiya, H., Tak, L., Singh, N. K., Majeed, M. A., & Mahajan, V. (2021). Reliability of Smart Grid Including Cyber Impact: A Case Study. In R. Singh, A. Singh, A. Dwivedi, & P. Nagabhushan (Eds.), *Computational Methodologies for Electrical and Electronics Engineers* (pp. 163–174). IGI Global. https://doi.org/10.4018/978-1-7998-3327-7.ch013

Quiza, R., La Fé-Perdomo, I., Rivas, M., & Ramtahalsing, V. (2021). Triple Bottom Line-Focused Optimization of Oblique Turning Processes Based on Hybrid Modeling: A Study Case on AISI 1045 Steel Turning. In L. Burstein (Ed.), *Handbook of Research on Advancements in Manufacturing, Materials, and Mechanical Engineering* (pp. 215–241). IGI Global. https://doi.org/10.4018/978-1-7998-4939-1.ch010

Rahmani, M. K. (2022). Blockchain Technology: Principles and Algorithms. In S. Khan, M. Syed, R. Hammad, & A. Bushager (Eds.), *Blockchain Technology and Computational Excellence for Society 5.0* (pp. 16–27). IGI Global. https://doi.org/10.4018/978-1-7998-8382-1.ch002

Ramdani, N., & Azibi, M. (2018). Polymer Composite Materials for Microelectronics Packaging Applications: Composites for Microelectronics Packaging. In K. Kumar & J. Davim (Eds.), *Composites and Advanced Materials for Industrial Applications* (pp. 177–211). Hershey, PA: IGI Global. doi:10.4018/978-1-5225-5216-1.ch009

Ramesh, M., Garg, R., & Subrahmanyam, G. V. (2017). Investigation of Influence of Quenching and Annealing on the Plane Fracture Toughness and Brittle to Ductile Transition Temperature of the Zinc Coated Structural Steel Materials. *International Journal of Surface Engineering and Interdisciplinary Materials Science*, *5*(2), 33–42. doi:10.4018/IJSEIMS.2017070103

Robinson, J., & Beneroso, D. (2022). Project-Based Learning in Chemical Engineering: Curriculum and Assessment, Culture and Learning Spaces. In A. Alves & N. van Hattum-Janssen (Eds.), *Training Engineering Students for Modern Technological Advancement* (pp. 1–19). IGI Global. https://doi.org/10.4018/978-1-7998-8816-1.ch001

Rondon, B. (2021). Experimental Characterization of Admittance Meter With Crude Oil Emulsions. *International Journal of Electronics, Communications, and Measurement Engineering*, *10*(2), 51–59. https://doi.org/10.4018/IJECME.2021070104

Rudolf, S., Biryuk, V. V., & Volov, V. (2018). Vortex Effect, Vortex Power: Technology of Vortex Power Engineering. In V. Kharchenko & P. Vasant (Eds.), *Handbook of Research on Renewable Energy and Electric Resources for Sustainable Rural Development* (pp. 500–533). Hershey, PA: IGI Global. doi:10.4018/978-1-5225-3867-7.ch021

Sah, A., Bhadula, S. J., Dumka, A., & Rawat, S. (2018). A Software Engineering Perspective for Development of Enterprise Applications. In A. Elçi (Ed.), *Handbook of Research on Contemporary Perspectives on Web-Based Systems* (pp. 1–23). Hershey, PA: IGI Global. doi:10.4018/978-1-5225-5384-7.ch001

Sahoo, P., & Roy, S. (2017). Tribological Behavior of Electroless Ni-P, Ni-P-W and Ni-P-Cu Coatings: A Comparison. *International Journal of Surface Engineering and Interdisciplinary Materials Science*, *5*(1), 1–15. doi:10.4018/IJSEIMS.2017010101

Sahoo, S. (2018). Laminated Composite Hypar Shells as Roofing Units: Static and Dynamic Behavior. In K. Kumar & J. Davim (Eds.), *Composites and Advanced Materials for Industrial Applications* (pp. 249–269). Hershey, PA: IGI Global. doi:10.4018/978-1-5225-5216-1.ch011

Sahu, H., & Hungyo, M. (2018). Introduction to SDN and NFV. In A. Dumka (Ed.), *Innovations in Software-Defined Networking and Network Functions Virtualization* (pp. 1–25). Hershey, PA: IGI Global. doi:10.4018/978-1-5225-3640-6.ch001

Salem, A. M., & Shmelova, T. (2018). Intelligent Expert Decision Support Systems: Methodologies, Applications, and Challenges. In T. Shmelova, Y. Sikirda, N. Rizun, A. Salem, & Y. Kovalyov (Eds.), *Socio-Technical Decision Support in Air Navigation Systems: Emerging Research and Opportunities* (pp. 215–242). Hershey, PA: IGI Global. doi:10.4018/978-1-5225-3108-1.ch007

Samal, M. (2017). FE Analysis and Experimental Investigation of Cracked and Un-Cracked Thin-Walled Tubular Components to Evaluate Mechanical and Fracture Properties. In P. Samui, S. Chakraborty, & D. Kim (Eds.), *Modeling and Simulation Techniques in Structural Engineering* (pp. 266–293). Hershey, PA: IGI Global. doi:10.4018/978-1-5225-0588-4.ch009

Samal, M., & Balakrishnan, K. (2017). Experiments on a Ring Tension Setup and FE Analysis to Evaluate Transverse Mechanical Properties of Tubular Components. In P. Samui, S. Chakraborty, & D. Kim (Eds.), *Modeling and Simulation Techniques in Structural Engineering* (pp. 91–115). Hershey, PA: IGI Global. doi:10.4018/978-1-5225-0588-4.ch004

Samarasinghe, D. A., & Wood, E. (2021). Innovative Digital Technologies. In J. Underwood & M. Shelbourn (Eds.), *Handbook of Research on Driving Transformational Change in the Digital Built Environment* (pp. 142–163). IGI Global. https://doi.org/10.4018/978-1-7998-6600-8.ch006

Sawant, S. (2018). Deep Learning and Biomedical Engineering. In U. Kose, G. Guraksin, & O. Deperlioglu (Eds.), *Nature-Inspired Intelligent Techniques for Solving Biomedical Engineering Problems* (pp. 283–296). Hershey, PA: IGI Global. doi:10.4018/978-1-5225-4769-3.ch014

Schulenberg, T. (2021). Energy Conversion Using the Supercritical Steam Cycle. In L. Chen (Ed.), *Handbook of Research on Advancements in Supercritical Fluids Applications for Sustainable Energy Systems* (pp. 659–681). IGI Global. doi:10.4018/978-1-7998-5796-9.ch018

Sezgin, H., & Berkalp, O. B. (2018). Textile-Reinforced Composites for the Automotive Industry. In K. Kumar & J. Davim (Eds.), *Composites and Advanced Materials for Industrial Applications* (pp. 129–156). Hershey, PA: IGI Global. doi:10.4018/978-1-5225-5216-1.ch007

Shaaban, A. A., & Shehata, O. M. (2021). Combining Response Surface Method and Metaheuristic Algorithms for Optimizing SPIF Process. *International Journal of Manufacturing, Materials, and Mechanical Engineering, 11*(4), 1–25. https://doi.org/10.4018/IJMMME.2021100101

Shafaati Shemami, M., & Sefid, M. (2022). Implementation and Demonstration of Electric Vehicle-to-Home (V2H) Application: A Case Study. In M. Alam, R. Pillai, & N. Murugesan (Eds.), *Developing Charging Infrastructure and Technologies for Electric Vehicles* (pp. 268–293). IGI Global. https://doi.org/10.4018/978-1-7998-6858-3.ch015

Shah, M. Z., Gazder, U., Bhatti, M. S., & Hussain, M. (2018). Comparative Performance Evaluation of Effects of Modifier in Asphaltic Concrete Mix. *International Journal of Strategic Engineering, 1*(2), 13–25. doi:10.4018/IJoSE.2018070102

Sharma, N., & Kumar, K. (2018). Fabrication of Porous NiTi Alloy Using Organic Binders. In K. Kumar & J. Davim (Eds.), *Composites and Advanced Materials for Industrial Applications* (pp. 38–62). Hershey, PA: IGI Global. doi:10.4018/978-1-5225-5216-1.ch003

Shivach, P., Nautiyal, L., & Ram, M. (2018). Applying Multi-Objective Optimization Algorithms to Mechanical Engineering. In M. Ram & J. Davim (Eds.), *Soft Computing Techniques and Applications in Mechanical Engineering* (pp. 287–301). Hershey, PA: IGI Global. doi:10.4018/978-1-5225-3035-0.ch014

Shmelova, T. (2018). Stochastic Methods for Estimation and Problem Solving in Engineering: Stochastic Methods of Decision Making in Aviation. In S. Kadry (Ed.), *Stochastic Methods for Estimation and Problem Solving in Engineering* (pp. 139–160). Hershey, PA: IGI Global. doi:10.4018/978-1-5225-5045-7.ch006

Siero González, L. R., & Romo Vázquez, A. (2017). Didactic Sequences Teaching Mathematics for Engineers With Focus on Differential Equations. In M. Ramírez-Montoya (Ed.), *Handbook of Research on Driving STEM Learning With Educational Technologies* (pp. 129–151). Hershey, PA: IGI Global. doi:10.4018/978-1-5225-2026-9.ch007

Sim, M. S., You, K. Y., Esa, F., & Chan, Y. L. (2021). Nanostructured Electromagnetic Metamaterials for Sensing Applications. In M. Bhat, I. Wani, & S. Ashraf (Eds.), *Applications of Nanomaterials in Agriculture, Food Science, and Medicine* (pp. 141–164). IGI Global. https://doi.org/10.4018/978-1-7998-5563-7.ch009

Singh, R., & Dutta, S. (2018). Visible Light Active Nanocomposites for Photocatalytic Applications. In K. Kumar & J. Davim (Eds.), *Composites and Advanced Materials for Industrial Applications* (pp. 270–296). Hershey, PA: IGI Global. doi:10.4018/978-1-5225-5216-1.ch012

Skripov, P. V., Yampol'skiy, A. D., & Rutin, S. B. (2021). High-Power Heat Transfer in Supercritical Fluids: Microscale Times and Sizes. In L. Chen (Ed.), *Handbook of Research on Advancements in Supercritical Fluids Applications for Sustainable Energy Systems* (pp. 424–450). IGI Global. https://doi.org/10.4018/978-1-7998-5796-9.ch012

Sözbilir, H., Özkaymak, Ç., Uzel, B., & Sümer, Ö. (2018). Criteria for Surface Rupture Microzonation of Active Faults for Earthquake Hazards in Urban Areas. In N. Ceryan (Ed.), *Handbook of Research on Trends and Digital Advances in Engineering Geology* (pp. 187–230). Hershey, PA: IGI Global. doi:10.4018/978-1-5225-2709-1.ch005

Stanciu, I. (2018). Stochastic Methods in Microsystems Engineering. In S. Kadry (Ed.), *Stochastic Methods for Estimation and Problem Solving in Engineering* (pp. 161–176). Hershey, PA: IGI Global. doi:10.4018/978-1-5225-5045-7.ch007

Strebkov, D., Nekrasov, A., Trubnikov, V., & Nekrasov, A. (2018). Single-Wire Resonant Electric Power Systems for Renewable-Based Electric Grid. In V. Kharchenko & P. Vasant (Eds.), *Handbook of Research on Renewable Energy and Electric Resources for Sustainable Rural Development* (pp. 449–474). Hershey, PA: IGI Global. doi:10.4018/978-1-5225-3867-7.ch019

Sukhyy, K., Belyanovskaya, E., & Sukhyy, M. (2021). *Basic Principles for Substantiation of Working Pair Choice*. IGI Global. doi:10.4018/978-1-7998-4432-7.ch002

Suri, M. S., & Kaliyaperumal, D. (2022). Extension of Aspiration Level Model for Optimal Planning of Fast Charging Stations. In A. Fekik & N. Benamrouche (Eds.), *Modeling and Control of Static Converters for Hybrid Storage Systems* (pp. 91–106). IGI Global. https://doi.org/10.4018/978-1-7998-7447-8.ch004

Tallet, E., Gledson, B., Rogage, K., Thompson, A., & Wiggett, D. (2021). Digitally-Enabled Design Management. In J. Underwood & M. Shelbourn (Eds.), *Handbook of Research on Driving Transformational Change in the Digital Built Environment* (pp. 63–89). IGI Global. https://doi.org/10.4018/978-1-7998-6600-8.ch003

Terki, A., & Boubertakh, H. (2021). A New Hybrid Binary-Real Coded Cuckoo Search and Tabu Search Algorithm for Solving the Unit-Commitment Problem. *International Journal of Energy Optimization and Engineering*, *10*(2), 104–119. https://doi.org/10.4018/IJEOE.2021040105

Tüdeş, Ş., Kumlu, K. B., & Ceryan, S. (2018). Integration Between Urban Planning and Natural Hazards For Resilient City. In N. Ceryan (Ed.), *Handbook of Research on Trends and Digital Advances in Engineering Geology* (pp. 591–630). Hershey, PA: IGI Global. doi:10.4018/978-1-5225-2709-1.ch017

Ulamis, K. (2018). Soil Liquefaction Assessment by Anisotropic Cyclic Triaxial Test. In N. Ceryan (Ed.), *Handbook of Research on Trends and Digital Advances in Engineering Geology* (pp. 631–664). Hershey, PA: IGI Global. doi:10.4018/978-1-5225-2709-1.ch018

Valente, M., & Milani, G. (2017). Seismic Assessment and Retrofitting of an Under-Designed RC Frame Through a Displacement-Based Approach. In V. Plevris, G. Kremmyda, & Y. Fahjan (Eds.), *Performance-Based Seismic Design of Concrete Structures and Infrastructures* (pp. 36–58). Hershey, PA: IGI Global. doi:10.4018/978-1-5225-2089-4.ch002

Vargas-Bernal, R. (2021). Advances in Electromagnetic Environmental Shielding for Aeronautics and Space Applications. In C. Nikolopoulos (Ed.), *Recent Trends on Electromagnetic Environmental Effects for Aeronautics and Space Applications* (pp. 80–96). IGI Global. https://doi.org/10.4018/978-1-7998-4879-0.ch003

Vasant, P. (2018). A General Medical Diagnosis System Formed by Artificial Neural Networks and Swarm Intelligence Techniques. In U. Kose, G. Guraksin, & O. Deperlioglu (Eds.), *Nature-Inspired Intelligent Techniques for Solving Biomedical Engineering Problems* (pp. 130–145). Hershey, PA: IGI Global. doi:10.4018/978-1-5225-4769-3.ch006

Verner, C. M., & Sarwar, D. (2021). Avoiding Project Failure and Achieving Project Success in NHS IT System Projects in the United Kingdom. *International Journal of Strategic Engineering*, *4*(1), 33–54. https://doi.org/10.4018/IJoSE.2021010103

Verrollot, J., Tolonen, A., Harkonen, J., & Haapasalo, H. J. (2018). Challenges and Enablers for Rapid Product Development. *International Journal of Applied Industrial Engineering*, *5*(1), 25–49. doi:10.4018/IJAIE.2018010102

Wan, A. C., Zulu, S. L., & Khosrow-Shahi, F. (2021). Industry Views on BIM for Site Safety in Hong Kong. In J. Underwood & M. Shelbourn (Eds.), *Handbook of Research on Driving Transformational Change in the Digital Built Environment* (pp. 120–140). IGI Global. https://doi.org/10.4018/978-1-7998-6600-8.ch005

Yardimci, A. G., & Karpuz, C. (2018). Fuzzy Rock Mass Rating: Soft-Computing-Aided Preliminary Stability Analysis of Weak Rock Slopes. In N. Ceryan (Ed.), *Handbook of Research on Trends and Digital Advances in Engineering Geology* (pp. 97–131). Hershey, PA: IGI Global. doi:10.4018/978-1-5225-2709-1.ch003

You, K. Y. (2021). Development Electronic Design Automation for RF/Microwave Antenna Using MATLAB GUI. In A. Nwajana & I. Ihianle (Eds.), *Handbook of Research on 5G Networks and Advancements in Computing, Electronics, and Electrical Engineering* (pp. 70–148). IGI Global. https://doi.org/10.4018/978-1-7998-6992-4.ch004

Yousefi, Y., Gratton, P., & Sarwar, D. (2021). Investigating the Opportunities to Improve the Thermal Performance of a Case Study Building in London. *International Journal of Strategic Engineering*, *4*(1), 1–18. https://doi.org/10.4018/IJoSE.2021010101

Zindani, D., & Kumar, K. (2018). Industrial Applications of Polymer Composite Materials. In K. Kumar & J. Davim (Eds.), *Composites and Advanced Materials for Industrial Applications* (pp. 1–15). Hershey, PA: IGI Global. doi:10.4018/978-1-5225-5216-1.ch001

Zindani, D., Maity, S. R., & Bhowmik, S. (2018). A Decision-Making Approach for Material Selection of Polymeric Composite Bumper Beam. In K. Kumar & J. Davim (Eds.), *Composites and Advanced Materials for Industrial Applications* (pp. 112–128). Hershey, PA: IGI Global. doi:10.4018/978-1-5225-5216-1.ch006

About the Contributors

Azad Kumar is currently serving as assistant professor in the Department of Chemistry, Siddharth University, Kapilvastu, Siddharthnagar. Dr. Azad has obtained B.Sc. Hons (2007), M.Sc. (2009), M.Phil. (2010) in Chemistry from Dayalbagh Educational Institute, Dayalbagh Agra, and Ph.D. (2018) in Chemistry from Babasaheb Bhimrao Ambedkar University, Lucknow. Dr. Azad is also a member of several scientific societies. His research and teaching interests include theory and application of nanomaterial, hybrid composite, and polymer. He has published 20 research papers in peer-reviewed & refereed journals of repute. He has served as editorial member of some journals. He has teaching experience of 10 years in various institutes.

***

Nuray Bekoz Ullen is currently working in the Department of Metallurgical and Materials Engineering, Istanbul University-Cerrahpaşa (IUC), Turkey. She obtained her BSc, MSc and PhD degree at Department of Metallurgical and Materials Engineering from Istanbul University, Turkey in 2003, 2006 and 2011, respectively. Her main research interest includes powder metallurgy, metal foams, coating technology, surface treatments, material characterizations, machinability and welding. In recent years, she has focused on improving the surface properties of dense porous materials. She has collaborated actively with researchers in several other disciplines of mechanical, chemistry and physics. She has taught on engineering materials and deformation mechanisms, highly porous material production techniques and characterization, failure of materials. She had role as a consultant in master theses in the fields of material characterization. She has published several research articles on material production, characterization and processing.

Chinmoy Bhattacharya is currently working as an Associate Professor in the Department of Chemistry and also serving as Associate Dean (Faculty Welfare) at the Indian Institute of Engineering Science & Technology (IIEST), Shibpur, West Bengal. Completed M.Sc. from Calcutta University (Presidency College) in 2000 with Physical Chemistry specialization and obtained Ph.D. degree in the area of Photoelectrochemistry from Bengal Engineering & Science University, Shibpur (presently, IIEST, Shibpur) in 2006 under the supervision of Professor Jayati Datta. His area of interest is the use of solar energy either to generate 'green' fuels (like H 2 /O 2) through water splitting or treatment of water-soluble wastes for a cleaner environment. He has published more than 45 papers in different international journals and contributed to three book chapters; he completed 8 R&D projects sponsored by various Governmental agencies. He has guided 7 PhD students; currently another 4 scholars are working with him. He has received different awards, like, "Global Alumni Association of BESU: Young Faculty Research Award (YFRA) 2016", "DST -BOYSCAST Fellowship (2010-2011) for working with Professor Allen J. Bard of The University of Texas at Austin, USA for 01 year", "Professor S.R. Palit Memorial Award by Indian Chemical Society in 2004 & 2007".

Amit Kumar Dutta is currently working as an Assistant Professor in the Department of Chemistry, Bangabasi Morning College (Affiliated to University of Calcutta). He has Completed M.Sc. from University of Calcutta in 2007 with Inorganic Chemistry specialization and obtained Ph.D. degree in the area of Nano-catalysis from Bengal Engineering and Science University, Shibpur (presently, IIEST, Shibpur) in 2012 under the supervision of Prof. Bibhutosh Adhikary. His area of research interest is the use of semiconducting nano-materials in photo-catalysis and bio-sensing application. He has published 35 research papers in different peer-reviewed journals and 3 book chapters. He has completed one UGC project and currently one research scholar is working with him. He has awarded Junior and Senior Research Fellowship (JRF/SRF) through joint CSIR–UGC (NET) and Research Associate Fellowship (CSIR-RA).

Gizem Karabulut is currently working as a Research Assistant in the Department of Metallurgical and Materials Engineering at Istanbul University-Cerrahpasa. She obtained her BSc degree at the Department of Metallurgical and Materials Engineering from Istanbul University-Cerrahpasa as an honor student in 2020. She completed her MSc education in 2023 and continues PhD.

Sukumar Kundu is Assistant Professor, Department of Metallurgy and Materials Engineering, IIEST, Shibpur.

Shahzad Naseem is a National Professor of Physics working at Centre of Excellence in Solid State Physics, University of the Punjab, Lahore, Pakistan. Prof. Naseem has supervised 292 research theses (28 PhD and 264 MS/MPhil) in the fields of PV, LCD, SC, Spintronics etc. He has published more than 500 IF papers with an impact factor of more than 950 and citations of around 9000.

Himanshu Sekhar Sahoo, received B.Sc degree in Chemistry from Utkal University in 2019 and M.Sc degree in Chemistry from Indian Institute of Engineering Science and Technology (IIEST), Shibpur in 2021. He is currently pursuing Ph.D. from IIEST, Shibpur under the supervision of Dr. Chinmoy Bhattacharya. His research interests include the metal oxide-based semiconductor for photoelectrochemical (PEC) applications.

Victor Shikuku is in Physical, Materials and Environmental Chemistry. Special interests in Drinking and Wastewater treatment processes, development of materials for water purification, and sustainable development

Jitendra Kumar Singh is currently working as Assistant Professor at Hanyang University, South Korea. He was born in Varanasi, India, and received his Ph.D. from the Department of Chemistry, Indian Institute of Engineering Science and Technology (IIEST), Shibpur-Howrah, West Bengal, India in 2017. He has worked as a project associate in the Corrosion and Surface Engineering division, CSIR-National Metallurgical Laboratory (CSIR-NML), Jamshedpur, India from April 2009 to October 2014, and as a researcher at the Department of Architectural Engineering, Hanyang University, ERICA campus, South Korea from August 2015 to March 2022. His research interests include the Synthesis of nanocomposite coating, Dual metal/polymer coating, Thermal Spray Technology, Hot-dip galvanizing, Corrosion, Inhibitor, Electrochemistry, Synthesis of novel materials for corrosion protection of metals and alloys, Conservation of metallic artifacts, Concrete, fabrication and durability performance of sensor and semi-conductor, advanced materials, the process to reduce the porosity of coating, energy storage materials, phase change materials. He has published around 100 research papers including SCI/SCIE, patent, copyright, national and international conferences and book chapters.

Index

P

S

W